The Scale Insects of the Tropical South Pacific Region

Part 1
THE ARMOURED SCALES (DIASPIDIDAE)

D.J. Williams and Gillian W. Watson

C·A·B International Institute of Entomology

Published by
C·A·B International
Wallingford
Oxon OX10 8DE
UK

Tel: Wallingford (0491) 32111
Telex: 847964 (COMAGG G)
Telecom Gold/Dialcom: 84:CAU001
Fax: (0491) 33508

British Library Cataloguing in Publication Data
Williams, D.J. (Douglas John), *1924–*
  The scale insects of the tropical South Pacific region.
  Pt. 1: The armoured scales (Diaspididae)
  1. South Pacific region. Tropical regions. Scale insects
  I. Title II. Watson, Gillian W.
595.7'52

ISBN 0–85198–608–0

D.J. Williams & Gillian W. Watson
CAB International Institute of Entomology
c/o British Museum (Natural History)
Cromwell Road
London SW7 5BD
UK

© C·A·B International, 1988. All rights reserved. No part of this publication may be reproduced in any form or by any means, electronically, mechanically by photocopying, recording or otherwise, without the prior permission of the copyright owner.

Printed in the UK by The Cambrian News Ltd., Aberystwyth.

Contents

Abstract	3
Introduction	5
Acknowledgements and depositories	5
History	7
Economic importance	7
Morphology	11
Classification	16
Systematics	16
Key to genera of tropical South Pacific Diaspididae	19
Genera and species	22
Systematic list of host-plants of tropical South Pacific Diaspididae	249
Index to plant genera in plant families	273
References	277
Index	287

Abstract

This taxonomic account of the armoured scales (Diaspididae) of the tropical South Pacific region is the first of three parts planned for all the important scale insects (Coccoidea) of the area. Keys to 37 genera and 124 species, including one new genus and 38 new species, are provided, accompanied by detailed descriptions and illustrations. Lectotypes of 12 species are designated. The main purpose of the work is to provide agricultural staff who are concerned with pest control and quarantine inspection in the South Pacific area with a reliable means of identification of species. Normally identification of scale insects has been left to specialists, but it is hoped that this work will encourage non-specialists to identify and study this destructive group. The area covered is Melanesia and Polynesia, from Irian Jaya in the west to Easter Island in the east; but Kiribati in Micronesia is also included. The work provides a record of the armoured scales of each territory and of the islands within each territory; knowledge that is essential for effective quarantine inspection and for export of plant produce. Records are based on material examined from all the major world collections of Pacific scale insects and on the economic literature. The work covers all the major pest species of the area including *Aspidiotus destructor* Signoret, *Parlatoria cinerea* Hadden, *P. pergandii* Comstock and *Unaspis citri* (Comstock), and concludes with a comprehensive list of the host-plants and their associated armoured scales.

New taxa described comprise the new genus *Fijifiorinia* and the new species *Allantomytilus dacryoides, Andaspis maai, A. spinosa, Aspidiotus macfarlanei, A. maddisoni, A. musae, A. pacificus, Aulacaspis martini, Chionaspis broughae, C. comys, C. freycinetiae, C. keravatana, C. pandanicola, C. rhaphidophorae, Diaspis casuarinae, Fijifiorinia astronidii, F. oconnori, Fiorinia biakana, F. coronata, F. fijiensis, F. reducta, Genaparlatoria araucariae, Lepidosaphes elmerilleae, L. eurychlidonis, L. geniostomae, L. karkarica, L. pometiae, L. securicula, L. stepta, Lopholeucaspis baluanensis, Oceanaspidiotus nendeanus, Pseudaulacaspis coloisuvae, P. leveri, P. multiducta, P. papulosa, P. ponticula, Schizentaspidus silvicola,* and *Silvestraspis ficaria*. One new combination, *Oceanaspidiotus pangoensis* (Doane & Hadden) is proposed. Four new synonymies are proposed: *Aulacaspis major* Rutherford and *A. rutherfordi* Morrison with *A. tegalensis* (Zehntner), *Chionaspis inday* Banks with *Pseudaulacaspis cockerelli* (Cooley), and *Mytilaspis fasciata* Green with *Lepidosaphes rubrovittata* Cockerell.

Key words

Coccoidea	Diaspididae	scale insects	armoured scales	South Pacific
systematics	generic keys	specific keys	morphology	plant pests
quarantine	host-plants			
Aspidiotus destructor	*Parlatoria cinerea*	*Parlatoria pergandii*		*Unaspis citri*

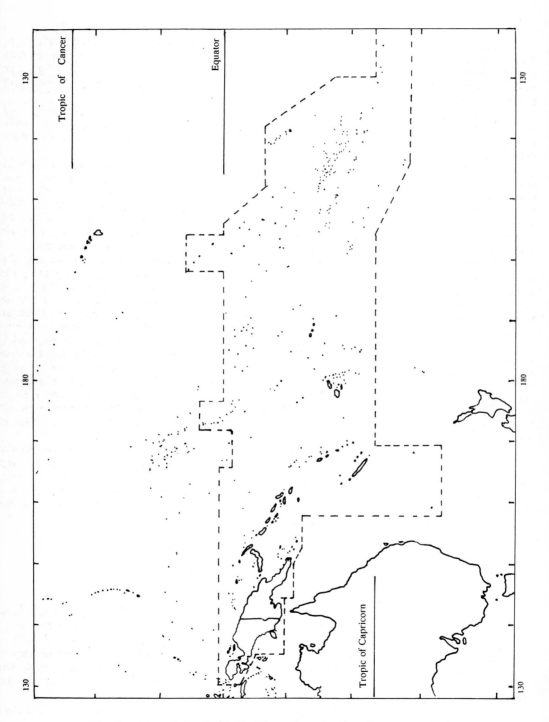

Fig. 1. Map of the South Pacific region showing the area studied.

Introduction

This work is a systematic treatment of the armoured scales of the tropical South Pacific area. The armoured scales, or Diaspididae, are only one of about 20 families of the homopterous superfamily Coccoidea, commonly known as scale insects, coccids or coccoids. Following other workers on the group, the present authors use the term scale insect for the whole group including armoured scales, mealybugs, soft scales, lac insects and cochineal insects.

Coccoidea are sap-sucking insects related to Aphidoidea (aphids), Aleyrodoidea (whitefly) and Psylloidea (jumping plant lice), collectively known as Sternorrhyncha, having the labium apparently arising from the sternum. Auchenorrhyncha, the other large homopterous division, was so named because the labium appears to arise from the neck.

Armoured scales occur in every part of the world where plants grow. Adult females are always sessile; first instars or crawlers, possessing legs, can locate feeding sites, and can be blown considerable distances by wind. The most common form of dispersal of species is on pieces of vegetable material transported by man. Scale insects are usually fixed to the plant by their mouthparts or secretions and, without effective quarantine regulations, they can easily be dispersed to new areas. Cohic (1959) criticised the lack of a quarantine station in the Wallis Is when an air service with Fiji opened in 1950. Many pests were unknown in the Wallis Is before that time.

The area covered in this work (Fig. 1) is Melanesia and Polynesia, and includes most of the territories presently within the South Pacific Commission area. Beardsley (1966) revised the scale insects of Micronesia, and this area (defined by Gressit, 1954) is excluded except for Kiribati, a territory included in the recent UNDP/FAO survey. Irian Jaya was formerly within the South Pacific Commission area. Geographically and ecologically, the scale insect fauna of Irian Jaya is similar to that of Papua New Guinea, and the whole island is here included. This decision was also influenced by an opportunity of borrowing from the Bernice P. Bishop Museum some collections which included important species from Irian Jaya. Lord Howe I. and Norfolk I. are included simply because their faunas are often omitted from studies of Australian insects. The Kermadec Is are the only tropical islands within the territory of New Zealand, but a study of the scale insects shows that many species have been introduced from there, and the islands are excluded from the study area.

The specimens examined number about 6000 on about 4500 microscope slides. The research was carried out at the CAB International Institute of Entomology in the British Museum (Natural History), where the Museum's collections were freely available.

This work should provide a reliable means of identification of scale insect species in the South Pacific area. It should also provide agricultural staff, concerned with pest control and quarantine inspection in the area, with a record of the armoured scale insect fauna of each territory and of the island groups within a territory. This knowledge is essential for effective quarantine within the region and for export of produce to other regions. It is hoped also that the work may stimulate further study of these insects by entomologists in the area.

Acknowledgements and depositories

Foundations for this work were planned many years ago after E.S. Brown, formerly Commonwealth Pool of Entomologists, collected extensively in the Solomon Islands in 1954-56, and after B.A. O'Connor, when Government Entomologist, Fiji, collected for the senior author during the 1950's.

The work, however, would not have been possible at present without generous financial support from the UK Overseas Development Administration Research Scheme R4216 for a two year period. The Natural Resources and Environment Department of the Overseas Development Administration also supplied funds towards equipment, travel and publication costs.

For extra financial support towards publication costs, the authors acknowledge with gratitude the Australian and Pacific Science Foundation.

Illustrations of 25 species, originally prepared by G.F. Ferris, are reproduced from the Atlas of Scale Insects of North America, Volumes I, II and IV, by kind permission of the publishers, Stanford University Press, Copyright 1937, 1938 and 1942, by the Board of Trustees of the Leland Stanford Junior University, California. In this connection, R.J. Gill, California Department of Food and Consumer Services, Sacramento, and R.O. Schuster, R.M. Bohart Museum of Entomology, University of California, Davis, kindly sorted original illustrations; also 22 others prepared by G.F. Ferris and H.L. McKenzie, published in the journal Microentomology; and sent them on loan. Permission to borrow these from the R.M. Bohart Museum of Entomology is gratefully acknowledged. The Trustees of the British Museum (Natural History) gave permission to use figures 6, 10 and 84, and the Registrar, Royal Entomological Society of London, has allowed use of figure 86, all formerly prepared by the senior author.

Much of the material examined in recent years was collected by P.A. Maddison, DSIR, Auckland, when on a tour of duty for UNDP/FAO. Dr Maddison has given much encouragement and help on many matters concerning this material.

During a recent visit to the Systematic Entomology Laboratory, USDA, Beltsville, Maryland, by one of us (D.J.W.), D.R. Miller allowed access to the collections of the United States National Museum (USNM) and gave much advice on scale insect problems. Thanks are also due to S. Nakahara of the same organisation, and V. Blackburn, Animal and Plant Health Inspection Service, Beltsville, for useful discussions on many species found throughout the Pacific and southern Asian regions. As a result of this visit, the USNM kindly sent specimens on loan.

J.W. Beardsley, University of Hawaii at Manoa, Honolulu, has sent specimens for study from Micronesia for comparison. S. Takagi, Hokkaido University, Sapporo, D. Matile-Ferrero, Muséum National d'Histoire Naturelle, Paris, and H. Strümpel, Zoologisches Museum, Universitat Hamburg, have sent on loan other specimens in connection with this study.

The collections of the Bernice P. Bishop Museum, Honolulu, were freely made available and thanks are due to the Collection Manager for the loan of a considerable amount of material.

Collections at the Department of Food and Consumer Services, Sacramento, and the R.M. Bohart Museum of Entomology, Davis, were kindly made available by R.J. Gill and R.O. Schuster respectively; due acknowledgement for the loan of material is given here.

Y. Ben-Dov, The Volcani Center, Israel, kindly examined all the specimens of *Odonaspis* discussed in this work and gave useful advice.

Before ending her duties at the University of Papua New Guinea, Elaine Brough collected extensively in the south-eastern provinces including the Trobriand Islands. This unique collection was handed over for study.

Without necessary field work, the collections used for this study would not have been available. Collectors' names have not been added in the text except in the descriptions of new species, but it seems fitting to mention them here.

B.N. Alalailima, J.H. Ardley, H.L. Autar, G. Baker, R.S. Baker, J. Balcuinas, J.H. Barrett, G. Bengtsson, M. Bigger, G. Brittin, M. le Bronnec, E.J. Brough, E.S. Brown, F. Brunck, P.A. Buxton, J. Casey, A. Catley, A. Charles, M.J.W. Cock, T.D.A. Cockerell, P. Cochereau, J. Clapham, G. Compere, W.L. Conroy, W. Cottier, G.C. Daft, P.S. Dale, L.L. Deitz, C. Dick, R.W. Doane, J.S. Dugdale, L.J. Dumbleton, G.S. Dun, C.A. Edwards, E.D. Edwards, A.C. Eyles, G. Fabres, W. Fletcher, J.L. Froggatt, R.A. Fullerton, W.C. Gagne, S. George, W.T. Goodwin, Grahams, G. Grandison, P.S. Green, P.J.M. Greenslade, W. Greenwood, J.L. Gressitt, T. van Haaren, M. Hand, K.L.S. Harley, A.G. Harrison, K.M. Harrow, D.W. Heatherington, P. Herbert, T. Hola, R. Holway, G.H. Hopkins, N. Howcroft, J.M. Hoy, J.W. Ismay, G. Jackson, F.P. Jepson, P.T. Joseph, M. Jowett, J. Keogh, K.R.S. Kamath, E. Kanjivi, M. Kirby, C.H. Knowles, A. Koebele, N.L.H. Krauss, K. Kumar, L.S. Kuniata, T. Langi, R.A. Lever, W. Leibregis, J.A. Litsinger, T. Maa, R. Macfarlane, E. McKenzie, P.A. Maddison, T. Mala, A.H. Mann, P.D. Manser, A. Maoma, J.H. Martin, M. McQuillan, D. Meadows, C.W. Meister, B. Meyers, M. Monnot, E. Morton, R. Morwood, L.A. Mound, E.P. Mumford, B.A. O'Connor, D.F. O'Sullivan, R.W. Paine, S. Pennycook, W.H. Pierce, J.S. Pillai, J.L. Reboul, M. Rennie, J.B. Risimeri, H. Roberts, Schaffer, J. Scott, D.E. Shaw, F.J. Simmonds, J. Simpson, R. Singh, E.S.C. Smith, B. Stride, J.A. Sutherland, G. Swaine, O.H. Swezey, J.J.H. Szent-Ivany, N. Takau, P.L. Tauiliili, T.H.C. Taylor, M. Tenang, B.M. Thistleton, W. Thomas, S. Tiko,

D. Tomlinson, N.I. Uroe, J.A. Uluinaceva, Varvalui, R. Veitch, R. Viner, Captain Wilkes, A.K. Walker, F.L. Washburn, J.C. Watt, J. White, L.A. Whitney, W. Wildin, F.X. Williams, G. Young, B. Zelazny, E.C. Zimmerman.

The technical assistance of Amitha Godage, and the advice and liaison with the printers by Clive Betts, is greatly appreciated.

The authors acknowledge with thanks advice and criticism of the manuscript by Jennifer M. Cox, British Museum (Natural History), London.

Abbreviations of depositories

ARSDC	Agricultural Research Station, Dodo Creek, Solomon Is.
BMNH	British Museum (Natural History), London, UK.
BPBM	Bernice P. Bishop Museum, Honolulu, Hawaii.
DSIR	New Zealand Arthropod Collection, DSIR, Auckland, New Zealand.
UCD	R.M. Bohart Museum of Entomology, University of California, Davis, USA.
USNM	United States National Museum of Natural History, Washington, D.C., USA.

History

The catalogue of the Diaspididae of the world by Borchsenius (1966) includes about 1700 species, and the total now probably exceeds this by another 200. Only seven species of Diaspididae were recorded from the South Pacific area by Fernald (1903) in a world catalogue. Even this figure had only increased to 35 when Dumbleton (1954) published his list of Pacific insect pests. The 120 species from the area discussed in the present work still probably represent a small proportion of the total.

One of the earliest records was made by Koebele (1893) who found *Unaspis citri* (Comstock) in Fiji. This scale was preyed upon by the ladybird *Anisorcus* sp., and Koebele attempted to take the beetle for use against the scale insect in California.

Outbreaks of pests often lead to surveys of territories to ascertain the extent of damage, and this usually results in further collecting. After Doane (1908) reported on the damage by *Aspidiotus destructor* Signoret to coconut in the Society Islands, there followed increasing reports of scale insect damage to coconut in other areas leading to important collecting and recording of scale insects. The appointment of government entomologists in some territories promoted further interest in the insect faunas. Careful collecting by these entomologists has increased our knowledge of the scale insects of the South Pacific area, and much of this material was sent to the CAB International Institute of Entomology (CIE) for identification. In recent years the most extensive survey of agricultural pests in the South Pacific islands was made by P.A. Maddison for the UNDP/FAO during the 1970's. Most of the scale insect material collected during this survey was sent to CIE, and the records are included in this study.

Economic Importance

Although food habits may have changed in many South Pacific islands as a result of imported foods and the introduction of a cash economy, the traditional food pattern is still maintained in many areas. There is no evidence of scale insect damage to local crops before European influence. The introduction of new plants and forest trees, without strict quarantine measures, has greatly increased the number of scale insect pests on both local food crops and cash economy crops.

Mealybug and soft scale pests of coffee and cacao occur in some areas, but there is little evidence of armoured scale insect damage to these crops. Any records of damage are listed in the text after each species, and some records of the more important species or potential pests are discussed here.

Aspidiotus destructor Signoret

Since the report by Doane (1908) of damage to coconuts in the Society Islands, this species has been reported from most territories in the South Pacific area. It is polyphagous and can be reintroduced from many plant species even after effective control on coconuts. Doane (1908) discussed the first incidence of biological control in the area, when *Aspidiotiphagus citrinus* Craw (now *Encarsia citrina*) was found to be controlling the scale insect with good success. Simmonds (1921) has indicated that the scale insect was probably introduced to the Society Islands in 1894. Jepson (1913) first discussed *A. destructor* in Fiji on bananas, after Australia had refused to accept infested fruit in 1912. Paine (1935) stated that it was first found in the Rewa district on bananas, but it was not until 1916 that it was recorded as a coconut pest in Fiji. Taylor (1935) reported that effective control was achieved there by the introduction of the coccinellid *Cryptognatha nodiceps* Marshall from Trinidad.

Only recently, a serious outbreak on coconuts in American Samoa has been brought under control by the ladybirds *Chilochorus nigritus* (Fabricius) and *Pseudoscymnus anomalus* Chapin, introduced from Vanuatu (Ikin, 1984; Macfarlane, 1986).

Abgrallaspis cyanophylli (Signoret)

Lever (1945b) recorded this species doing damage to *Psidium guajava* in Fiji. In recent years Greve & Ismay (1983) have reported damage to tea in Western Highlands Province (W.H.P.), Papua New Guinea, when a severe infestation caused chlorosis of most leaves.

Andaspis numerata Brimblecombe

This species was found in W.H.P., Papua New Guinea, on the woody stems of recently pruned bushes of tea. According to Greve & Ismay (1983) it was associated with the symbiotic fungus *Septobasidium* sp., which formed a net over the scale, finally obscuring it completely. Any scale insect is difficult to control under these circumstances.

Aonidiella aurantii (Maskell)

This citrus pest was discussed by O'Connor (1969) as occurring in, and needing prevention from spreading within the South Pacific area. The present distribution is given under the species, but it probably occurs almost everywhere citrus is grown worldwide. *A. aurantii* infests all aerial parts of the tree, causing spotting on the leaves and fruit, and may eventually kill the plant. Some references to reports of damage are given under the species.

Aspidiella hartii (Cockerell)

The yams *Dioscorea alata* and *D. esculenta* are particularly infested by this species, especially when in storage. Cohic (1958a) has discussed how yam tubers in storage in the Loyalty Is were encrusted with the scale, making the tubers fibrous; he describes similar damage in the Wallis Is (Cohic, 1959). The species is sometimes a pest on Zingiberaceae, and Swaine (1971) stated that in Fiji it is occasionally found on ginger rhizomes that have been grown in the same field as yams.

Aulacaspis madiunensis (Zehntner)

According to Williams *et al.* (1969) this species can cause damage to certain Javan sugarcane varieties in Queensland. It infests grasses as well as sugarcane in southern Asia, Australia, East Africa and South Africa, but has not been reported previously from the South Pacific area. A record herein from Tuvalu on *Pandanus*, an unusual host plant, indicates that it could be more widespread in the region and could infest sugarcane.

Aulacaspis tegalensis (Zehntner)

This sugarcane pest was also discussed by Williams *et al.* (1969) as sometimes causing severe damage when in large numbers, resulting in loss of yield in Mauritius. The species is known throughout southern Asia, the Malagasian area and East Africa. It may have been present for some time in Papua New Guinea, and could be introduced to other parts of the South Pacific area.

Chrysomphalus aonidum (Linnaeus)
Known as the Florida red scale, this species has a wide distribution in the South Pacific area. Although it is polyphagous, it is usually known as a pest of citrus. In New Caledonia, Cohic (1950a) discussed how, at the end of the wet season, colonies spread from the leaves of citrus to the twigs and small branches, and became so dense as to kill the trees. Cohic (1958a) later stated that the species was one of the greatest pests of citrus in New Caledonia, where its list of host-plants is particularly long. In Papua New Guinea, Szent-Ivany & Stevens (1966) reported it as severely damaging seedlings of *Pinus* sp.

Fiorinia fioriniae (Targioni)
This polyphagous and widespread species is not usually recorded as doing much damage, but in the South Pacific area it is often injurious to avocados. In New Caledonia, Cohic (1958a) stated that it is always common on avocados and causes serious defoliation.

Lepidosaphes beckii (Newman)
Well-recorded as one of the most important citrus pests of the world, *L. beckii* often forms heavy infestations in hot and humid parts of the South Pacific region. According to Cohic (1955), Tahiti is an ideal area for its spread on citrus; it is found there also on other Rutaceae, especially *Murraya exotica*. Dense colonies build up on all aerial parts of the plant, but it prefers the inner parts of the canopy, weakening the branches and eventually causing desiccation; the leaves look chlorotic, and the fruit becomes unsaleable.

Lopholeucaspis cockerelli (Charmoy)
This is a species becoming more widespread throughout the South Pacific area, and is often found on citrus. Although normally reported on angiosperms, it has attacked *Pinus caribaea* in Fiji, where large numbers build up on the leaves, causing some distortion.

Morganella longispina (Morgan)
Cohic (1955) has reported this tropicopolitan and polyphagous species to be common in French Polynesia; in Tahiti it mainly damages grapefruit, lemon and fig. It completely covers the trunk, branches, leaves and fruit. The feeding punctures promote the development of tumorous cankers on the twigs and branches.

Oceanaspidiotus araucariae (Adachi & Fullaway)
Originally described from Hawaii, this species has been reported recently from French Polynesia, Florida and Puerto Rico, always confined to *Araucaria* spp. At present it is localised in New Caledonia, but in the Wallis Is, Cohic (1959) reports that it is a serious pest on *A. columnaris* (now *A. cookii*) on the extremities of the branches, causing desiccation. It has not been reported from Papua New Guinea, but if introduced there it could be damaging to *Araucaria* species.

Parlatoria cinerea Hadden and *P. pergandii* Comstock
These species are known as chaff scales, and both are common on citrus throughout the South Pacific area. They have caused considerable damage to *Citrus sinensis* in Rarotonga, Cook Is, where Walker & Deitz (1979) reported *P. cinerea* accounting for 94% of the armoured scale infestation of the bark and 60% on the stems, *P. pergandii* making up most of the remaining population. These scale insects in Rarotonga are also associated with gumming and flaking of the bark.

Pseudaulacaspis pentagona (Targioni)
Commonly known as the peach scale, this cosmopolitan and polyphagous species has become widespread in the South Pacific area. In Fiji, Lever (1946) recorded it as a pest of *Passiflora*

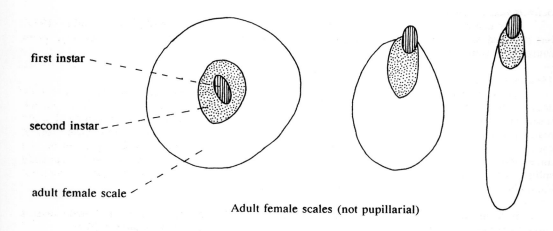
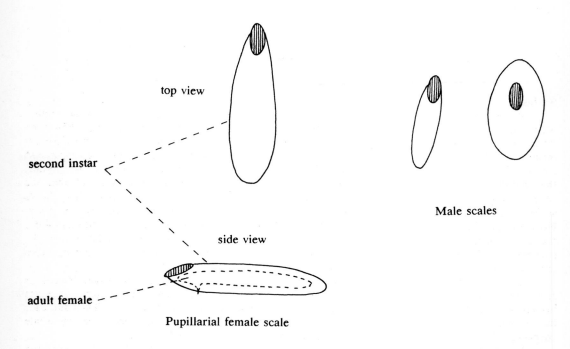

Fig. 2. Scale covers.

quadrangularis and of okra, *Hibiscus esculentus*. Cohic (1958a) reported it as a serious pest in New Caledonia on most introduced plants.

Selenaspidus articulatus (Morgan)

This is a common tropicopolitan species found on numerous species of plants. Swaine (1971) has reported severe local damage to citrus in Fiji, where the build up on leaves and fruit often causes desiccation and death of the tree.

Unaspis citri (Comstock)

As mentioned in the introduction to this section, *U. citri* has been known in the South Pacific area since the last century. It is now a widespread major pest in the area. Although the female insect is brown, the species is known as the snow scale because of the unusual prevalence of white male scales. It is usually a pest of citrus, and in the Solomon Is, Lever (1933) stated that it caused enough damage to be worth spraying. Heavy infestations produce yellow patches on the undersides of leaves. On the trunk and main limbs the bark becomes hard and dark, and subsequently splits. Such damage is probably common, and has been reported from Fiji by Swaine (1971) and from Papua New Guinea (Anon, 1969; Brough, 1986).

Students in the South Pacific area interested in the biology of the armoured scale insects should consult the works by Beardsley & Gonzalez (1975) and Miller & Kosztarab (1979).

It is controversial whether biological control is more successful on islands rather than in continental areas (Greathead, 1971). Examples of successful biological control may be found in Wilson (1960), Rao *et al.* (1971) and Walker & Deitz (1979). In this context, armoured scales are parenchyma feeders (Baranyovits, 1953) and do not produce honeydew, so large numbers may be bred for biological control purposes on tubers and melons without honeydew contamination. In Africa there are associations of ants and armoured scales for other reasons (Ben-Dov & Matile-Ferrero, 1984), but such associations have not yet been found in the South Pacific area.

Morphology

Scales

All of the adult female Diaspididae discussed in this work produce a fibrous scale, incorporating the dorsal exuviae of the previous instars. The actual scale formation has been discussed at some length by Dickson (1951), and the steps in scale formation through the different instars have been given by Stoetzel (1976). So far as is known, there are always three instars in the female. The first, or crawler instar, is the only one that is mobile. After the first moult, the second instar incorporates the dorsal exuviae of the first instar in its covering, and after the second moult the adult female incorporates the dorsal exuviae of both the previous instars in its scale. This scale must be lifted off before the adult females are prepared for microscope study. Normally there are three forms of scale, and these give a clue to the generic group. Although there are exceptions, rotund species produce a circular scale (Fig. 2), with the exuviae central or subcentral. The more oval or elongate the species, the more oval or elongate the scale becomes (Fig. 2), with the exuviae of the previous instars at the edge or at one end.

In some genera the adult female remains within the exuviae of the second instar instead of secreting a scale covering. These are known as pupillarial forms. The second instar in these insects becomes heavily sclerotized, and the adult female within remains membranous. The adult uses the second instar to protect the eggs in the same way that non-pupillarial forms use the scale covering. Adult pupillarial females are difficult to remove intact for study.

Males, when present, have five instars. The scale covering is always smaller than that of the female (Fig. 2). After the first moult, the second instar secretes a fibrous scale incorporating the exuviae of the first instar. The second instar is the last feeding stage; subsequently the male

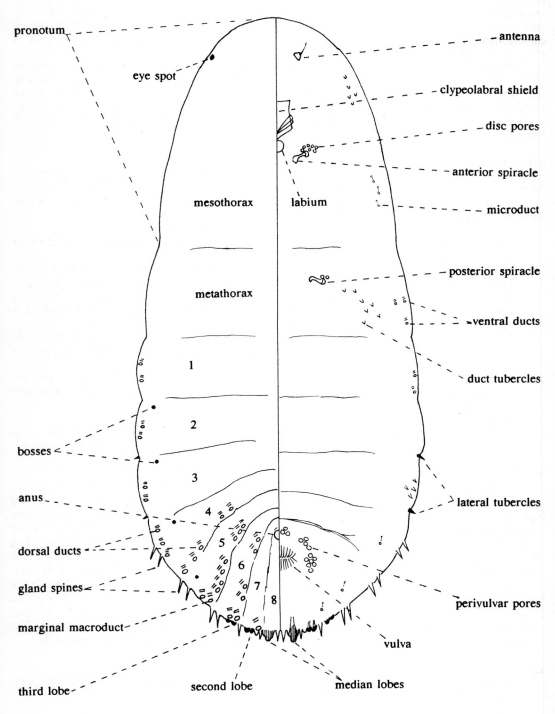

Fig. 3. Generalized drawing of adult female.

passes through two pupal instars before emerging as an adult. The adult is without functional mouthparts and usually has a single pair of wings, but wingless forms are known. Males have been little studied, but as Ghauri (1962) has shown, they are useful in establishing a basis for higher classification. More recently Howell (1980) and Takagi (1969, 1980) have shown the importance of studying second instar males in the systematics of the Diaspididae.

At present, however, the classification of the Diaspididae is based mainly on the characters of the adult female, the instar that is usually collected and identified: it is, therefore, the only instar discussed in this work, with a few exceptions. Although the family Diaspididae can be recognized in life by the scale covering, identification of species can only be achieved by a study of minute characters on slide-mounted specimens. Preparing adult females on microscope slides is often a tedious process, but a well prepared specimen makes identification easier.

Preparation

Various methods of preparation have been discussed in the literature, and one suggested by McKenzie (1956) has gained wide acceptance. The following method has been in use at CIE for many years and has proved to be successful; care is necessary, however, to avoid skin contact with acid fuchsin or xylene, or inhalation of xylene vapour.

1. After removing the specimens from under the scale, place in a 10% solution of KOH. Heat for about 10 minutes but do not allow to boil.
2. Remove to distilled water and, where necessary, express the body contents by gentle pressure. Occasionally waxy droplets may remain in the body; which can be removed by first placing in 95%-100% alcohol; then transferring to carbol xylene for about 10 minutes. After the specimens have cleared, place in alcohol again to remove the carbol xylene. Carbol xylene is composed of: xylene - 3 parts, carbolic acid crystals - 1 part.
3. Transfer from the water, or if the latter method has been used, from the alcohol, to an acid alcohol bath made up of: glacial acetic acid - 20 parts, 50% alcohol - 80 parts. Leave for a few minutes.
4. Stain for about an hour in: acid fuchsin - 0.5 g, 10% HCl - 25 ml, distilled water - 300 ml.
5. Transfer from the stain to 95% alcohol for a very few minutes to remove surplus stain; then transfer to absolute alcohol.
6. Transfer to clove oil for 20 minutes.
7. Place the specimen on a slide and remove surplus clove oil by means of fine filter paper. Put a drop of Canada balsam on the specimen and gently lower the cover slip by its own weight.

Body

A typical elongate species as prepared on a slide is shown in Fig. 3. It is customary in scale insect illustrations to draw the dorsal surface on the left and the ventral surface on the right, the system adopted throughout this work. The head and thorax are fused, there being in most species little indication of segmentation. There are often, however, some signs of segmentation between the mesothorax and and metathorax, and it is usual to regard the head and the two succeeding thoracic segments as the prosoma, and the metathorax and posterior segments as the postsoma. The numbering of the abdominal segments in the accompanying diagram is generally accepted by most students of scale insects. Segmentation of the posterior segments is more difficult to define because this area becomes sclerotized, and the intersegmental lines often disappear. It is usual, therefore, to give this area the term pygidium, that is the segments posterior to and including segment 4. Marginal segmental setae, however, determine the segments in most species. The abdominal segments anterior to the pygidium are often known as the free abdominal segments.

Eyes

Eyes are sometimes conspicuous as simple spots, but occasionally they are modified into spurs, directed anteriorly or posteriorly.

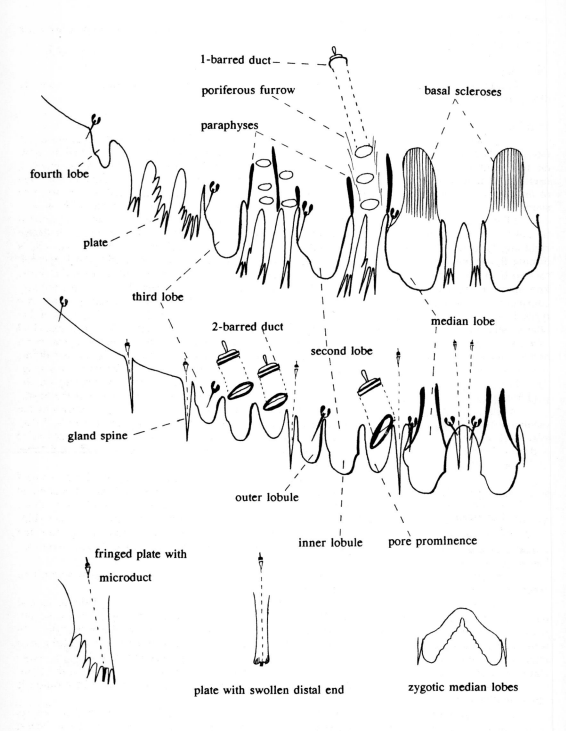

Fig. 4. Details of pygidial margins.

Antennae

The antennae are usually simple tubercles with 1 or more setae. In some genera they are close together and become elongate.

Spiracles

Spiracles are represented by only the anterior and posterior pairs on the thorax. Sometimes disc pores are associated with the spiracular openings.

Pygidium

The scale covering is produced mainly from secretions from various ducts on the pygidium, and from excretions from the anus. As Dickson (1951) has pointed out, the margin of the pygidium plays an important part in the formation of the scale by helping the insect to rotate while pivoting itself by the mouthparts. A knowledge of the characters on the pygidium is fundamental to the study of armoured scale insects, and these characters are shown in detail in Figs 3 and 4. In properly stained specimens, the more heavily sclerotized the structure, the more heavily stained it becomes.

Filamentous material is extruded through ducts, and each duct orifice is usually sclerotized in some way. In some genera the duct orifice lies parallel to the margin and in others it lies at right angles to it. The inner end of the duct usually ends in a conspicuous transverse bar with a minute filamentous extension. There are two main types of these ducts, one with the inner end formed by a single bar, commonly known as a 1-barred duct, and the other with two parallel bars, known as a 2-barred duct (Fig. 4). When the ducts are large, these bars are particularly conspicuous in properly stained specimens, and even in smaller ducts the separation into the two types is not difficult. In many genera the marginal ducts of the pygidium are larger than the others, and are often called marginal macroducts. Sometimes the duct orifice lies on a marginal lobe-like structure, or pore prominence. All other ducts on the dorsum are here called dorsal ducts and are often present in submarginal and submedian groups, segmentally arranged. On the pygidium, the submarginal rows sometimes appear to lie in what is normally called a poriferous furrow (Fig. 4). The arrangement has particular specific significance. Dorsal ducts may be present on the pygidium only, or they may extend to the anterior segments, sometimes as far forward as the head.

Arranged on the margin are projections known as lobes. A pair at the apex are known as the median lobes. They may be separate, or joined at the base, when they are referred to as being zygotic (Fig. 4); or they may be 'yoked' at the base by an arched sclerotized bar. Second, third and even fourth lobes may be present. In genera with 2-barred ducts, the second and third lobes each may be divided into inner and outer lobules.

Between the lobes, and anteriorly around the margins, are frequently spine-like or fringed structures. The spine-like structures are known as gland spines, each having a microduct opening at its apex. The fringed processes are known as plates, and may or may not possess a microduct (Fig. 4). Other plates may not be fringed, but are more truncate, and some may have swollen apices (Fig. 4). In some species the plates or gland spines decrease in size on the free abdominal segments and thorax, and may be replaced by small sclerotized tubercles, each with a minute duct. These are known as duct tubercles (Fig. 3).

The bases of the median lobes sometimes extend anteriorly to form 2 definite sclerotized areas, each known as a basal sclerosis (Fig. 4).

Paraphyses are elongate sclerotized structures arising from the dorsal angles of the lobes and on other parts of the margins (Fig. 4). The term is also given to similar structures on the ventral surface of the lobes. On the dorsum they seem to act as supports to the poriferous furrows. In some genera they are conspicuous and in others they are small, but their presence or absence is important in the identification of a genus. Occasionally the paraphyses are modified into large clavate processes (Fig. 51), when the round or oval inner end may become detached from the stem.

The anus is simple and variously placed on the midline of the pygidium, where its position is often significant.

In most species the vulva (Fig. 3) is conspicuous, but in others it is not so evident, and is represented by a mere transverse or curved slit. Many species have disc pores anterior and lateral to the vulva, often in groups. These are perivulvar pores. Each disc pore has five minute openings or loculi; such pores are sometimes known as quinquelocular pores. It is evident that if a specimen has perivulvar pores it is definitely mature, but a specimen without a vulva is immature.

Small disc-like scars are sometimes present on the dorsal submargins. These are commonly known as bosses (Fig. 3). They have some significance in distinguishing species, but at present their function is unknown.

There may be other terms used in the text for characters belonging to one or two species, but they should be apparent by reference to the appropriate illustration.

Classification

It has already been mentioned that species having the second and third lobes bilobed never have 1-barred ducts. Species with plates may have either 1-barred or 2-barred ducts, but never both types. Such combinations of characters have been used with success in defining groups of genera and even subfamilies and tribes: thus Ferris (1942) divided the family into two subfamilies that have since been raised to family status. It is Ferris's subfamily, the Diaspidinae, that is here regarded as the family Diaspididae. Ferris divided the Diaspidinae into four tribes. One of these, the Odonaspidini, with *Odonaspis* (Fig. 98) as the type-genus, lacks plates or gland spines and usually has but a single median lobe. The tribe Diaspidini thus contained species with 2-barred ducts, and the tribe Aspidiotini, species with 1-barred ducts. These tribes have been elevated to subfamily rank by certain authors, and new tribes have been suggested for other genera with common characteristics. In a catalogue of armoured scale insects, Borchsenius (1966) recognized 5 subfamilies, 12 tribes and 23 subtribes, whereas Takagi (1969) recognised only 7 tribes. There is still no general agreement on a system of classification, but normally the tribal distinctions mentioned above can be separated early in any key to genera. This method was used by Beardsley (1966) in his key to the Diaspididae of Micronesia. The same method is adopted here because the present work may be used by agricultural and quarantine entomologists who may be unfamiliar with works on classification. In the following key to genera, reference should be made to the terms already discussed. Identification will be made easier by reference to the illustrations of species.

Systematics

In the time available for this work it has only been possible to study what are here regarded as the most important species from the area. Some other species are available, but many are on endemic plants such as *Nothofagus*, and on plant species in specialised environments. Some interesting species are represented by material that, at present, is inadequate for critical study. These may form the basis of a later supplement.

Apart from the cosmopolitan species that have been introduced in modern times, most species seem to be related to the fauna of the Oriental region. There are certainly endemic species of Diaspididae in the remoter islands of Samoa, Fiji and New Caledonia. The scale insect fauna of New Caledonia will probably be studied by D. Matile-Ferrero, Muséum National d'Histoire Naturelle, Paris, but even so a few important species from New Caledonia are discussed in the present work. Further collecting in Papua New Guinea and Irian Jaya will almost certainly reveal an interesting endemic fauna. Most of the species found in these territories are related to many in southern Asia, but a new species of *Allantomytilus* described here from Irian Jaya is related to the only other known species, from Australia. The scale insect fauna of the whole island of New Guinea should be considered by entomologists working further east. It seems only a matter of time before many species will disperse eastwards if the host-plants are available, and some will undoubtedly be intercepted at plant inspection stations.

Many species discussed in this work are illustrated here for the first time. One of the limiting factors of any work on scale insect taxonomy is the time taken to illustrate the species. Photographs are of limited value because even when specimens are flattened on microscope slides, it is almost impossible to photograph a single surface. Some of the cosmopolitan species have been illustrated many times in various publications. An opportunity arose to borrow some original drawings made by G.F. Ferris and H.L. McKenzie, already published, and these are reproduced here. Although many of these drawings should possibly be revised, they are still of value, and many workers on scale insects have used them to advantage. It is hoped they may be useful in the South Pacific area, where some literature may be scarce.

The economic literature of the South Pacific area is voluminous, and species names may have been listed, based on earlier identifications, which may need modification. It is hoped that all published names that have been mentioned in the literature are listed in the present index. A few species not studied, but needing special comment, are discussed here.

Aonidia longa Lindinger

This is a species described by Lindinger (1911: 172) from New Caledonia, on *Podocarpus gnidioides*. Through the kindness of H. Strümpel, Zoologisches Institut und Zoologisches Museum, Universität Hamburg, it has been possible to see a few specimens of this interesting pupillarial species. It does not belong to *Aonidia* as presently understood, but the specimens available are not adequate for illustration and further description.

Aonidia rageaui Cohic

Cohic (1958a) listed this species from New Caledonia as a *nomen nudum*.

Aspidiotus suvaensis Green & Laing

This species was listed by Lever (1945a, 1947) from Fiji on *Rhizophora mangle*. The name is a *nomen nudum*, but specimens are available with identical data under different manuscript names. Although this seems to be a distinct species, the specimens are not good enough for critical study.

Cryptophyllaspis ruebsaameni Cockerell

Cockerell (1902b) described this species from the Bismarck Archipelago on the leaves of *Codiaeum* sp., causing small subcylindrical galls. A few original specimens are available for study, but they are in poor condition. The species seems to be a component of *Abgrallaspis* or a genus close to it.

Galeraspis sp.

Specimens are available identified as this genus from Fiji, on *Flagellaria* sp. The name was listed by Hinckley (1965), but is excluded from this work because the available material is not good enough to illustrate.

Leucaspis bugnicourti Cohic

The description and illustration of this species by Cohic (1958b) are adequate for identification. The species was described from New Caledonia, Nouméa on *Ficus* sp. No specimens have been available for study, but the species is pupillarial and seems to be related to *Fijifiorinia* gen. n., described herein, in possessing median lobes that are contiguous for most of their length. It differs from *Fijifiorinia* in possessing supplementary perivulvar pores on three segments anterior to the normal perivulvar pores, and in having a reddish brown scale instead of a black one. The species does not belong to *Leucaspis* as presently understood, and cannot be included under this genus in a key to genera. Further research on this species is needed.

Pseudaulacaspis dubia (Maskell)

Lever (1945a) listed this species from Fiji on *Adiantum* sp. as *Chionaspis dubia*, and later Hinckley (1965) listed it as *Phenacaspis dubia*. Specimens are available upon which the records are based, but the species is different from *P. dubia* described from New Zealand. The specimens need further study, and the species is omitted from the present work.

Most of the following records of species are accompanied by synonymy, description, material examined and comments. Synonymy has been kept to a minimum and represents only the important names that have been used in the South Pacific area. A fuller synonymy is given by Borchsenius (1966). The records listed under each territory refer to the major islands, with the earliest date of the material examined. Host-plant records are listed alphabetically after each territory. Provinces in Papua New Guinea and the Solomon Islands are abbreviated as follows.

Papua New Guinea
C.P.	Central Province (including National Capital District)
Chimbu P.	Chimbu Province
E.H.P.	Eastern Highlands Province
E.P.	Enga Province
G.P.	Gulf Province
Madang P.	Madang Province
Manus P.	Manus Province
Milne B.P.	Milne Bay Province
Morobe P.	Morobe Province
N.I.P.	New Ireland Province
N.P.	Northern Province
N.S.P.	North Solomons Province
S.H.P.	Southern Highlands Province
W.N.B.P.	West New Britain Province
W.S.P.	West Sepik Province
W.H.P.	Western Highlands Province
W.P.	Western Province

Solomon Islands (Names kindly supplied by R. Macfarlane, South Pacific Commission)
C.I.P.	Central Islands Province
G.P.	Guadalcanal Province
M.P.	Malaita Province
M.U.P.	Makira Ulawa Province
T.P.	Temotu Province
W.P.	Western Province
Y.P.	Ysabel Province

Records from the literature of damage, localities and host-plants, if listed under comments, are in addition to the records listed under material examined.

Lettering used in the figures
- A. Adult female, general aspect.
- B. Pygidium of adult female.
- C. Antenna of adult female.
- D. Anterior spiracle of adult female.
- E. Posterior spiracle of adult female.
- F. Pygidium of second instar female.
- G. Dorsal detail of pygidial margin of adult female.
- H. Ventral detail of pygidial margin of adult female.
- I. Antenna and cephalic margin of first instar.
- J. Habit.

Key to genera of tropical South Pacific Diaspididae

1   Pygidium with a tuft of 6 spiniform processes at apex...................... *Froggattiella* Leonardi
--   Pygidium without a tuft of 6 spiniform processes at apex 2

2   Pygidium with 7 long bifid processes on margin, these at least one third length of pygidium (pupillarial) .... *Agrophaspis* Borchsenius & Williams
--   Pygidium without 7 long bifid processes on margin (sometimes pupillarial) 3

3   Pygidium without plates or gland spines. Pygidial lobes represented by a single median lobe only. Ducts usually small, in no fixed arrangement, abundant on both dorsal and ventral surfaces .... *Odonaspis* Leonardi
--   Pygidium normally with plates or gland spines. Pygidial lobes usually present; rarely with median lobes united into a single lobe. Ducts usually present in dorsal segmental rows or series. If present on venter, they are not abundant 4

4   Dorsal ducts 1-barred. Second and third lobes never bilobed. Plates present, fringed; gland spines absent 5
--   Dorsal ducts 2-barred. Second and third lobes, when present, often bilobed. Gland spines present, but sometimes replaced by plates 21

5   Dorsal surface of pygidium with a conspicuous mosaic or areolate appearance 6
--   Dorsal surface of pygidium without this mosaic appearance 7

6   Pygidial margin with 2 pairs of elongate paraphyses, each with anterior circular knob .... *Duplaspidiotus* MacGillivray
--   Pygidial margin without knobbed paraphyses .... *Pseudaonidia* Cockerell

7   Pygidial margin with paraphyses arising from bases of lobes, or if some lobes are absent, then from the normal position of these lobes. Sometimes paraphyses minute. Occasionally paraphyses also extending around margins of pygidium lateral to lobes 8
--   Pygidial margin without paraphyses 15

8   Prosoma of mature female expanded postero-laterally and often almost enclosing pygidium, giving the body a reniform shape .... *Aonidiella* Berlese & Leonardi
--   Prosoma not expanded postero-laterally, body never reniform 9

9   Paraphyses arising from outer basal angles of median lobes elongate, each with a distinct knob at anterior end .... *Clavaspis* MacGillivray
--   Paraphyses, if present at outer basal angles of median lobes, not knobbed, usually fusiform 10

10   Median lobes only present. Plates absent between median lobes, but present lateral to lobes; of unusual form, all same length and elaborately fringed (note: the paraphyses in this genus are in the first and second spaces and are very small) .... *Morganella* Cockerell
--   Median, second and third lobes present, either graded in size, or second and third lobes reduced to minute points. Plates present between median lobes, and lateral to lobes; not elaborately fringed 11

11   Plates between lobes apically chelate, the fingers of the claw slightly sclerotized and connected by a membrane .... *Furcaspis* Lindinger
--   Plates not chelate 12

12	All paraphyses shorter than median lobes	13
--	Most paraphyses longer than median lobes	14

13 Anus large, the space between apex of pygidium and anus not more than twice diameter of anus. Second and third lobes represented by small points. Paraphyses robust .... *Hemiberlesia* Cockerell
-- Anus, even if large, situated about one third or one quarter length of pygidium from apex, the space always longer than twice diameter of anus. Second and third lobes developed to some extent, not represented by mere points. Paraphyses very small .... *Abgrallaspis* Balachowsky

14 Paraphyses associated only with the lobes .... *Chrysomphalus* Ashmead
-- Paraphyses associated with the lobes and numerous around pygidial margins to fourth abdominal segment .... *Lindingaspis* MacGillivray

15 Prosoma of mature female expanded postero-laterally to almost enclose pygidium .... *Aspidiotus* Bouché (in part)
-- Prosoma of mature female not expanded postero-laterally 16

16 Prosoma with a marked constriction between mesothorax and metathorax 20
-- Prosoma without such a constriction 17

17 Pygidial lobes represented by 2 pairs. Margins of pygidium crenulate. Ducts on ventral surface of pygidium numerous, almost same size as dorsal ducts .... *Aspidiella* Leonardi
-- Pygidial lobes represented by 3 or 4 pairs. Margins of pygidium not noticeably crenulate. Ventral ducts of pygidium, if present, much narrower and smaller than dorsal ducts 18

18 Plates lateral to third lobe dentate, edges sclerotized, with fleshy processes. Innermost poriferous furrow reaching to anus or beyond. Macroducts extending well anteriorly to anus. Dorsal marginal setae on lobes lanceolate (thick and dorsoventrally flattened). Median lobes with narrow membranous strip at base, appearing hinged. Pygidial recess always absent .... *Octaspidiotus* MacGillivray
-- Plates lateral to third lobes not dentate or sclerotized, fringed or with slender processes. Innermost poriferous furrow usually ending before anus. Macroducts not extending very far anteriorly to anus. Dorsal marginal setae slender to thick, not flattened. Median lobes lacking membranous strip at base, not appearing hinged. Pygidial recess sometimes present 19

19 Macroducts fairly slender, with sclerotized rims. Lobes numbering 3 or 4 pairs. Apical recess absent or very slight. Perivulvar pores present or absent. Plates between median lobes often simple; all plates beyond outermost lobes with simple or no fringe. Prosoma of mature female expanded, but not postero-laterally, sometimes becoming moderately sclerotized at maturity .... *Oceanaspidiotus* Takagi
-- Macroducts fairly wide, rims not usually sclerotized. Lobes numbering 3 pairs. Apical recess absent to well developed. Perivulvar pores present. Plates between median lobes usually fringed; at least some plates beyond third lobes with well-developed fringe. Prosoma of mature female not greatly expanded, normally remaining membranous .... *Aspidiotus* Bouché (in part)

20 With a marked constriction between metathorax and abdominal segment 1, this in addition to constriction between mesothorax and metathorax. Third lobes similar shape to median and second lobes .... *Schizentaspidus* Mamet
-- Without a marked constriction between metathorax and abdominal segment 1. Third lobes spur-shaped .... *Selenaspidus* Cockerell

21	Plates present. Gland spines absent	22
--	Plates absent. Gland spines present	25
22	Plates not fringed, apices swollen. Thorax expanded postero-laterally (pupillarial) .... *Silvestraspis* Bellio	
--	Plates fringed. Thorax not expanded postero-laterally (pupillarial or not pupillarial)	23
23	Orifices of marginal ducts with simple sclerotized rims. With supplementary groups of disc pores on 2 segments preceding normal perivulvar pores (pupillarial) .... *Lopholeucaspis* Balachowsky	
--	Orifices of marginal ducts with semi-lunate sclerotized rims (not pupillarial)	24
24	Prosoma at maturity heavily sclerotized .... *Genaparlatoria* MacGillivray	
--	Prosoma at maturity membranous .... *Parlatoria* Targioni	
25	Gland spines present between median lobes, sometimes minute	26
--	Gland spines absent from between median lobes	30
26	Median lobes separated by a space equal to at least twice width of a single lobe .... *Allantomytilus* Leonardi	
--	Median lobes close together, separated by a space no more than width of a single lobe	27
27	Median lobes normally parallel, each with inner and outer edges about same length. Paraphyses on lobes vertical and normal	28
--	Median lobes usually triangular, with inner margin shorter than outer margin. Paraphyses expanded, club-shaped or transverse	29
28	Body almost circular or turbinate .... *Carulaspis* MacGillivray	
--	Body fusiform .... *Lepidosaphes* Shimer	
29	Marginal ducts large, conspicuously larger than dorsal ducts .... *Andaspis* MacGillivray	
--	Marginal ducts small, same size as dorsal ducts .... *Howardia* Berlese & Leonardi	
30	Dorsum of pygidium with a 'lattice work' or coarse reticulate pattern .... *Ischnaspis* Douglas	
--	Dorsum of pygidium without this pattern	31
31	Median lobes separated at base, although at times the bases may be close together	32
--	Median lobes yoked at base by an internal sclerosis	33
32	Body turbinate .... *Diaspis* Costa	
--	Body fusiform .... *Unaspis* MacGillivray	
33	Pygidium with a series of marginal ducts only, sometimes few and indistinct	34
--	Pygidium with at least submarginal ducts in addition to marginal ducts	35
34	Median lobes of pygidium divergent, or if parallel, with a space between. A pair of setae present at base between median lobes (pupillarial) .... *Fiorinia* Targioni	
--	Median lobes of pygidium contiguous. Setae absent from between bases of median lobes (pupillarial) .... *Fijifiorinia* gen. n.	
35	With a pair of setae between median lobes .... *Pseudaulacaspis* MacGillivray	
--	Without a pair of setae between median lobes	36

36	Gland spines and macroducts absent anterior to first abdominal segment. Prosoma usually swollen, wider than abdomen; if head narrow, then with lateral tubercles .... *Aulacaspis* Cockerell
--	Gland spines or macroducts present anterior to first abdominal segment. Prosoma not swollen, without lateral tubercles 37
37	Median lobes contiguous for most of their length. Submarginal and submedian ducts never present on segment 6 and 7 .... *Pinnaspis* Cockerell
--	Median lobes separated by a space; if space is very narrow then there are submarginal or submedian ducts on segments 6 and 7 .... *Chionaspis* Signoret

Genera and species

Genus ABGRALLASPIS Balachowsky

Abgrallaspis Balachowsky, 1948: 306. Type-species *Aspidiotus cyanophylli* Signoret, by original designation.

The genus is closely related to *Hemiberlesia* Cockerell, but Balachowsky described *Abgrallaspis* originally for a few European species with paraphyses and anus smaller than in *Hemiberlesia*. Later Balachowsky (1956) redefined the genus, distinguishing it from *Hemiberlesia* mainly by the size and position of the anus, which in *Abgrallaspis* lies over twice its diameter from the apex, whereas in *Hemiberlesia* it lies less than its diameter from the apex. Komosińska (1969) revised the entire genus comprising 22 species. The only species present in the South Pacific area is *A. cyanophylli* (Signoret), normally a tropicopolitan and polyphagous species, but it is often a common greenhouse pest elsewhere.

Abgrallaspis cyanophylli (Signoret) (Fig. 5)

Aspidiotus cyanophylli Signoret, 1869b: 119.
Hemiberlesia cyanophylli (Signoret), Ferris, 1938: 237.
Abgrallaspis cyanophylli (Signoret), Balachowsky, 1948: 306.

Material examined
COOK IS. Aitutaki, 1977; Atiu, 1977; Mangaia, 1977; Rarotonga, 1975. On *Acalypha hispida, Artocarpus altilis, Cordyline terminalis, Eriobotrya japonica, Musa* sp., *Persea americana, Psidium guajava.*
FIJI. Taveuni, 1913; Vanua Levu, 1975; Viti Levu, 1945; Wakaya, 1946. On *Artocarpus altilis, Barringtonia* sp., *Cocos nucifera, Musa* sp., *Plumeria acutifolia, Psidium guajava.*
FRENCH POLYNESIA. Society Is, Tahiti, 1976. On *Capsicum ovatum.*
KIRIBATI. Tarawa, 1976. On *Coccoloba uvifera, Guettarda speciosa, Musa* sp.
PAPUA NEW GUINEA. C.P.: 1958. E.H.P.: 1959. E.N.B.P.: 1980. Morobe P.: 1979. W.H.P.: 1972. On *Camellia sinensis, Coleus* sp., *Elettaria cardamomum, Eugenia* sp., *Ficus* sp.
TONGA. Ha'apai Group, Foa, 1977, Nomuka, 1977; Tongatapu Group, 'Eua, 1977, Tongatapu, 1975; Vava'u Group, Pangaimotu, 1977, Vava'u, 1977. On *Annona squamosa, Bauhinia* sp., *Ceiba pentandra, Dioscorea alata, Hibiscus syriacus, Jatropha curcas, Manihot esculenta, Persea americana.*
TUVALU. Funafuti, 1976; Vaitupu, 1976. On *Artocarpus altilis.*
VANUATU. Efate, 1983. On *Musa* cv.

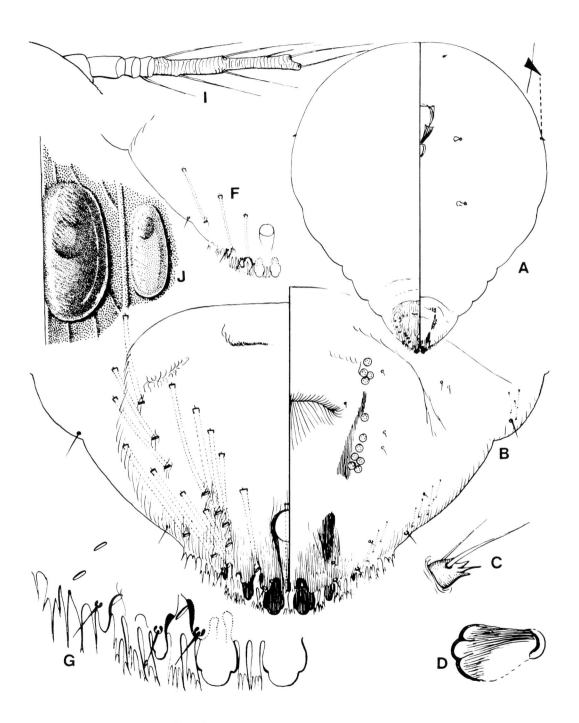

Fig. 5. *Abgrallaspis cyanophylli* (Signoret).

WESTERN SAMOA. Savai'i, 1975; Upolu, 1975. On *Acalypha hispida, Annona muricata, Artocarpus altilis, Cedrela toona, Cinnamomum zeylanicum, Clerodendrum* sp., *Macadamia tetraphylla, Musa* sp., *Persea americana, Piper methysticum, Psidium guajava, Swietenia macrophylla, Theobroma cacao.*

Comments
    In addition to the material listed above, *A. cyanophylli* has been recorded from Fiji, on *Hevea brasiliensis* and *Psidium* sp. (Lever, 1945a), and causing damage to *Cocos nucifera, Dioscorea* spp. (Dumbleton, 1954) and *P. guajava* (Lever, 1945b); from New Caledonia, on *Annona* sp., *A. squamosa, Cocos nucifera, Coffea* sp., *C. arabica, Mangifera indica* and *Musa sapientum* (Cohic, 1956 and 1958a, and Brun & Chazeau, 1980); from Papua New Guinea, W.H.P., on *Camellia sinensis*, causing chlorosis of most leaves, and E.N.B.P., on *Cocos nucifera* (Greve & Ismay, 1983); and from Tonga, causing damage to *C. nucifera* and *Dioscorea* spp. (Dumbleton, 1954). Borchsenius (1966) also mentions its occurrence in Samoa and Tahiti.

Genus **AGROPHASPIS** Borchsenius & Williams

Agrophasphis Borchsenius & Williams, 1963: 375. Type-species *Aonidia buxtoni* Laing, by original designation.

This genus is pupillarial, and was described for a single species from New Caledonia. The important characters are discussed under the species.

Agrophaspis buxtoni (Laing) (Fig. 6)

Aonidia buxtoni Laing, 1933: 676.
Agrophaspis buxtoni (Laing), Borchsenius & Williams, 1963: 374.

Description
    The adult female of this pupillarial species is easily recognizable by the plates that are modified into 7 long bifid processes at the posterior end of the pygidium, these processes about one third the length of pygidium. Few other recognition characters can be mentioned for this peculiar species, except that the body is circular and membranous throughout.
    The second instar has three pairs of lobes and fringed plates. Dorsal ducts apparently 2-barred, but second bar extremely thin.

Material examined
NEW CALEDONIA. New Caledonia, 1925. On 'switch grass'.

Comments
    This species seems to be endemic to New Caledonia. Cohic (1958a) stated it occurred on an indeterminate species of Cyperaceae, and was common at the base of *Carex* sp.; this was probably the same as the original host plant listed above.

Genus **ALLANTOMYTILUS** Leonardi

Allantomytilus Leonardi, 1898a: 45; Ferris, 1936: 24,34. Type-species *Mytilaspis maideni* Maskell, by monotypy.

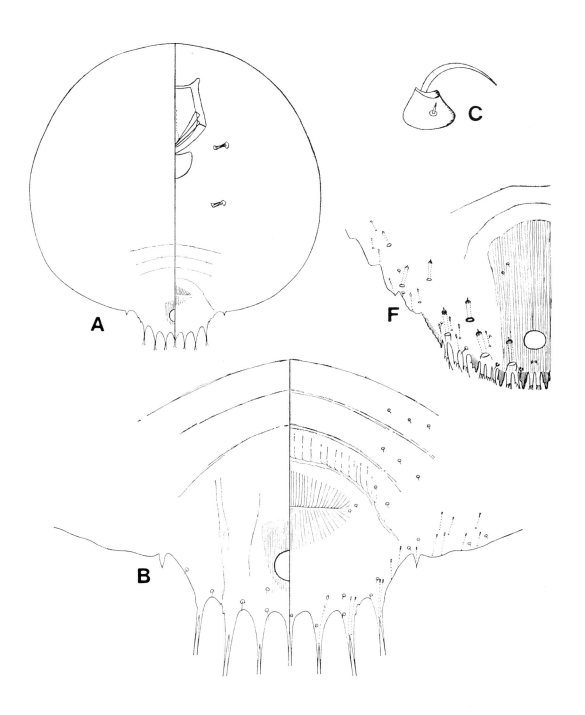

Fig. 6. *Agrophaspis buxtoni* (Laing).

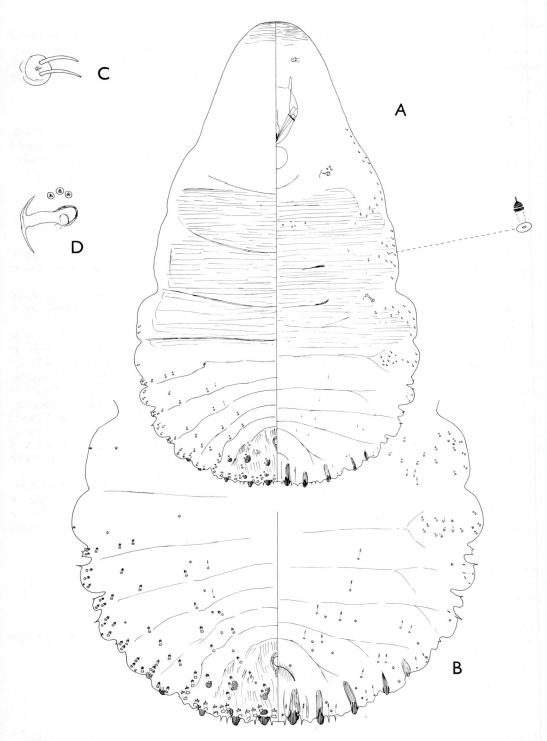

Fig. 7. *Allantomytilus dacryoides* sp. n.

Description
A genus with 2-barred ducts more or less scattered over the dorsum. With at least 4 pairs of lobes, not bilobed, but sometimes lobes obscured by overlapping of dorsal surface; all lobes, including median pair, set well apart. Gland spines short, those on margins of pygidium well separated, a pair wide apart between median lobes. Perivulvar pores absent. Head and thorax becoming sclerotized in mature specimens.

Comments
This genus was described for an Australian species which was illustrated by Ferris (1936). The new species seems to be congeneric and, because it is found in Irian Jaya, the genus may have a wider distribution in the South Pacific area than presently known. Both species have a peculiar scale that is convex and striped, affording them easy recognition in life.

Allantomytilus dacryoides sp. n. (Fig. 7)

Description
Scale of adult female elongate, convex, about 1.2 mm long, dark brown with white transverse ridges giving the scale a 'tiger-striped' appearance. Male scale smaller but similar to female scale.

Adult female, slide-mounted, up to 1.1 mm long, pyriform or teardrop-shaped, widest at about first abdominal segment; head sharply rounded; pygidium gently rounded; body becoming more heavily sclerotized on mesothorax, metathorax and head margin with maturity.

Pygidium with 4 recognizable lobes. Median lobes separated by a space equal to twice as wide as a single lobe, each lobe often with a notch on each side. Remaining lobes present as more or less rounded sclerotized projections from the margin, widely separated, made more conspicuous by the ventral paraphyses and anterior sclerotized areas continuous with the lobes; a fifth pair of smaller projections also present. Gland spines short; a pair present between median lobes, 1 lateral to each median lobe and 2 lateral to each second lobe. A few other gland spines, sometimes indistinct, on margins of segments 2 and 3. Marginal ducts largest between the lobes, numerous as far forward as metathorax. Submarginal ducts well separated from margins, present as far forward as abdominal segment 1; submedian ducts present as far forward as segment 2, there being rarely more than 3 or 4 submarginal and submedian ducts altogether on each side. Some ducts on pygidium associated with sclerotized patches around orifices. Anus situated near base of pygidium.

Ventral surface without perivulvar pores. Microducts present around free abdominal margins, in submedian abdominal areas and submarginally on thorax. Duct tubercles few, present laterally on segment 1 of abdomen. Small ducts, smaller than dorsal ducts, on margins of thorax and laterally on segment 1. Antennae each with 2 short blunt setae. Anterior spiracles each usually with 3 disc pores. A few sclerotized spicules located on median area of mesothorax.

Material examined
Holotype female. **IRIAN JAYA**. Biak, on leaves of tree, 22.v.1959 (BPBM).
Paratypes female. **IRIAN JAYA**. Same data as holotype. 9 (BPBM), 9 (BMNH).

Comments
The species differs from the type-species in having fewer dorsal ducts and fewer gland spines on the free abdominal segments.

Genus **ANDASPIS** MacGillivray

Andaspis MacGillivray, 1921: 275; Rao & Ferris, 1952: 17; Williams, 1963: 13. Type-species *Mytilaspis flava* var. *hawaiiensis* Maskell, by original designation.

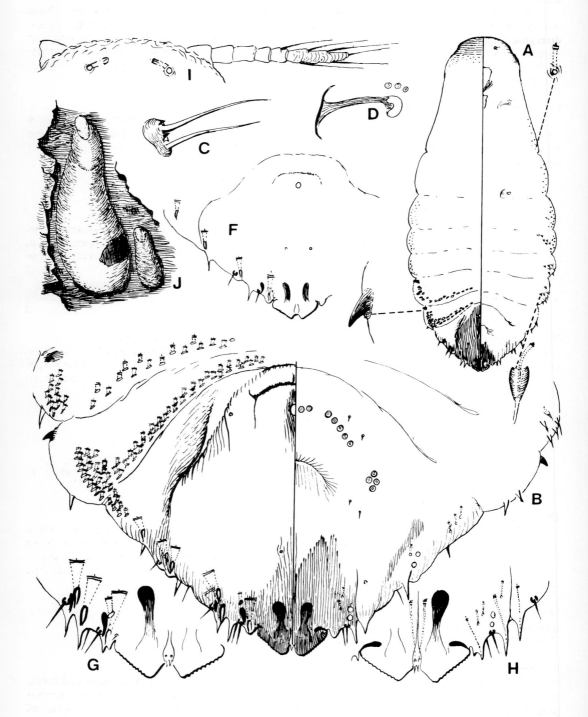

Fig. 8. *Andaspis hawaiiensis* (Maskell).

Description
   Closely related to *Lepidosaphes*, from which it differs in the form of the median lobes. These are usually prominent and close together, with small gland spines between; the inner edges short, diverging to long oblique outer margins. The ventral paraphyses on the median lobes are variously shaped, often transverse, and sometimes in addition, a club-shaped paraphysis arises from the basal angle or from the middle basal part of lobe. Second lobes usually reduced to points or lacking. Marginal macroducts enlarged, numbering 4-6.

Comments
   This genus is probably of southern Asian origin, with extensions into some temperate areas and Australia. Four species are here discussed from the South Pacific area; these include the type-species, the only one to have a wide distribution in the tropics. There are probably many more species in the South Pacific region.

Key to species of *Andaspis*

1	Marginal macroducts numbering 5 on each side. Eye spots modified into prominent spurs .... ***spinosa*** sp. n.	
--	Marginal macroducts numbering 6 on each side. Eye spots, if present, not modified into prominent spurs 2	
2	Submedian group of ducts absent from abdominal segment 6 .... ***hawaiiensis*** (Maskell)	
--	Submedian group of ducts present on abdominal segment 6 3	
3	Bosses present on abdomen. Median lobes separated for most of their length .... ***numerata*** Brimblecombe	
--	Bosses absent from abdomen. Median lobes separated for only about one quarter of their length .... ***maai*** sp. n.	

Andaspis hawaiiensis (Maskell) (Fig. 8)

Mytilaspis flava var. *hawaiiensis* Maskell, 1895, 47.
Lepidosaphes moorsi Doane & Ferris, 1916: 401.
Andaspis hawaiiensis (Maskell), MacGillivray, 1921: 292; Ferris, 1937a: 4; Zimmerman, 1948: 407; Balachowsky, 1954: 132.

Description
   This is a distinctive species with minute gland spines between prominent median lobes which are almost triangular, set close together, the short inner edges divergent and the lateral edges serrate, each lobe with a stout, club-shaped paraphysis extending into pygidium from inner angle. In addition to the 6 marginal macroducts on each side of the pygidium there is also a conspicuous row of ducts across segment 4 of the abdomen.

Material examined
COOK IS. Atiu, 1977. On *Albizia falcataria*.
WESTERN SAMOA. Upolu, 1913. On *Citrus* sp.

Comments
   Although this species was described originally from Hawaii, it was probably introduced there. It is now known to be tropicopolitan, extending into some temperate areas, and is found on numerous plant species. Specimens have not been reported yet from Micronesia. A few specimens of *Lepidosaphes moorsi* Doane & Ferris, collected on citrus at Apia in Western Samoa in 1913, have

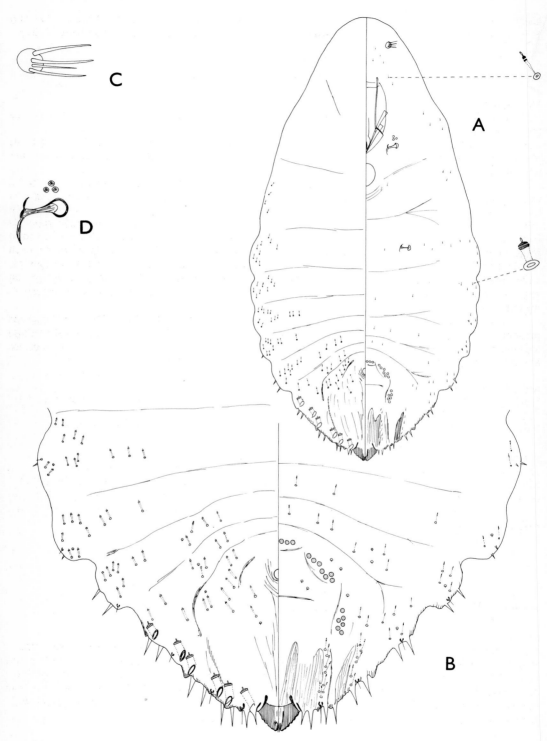

Fig. 9. *Andaspis maai* sp. n.

kindly been made available by R.O. Schuster, University of California, Davis. Dale (1959) and Maddison (1976) both record *L. moorsi* from Western Samoa damaging the leaves and fruit of citrus.

Andaspis maai sp. n. (Fig. 9)

Description

Female scale and exuviae pale brown, elongate, about 1.5 mm long. Male scale same colour as female scale, about 1.0 mm long.

Adult female on slide small, about 0.7 mm long, oval, widest at about first abdominal segment, membranous except for pygidium; lateral lobes of abdominal segments only moderately developed, without tubercles or lateral spurs.

Pygidium with median lobes well developed, projecting; outer margins sloping, giving the pygidium a pointed appearance; fused except for short apical cleft bearing 2 short gland spines at base, reaching apices of lobes. Outer edges of lobes straight or curved, dentate. Ventrally each lobe with a slender paraphysis curving antero-laterally from near apex, and a stouter paraphysis curving antero-medially from outer basal angle. Other lobes lacking, except for minute sclerotized projections. Gland spines lateral and anterior to median lobes well developed, in pairs on each segment as far forward as segment 4. Marginal macroducts 6 on each side. Dorsal ducts slender, wider than ventral microducts, almost as long as marginal macroducts; in submedian groups on segments 4 to 7, those on segment 7 reduced to 1 or 2; present also in marginal groups anteriorly to mesothorax.

Ventral surface of pygidium with perivulvar pores in 5 groups, totalling 35-40. Microducts present in elongate rows on pygidium and around margins to head; others in rows across abdominal segments, but many marginal ducts on thorax approaching dorsal ducts in width. Duct tubercles absent. Antennae each with 4 long setae. Anterior spiracles each with 2 or 3 disc pores.

Material examined

Holotype female. **PAPUA NEW GUINEA.** C.P., Port Moresby, on leaves of ornamental tree. 1.ix.1959 (*T. Maa*) (BPBM).
Paratypes female. **PAPUA NEW GUINEA.** Same data as holotype, 9 (BPBM), 9 (BMNH).

Comments

This species probably comes closest to *A. micropori* Borchsenius, described from China. It differs in possessing dorsal submedian ducts on the seventh segment, and having a shorter cleft between the median lobes, with the gland spines in the cleft reaching the apices of the lobes. In *A. micropori* the space between the median lobes is wider and longer, and the gland spines do not reach the apices of the lobes.

Andaspis numerata Brimblecombe (Fig. 10)

Andaspis numerata Brimblecombe, 1959: 393.
Andaspis dasi Williams, 1963: 14. [Synonymized by Williams 1980: 260].

Description

The most important characters of this species are the triangular median lobes, each with a long blunt paraphysis at its base, the lobes separated by a narrow space with a pair of minute gland spines between. There are 6 marginal macroducts on each side of the pygidium and the dorsal ducts are very slender, fairly numerous, and form an elongate group on segment 6. Bosses are present near the margins of abdominal segments 1 to 6, and there is a pair on each side of the prothorax. Lateral spurs on segments 1 to 3 of abdomen are present or absent.

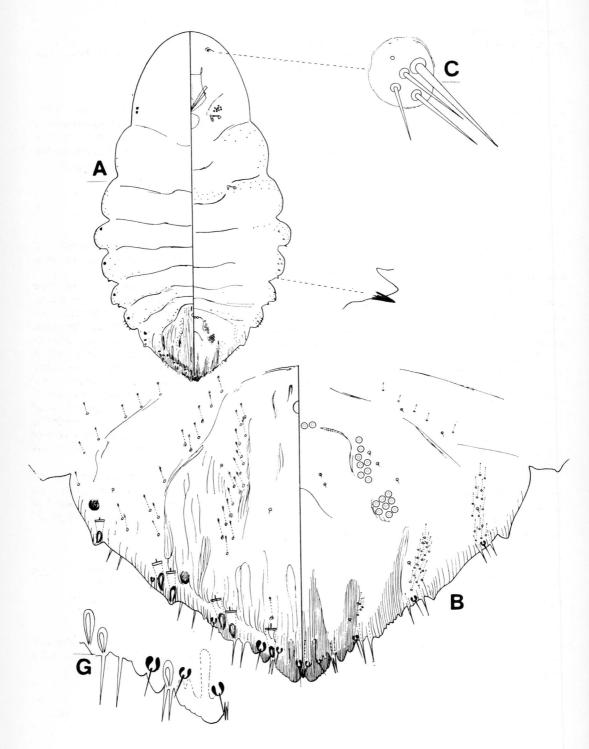

Fig. 10. *Andaspis numerata* Brimblecombe.

Material examined
PAPUA NEW GUINEA. W.H.P.: 1978. On stems of *Camellia sinensis*.
TONGA. Ha'apai Group, Lifuka, 1977; Tongatapu Group, Tongatapu, 1975. On *Hibiscus rosa-sinensis*.

Comments
Williams (1980) has commented on the variability of the lateral spurs of this species. They are sometimes absent entirely, reduced in number or represented by a full complement of 3 pairs. *A. numerata* was described from Queensland and, apart from the records given, it has been intercepted at quarantine from a wider area of the Pacific as listed by Williams (1980). It is often associated with the symbiotic fungus *Septobasidium*.

Andaspis spinosa sp. n. (Fig. 11)

Description
Scale of adult female elongate, dusty brown, exuviae yellow-brown. Male scale not seen.
Slide-mounted specimens up to 1.0 mm long, elongate-oval, widest at first abdominal segment, head and pygidium rounded, body membranous except for pygidium.
Pygidium with well-developed and prominent median lobes separated by a narrow space, each lobe wider than long, almost rectangular but outer edges oblique; with 1 or 2 notches at inner corner, and 3 or 4 notches at outer corner; ventral surface with a wide, blunt paraphysis extending from basal inner and outer angles. Second lobes minute, each with an inner and outer lobule, the outer lobule barely perceptible. Gland spines between median lobes small, scarcely exceeding lobes in length; represented by 2 between each median and second lobe, 1 lateral to each second lobe, and in pairs to about segment 2 of abdomen, except singly on third segment. Marginal macroducts numbering 5 on each side. Dorsal ducts slender; segment 6 on each side usually with a single submarginal duct and a small group of submedian ducts; segments 2 to 5 with ducts in an almost continuous transverse row on each, and marginal groups of ducts on segment 1 and metathorax. Anus situated at base of pygidium. Eye spots modified into relatively large thorn-like spines, projecting anteriorly. Spur-like lateral tubercles on lateral margins of segment 1. Sclerotized thickenings on margins of segments 2 and 4; small projections, each with one or two points, between segments 2 and 3, and between segments 3 and 4.
Ventral surface with perivulvar pores numbering 38-45 in 5 groups. Microducts present on head. Small, short ducts present on margins of mesothorax, and between spiracles and margins on metathorax. Duct tubercles present in small groups of 2-5 on margins of metathorax and segment 1 of abdomen; gland spines also present on margins of segment 1. Antennae each with 2 long setae. Anterior spiracles each with 2 or 3 disc pores.

Material examined
Holotype female. **PAPUA NEW GUINEA.** C.P., Port Moresby, on *Ficus* sp., 3.ix.1959 (*T. Maa*) (BPBM).
Paratypes female. **PAPUA NEW GUINEA.** Same data as holotype. 2 (BPBM), 3 (BMNH).

Comments
This species, with slender dorsal ducts resembling microducts, is different from all others of the genus so far described, in having the eyes modified into robust thorn-like spines directed anteriorly.

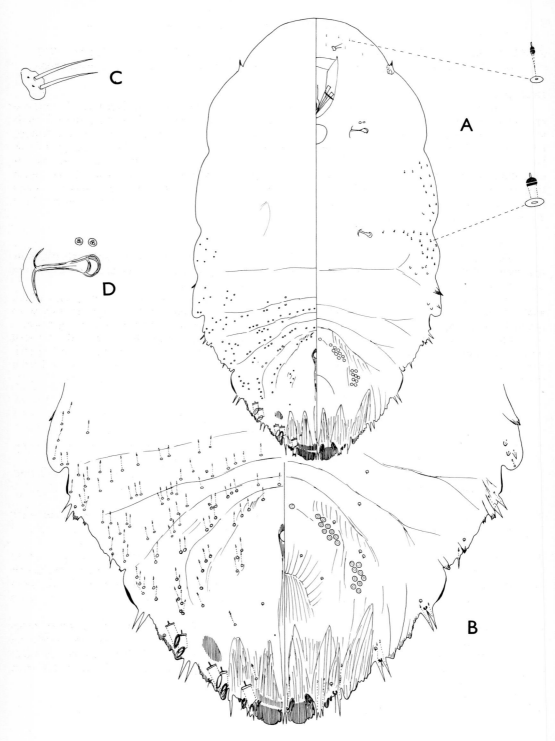

Fig. 11. *Andaspis spinosa* sp. n.

Genus **AONIDIELLA** Berlese & Leonardi

Aonidiella Berlese & Leonardi *in* Berlese 1895: 77; Ferris, 1938: 178; McKenzie, 1938: 1; Balachowsky, 1956: 22. Type-species *Aspidiotus aurantii* Maskell, by original designation.

Description
Prosoma, consisting of cephalothorax and abdominal segment 1 and sometimes segment 2, enlarged and sclerotized at maturity, produced postero-laterally into lobes between which the postsoma tends to retract, giving a more or less reniform outline. Ducts 1-barred. With 3 pairs of well-developed lobes, never bilobed; a fourth pair sometimes represented as small points. Paraphyses present as far laterally as third lobe, mostly shorter than the adjacent lobes. Plates as long as, or exceeding, lobes, those present between lobes always fringed; present as far laterally as fourth lobe or dorsal marginal seta of fifth segment. Anus smaller than a median lobe, situated at about the posterior fifth of the pygidium. Dorsal macroducts long and quite slender, arranged in 3 distinct poriferous furrows on each side. Perivulvar pores present or absent.

Comments
This genus has been synonymized with *Chrysomphalus* by several authors, although it is now regarded as distinct. Unlike in *Aonidiella*, the prosoma in *Chrysomphalus* never becomes sclerotized or reniform, and the paraphyses are normally longer than the adjacent lobes.

Aonidiella is a genus of considerable economic importance, containing several species which attack citrus and occasionally coconuts. The 5 species known from the South Pacific area were probably introduced on citrus fruit and plants, and have been spread between the islands by trade.

Specimens on which records of *A. citrina* (Coquillet) from Papua New Guinea and Fiji are based have been re-examined, and all appear to be *A. aurantii* (Maskell): however, material from Western Samoa on *Erythrina indica* (Reddy, 1970) has not been available. This species has been omitted from the following key.

Key to species of *Aonidiella*

1   Plates lateral to third lobe not fringed, each with 1 long fleshy process; abdominal segments 1 to 3 with a submarginal row or cluster of dorsal ducts either side .... *orientalis* (Newstead)
--   Plates lateral to third lobe fringed; abdominal segments lacking rows of dorsal ducts 2

2   Prevulvar scleroses and apophyses present; perivulvar pores absent .... *aurantii* (Maskell)
--   Prevulvar scleroses and apophyses absent; perivulvar pores present or absent 3

3   4 groups of perivulvar pores present; thoracic tubercles present; fourth lobes present as distinct points .... *eremocitri* McKenzie
--   0-2 groups of perivulvar pores present; thoracic tubercles present or absent; fourth lobes present as very small to distinct points 4

4   2 groups of perivulvar pores present; thoracic tubercles absent or very small; fourth lobes present as very small points .... *comperei* McKenzie
--   Perivulvar pores absent; thoracic tubercles present or absent; fourth lobes present as small to distinct points .... *inornata* McKenzie

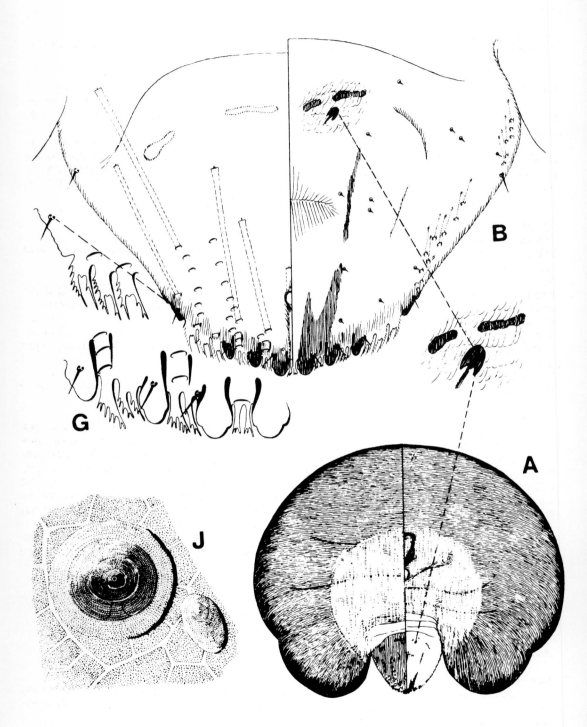

Fig. 12. *Aonidiella aurantii* (Maskell).

Aonidiella aurantii (Maskell) (Fig. 12)

Aspidiotus aurantii Maskell 1879: 199.
Aonidiella aurantii (Maskell), Berlese & Leonardi, *in* Berlese 1895: 83; McKenzie, 1937a: 324; 1938: 6; Ferris, 1938: 179; Balachowsky, 1956: 30.

Description
Scale of adult female circular, flat, translucent yellow-brown, with red-brown insect visible through it, exuviae more or less central. Male scale elongate-oval, paler than female scale, exuviae present towards one end.
Slide-mounted adult female 0.7-1.2 mm long, slightly wider than long; prosoma strongly reniform and heavily sclerotized at maturity; postsoma largely membranous, often retracted between prosomal lobes. Median lobes slightly longer than second and third lobes; fourth lobes slightly indicated. Plates all fringed. Thoracic tubercles absent. Perivulvar pores absent; prevulvar scleroses and apophyses present.

Material examined
COOK IS. Mangaia, 1977; Rarotonga, 1973. On *Citrus paradisi, Rosa indica.*
FIJI. Taveuni, 1977; Viti Levu, 1957. On *Broussonetia papyrifera, Capsicum frutescens, Citrus maxima, Erythrina lithosperma.*
NIUE. 1975. On *Citrus* sp.
PAPUA NEW GUINEA. C.P.: 1956. E.H.P.: 1958. E.N.B.P.: 1951. E.P.: 1959. M.B.P.: 1961. Morobe P.: 1958. S.H.P.: 1983. W.H.P.: 1954. On *Citrus* sp., *C. aurantifolia, C. aurantium, C. maxima, C. reticulata, Cucurbita pepo.*
SOLOMON IS. C.I.P.: Bellona, 1974; Russell Is, Mbanika, 1955. On *Citrus* sp., *C. limon.*
TONGA. Ha'apai Group, Foa, 1977, Lifuka, 1977, Tongatapu Group, Tongatapu, 1974. On *Artocarpus altilis, C. aurantium, C. limon, C. grandis, C. maxima, Ricinus communis.*
VANUATU. Espiritu Santo, 1923. On *Cocos nucifera.*
WESTERN SAMOA. Savai'i, 1975; Upolu, 1977. On 'akone', *Citrus limon,* [?]*Bambusa vulgaris.*

Comments
The California red scale occurs in all the major citrus-producing areas of the world on a wide range of host plants. If unchecked, it is a serious pest of citrus, attacking all the aerial parts of the tree. It kills leaves, twigs and branches, and stunts and pits the fruit so that it falls or is unmarketable. It is most prolific in hot dry conditions. In addition to the material available, it has been recorded from Fiji on *Musa* sp. and *M. sapientum* (Doane, 1909; Jepson, 1915), and on *Cocos nucifera* (Veitch & Greenwood, 1924); from French Polynesia damaging citrus (Reboul, 1976); from Irian Jaya on *Cycas rumphii* (Thomas, 1962a; Reyne, 1961); from New Caledonia on *Cycas* sp., *M. sapientum* (Cohic, 1956) and *Rosa* sp. (Cohic, 1958a); from Papua New Guinea on *Citrus sinensis* (Greve & Ismay, 1983); from Vanuatu damaging citrus (O'Connor, 1969) and *Cocos nucifera,* although the latter probably refers to *A. eremocitri* (Williams & Butcher, 1987); from Wallis Is on *Artocarpus incisa* (Gutierrez, 1981); and from Western Samoa on *Erythrina indica* (Reddy, 1970). Firman (1982) remarks that *A. aurantii* does not seem to be a serious problem in any of the Pacific countries.

Aonidiella comperei McKenzie (Fig. 13)

Aonidiella comperei McKenzie, 1937a: 327; 1938: 8.

Description
Scale of adult female circular and smooth, translucent yellow with yellow-brown insect visible through it, rather brittle; exuviae more or less central. Male scale smaller than female scale, oval, pale yellow, with acentric exuviae present.

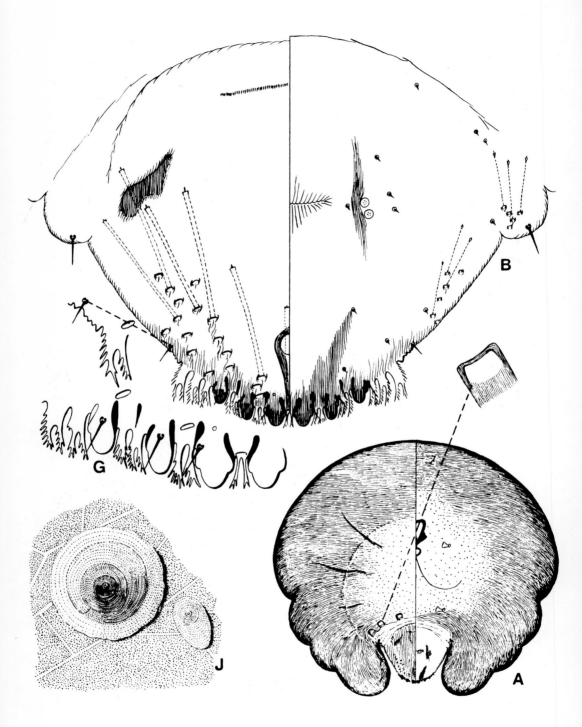

Fig. 13. *Aonidiella comperei* McKenzie.

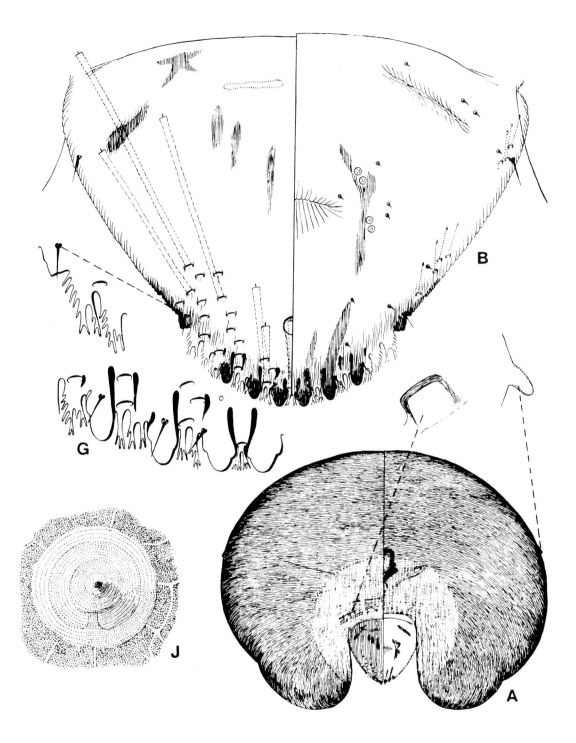

Fig. 14. *Aonidiella cremocitri* McKenzie.

Adult female, slide-mounted, 0.9-1.1 mm long, usually wider than long; prosoma strongly reniform and heavily sclerotized at maturity; postsoma less sclerotized than prosoma or membranous, often retracted between prosomal lobes. Median lobes usually larger than second and third lobes; fourth lobes represented by small to moderate-sized points. Plates all fringed. Thoracic tubercles small or absent. Perivulvar pores present in 2 groups, usually 1 or 2 at each postero-lateral position; prevulvar scleroses and apophyses absent.

Material examined
KIRIBATI. Butaritari, 1976; Tarawa, 1976. On *Morinda citrifolia, Musa* sp., *Pluchea odorata*.
PAPUA NEW GUINEA. E.H.P.: 1959. Morobe P.: 1979. On *Annesijoa* sp., *Diospyros* sp., *Ficus* sp.

Comments
A. comperei was described from Sri Lanka, and has also been collected from the USA, Malaysia and Sarawak. It has not been recorded as causing any damage in the South Pacific area.

This species is very close to *A. eremocitri* and *A. inornata*, its most distinctive characters being the absence or smallness of thoracic tubercles, the small points of the fourth lobes, and the restriction of perivulvar pores to 2 postero-lateral groups of only 1-2 pores each.

Aonidiella eremocitri McKenzie (Fig. 14)

Aonidiella eremocitri McKenzie, 1937b: 177; 1938: 9.

Description
Scale of adult female circular, translucent pale brown with yellow insect visible through it, exuviae more or less central. Male scale smaller than female scale, elongate-oval, with exuviae present towards one end.

Slide-mounted adult female 0.6-1.2 mm long, usually wider than long; prosoma strongly reniform and heavily sclerotized or membranous, often retracted between prosomal lobes. Median lobes slightly larger than, or same size as, second lobes; third lobes slightly smaller than second; fourth lobes represented by moderate-sized to relatively large points. Plates all fringed. Distinct sclerotized thoracic tubercles present on or near the margin on either side. Perivulvar pores present in 3-4 groups, up to 3 in each antero-lateral group and 1-4 in each postero-lateral group; prevulvar scleroses and apophyses absent.

Material examined
FIJI. Viti Levu, 1955. On *Barringtonia* sp.
PAPUA NEW GUINEA. W.N.B.: 1956. On *Cocos nucifera*.
SOLOMON IS. G.P.: Guadalcanal, 1950. W.P.: New Georgia Is, Rendova, 1934. On *Cocos nucifera*.
VANUATU. Espiritu Santo, 1923. On *C. nucifera, Citrus* sp.

Comments
A. eremocitri was originally described from Australia; it has also been collected from Sarawak, Malaysia and the USA on citrus, in addition to the countries given above. In the South Pacific area it has been recorded from Fiji on *Bischofia javanica* (Hinckley, 1965), and from the Solomon Is on citrus (Lever, 1968) in addition to the hosts given above. Williams & Butcher (1987) point out that damage to *Cocos nucifera* in Vanuatu attributed to *Aonidiella* was probably due to *A. eremocitri* rather than to *A. aurantii*.

This species is extremely close to *A. comperei* and *A. inornata*, the most useful distinguishing characters being the well-developed thoracic tubercles and fourth lobe points, and more than 2 groups of perivulvar pores.

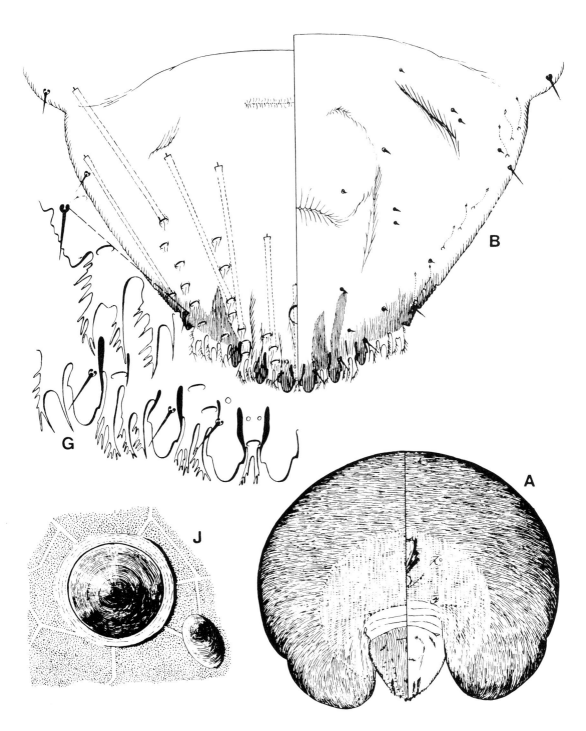

Fig. 15. *Aonidiella inornata* McKenzie.

Aonidiella inornata McKenzie (Fig. 15)

Aonidiella inornata, McKenzie 1938: 10.

Description
Scale of adult female circular, translucent pale brown, with outline of insect visible through it; margin descending abruptly on to plant at edge of insect; exuviae more or less central. Male scale smaller than female scale, elongate-oval, yellow-brown, with exuviae located towards one end.
Slide-mounted adult female 0.5-0.9 mm long, usually wider than long; prosoma less sclerotized than prosoma or membranous, often retracted between prosomal lobes. Median lobes rather larger than, or some size as, second lobes; third lobes slightly smaller than second; fourth lobes represented by small to relatively large points. Plates all fringed. Thoracic tubercles of varying sizes present or absent. Perivulvar pores, prevulvar scleroses and apophyses absent.

Material examined
FIJI. Taveuni, 1913; Viti Levu, 1949. On *Allemanda* sp., *Barringtonia* sp., *Bischofia javanica*, *Musa* sp., *Piper aduncum*, *Platanocephalus morindaefolius*, *Vitis vinifera*.
IRIAN JAYA. Biak, 1959. On ?.
KIRIBATI. Tarawa, 1976. On *Cassia* sp., *Euphorbia* sp., *Musa* sp., *Pandanus odoratissimus*, *Plumeria rubra*.
PAPUA NEW GUINEA. C.P.: 1959. E.H.P.: 1959. Morobe P.: 1979. On *Annesijoa* sp., cultivated shrub, *Elettaria cardamomum*, *Melaleuca* sp., ?*Polygonum* sp., *Salacea* sp.
VANUATU. Efate, 1983. On *Musa* sp., *Piper methysticum*.
WESTERN SAMOA. Upolu, 1975. On *Musa* sp.

Comments
A. inornata was first described from the Philippines, and in addition to the above countries, has been collected from the USA, Cayman Is, East Africa (Pemba), Thailand, Malaysia and Sarawak. In the South Pacific region it has been recorded from Fiji on *Allemanda cathartica* and *Veitchia joannis* (Hinckley, 1965) in addition to the hosts listed above. No damage has been reported.
In view of the variability in the presence or absence of perivulvar pores which can occur within one species, e.g. in *Oceanaspidiotus pangoensis*, there does not seem to be sufficient grounds to separate *A. inornata* as a distinct species. McKenzie (1938) does not mention thoracic tubercles in his description, a useful character to distinguish *A. comperei* from *A. eremocitri*; in material identified as *A. inornata*, they may be present or absent. The paratype specimens of *A. inornata* have small thoracic tubercles. It seems likely that specimens which have been identified as *A. inornata* may be a mixture of *A. comperei* and *A. eremocitri* without perivulvar pores. Further study is needed to decide whether *A. inornata* should be synonymised with one of these other two species.

Aonidiella orientalis (Newstead) (Fig. 16)

Aspidiotus orientalis Newstead, 1894: 26.
Aonidiella orientalis (Newstead), McKenzie, 1937a: 327; McKenzie, 1938: 12; Ferris, 1938: 180; Balachowsky, 1956: 42.

Description
Scale of adult female circular, flat, almost white to pale brown or yellow; yellow to dark brown exuviae more or less central. Male scale elongate-oval, yellow exuviae near one end.
Slide-mounted adult female 1.0-1.4 mm long; prosoma pyriform expanding to subcircular and becoming moderately sclerotized around margins at maturity; free abdominal segments forming small lobes on either side, membranous, often retracted into prosoma; pygidium quite well sclerotized dorsally. Median lobes distinctly larger than second lobes; fourth lobes represented by small point either side. Plates lateral to third lobes not fringed, each with long fleshy process present at

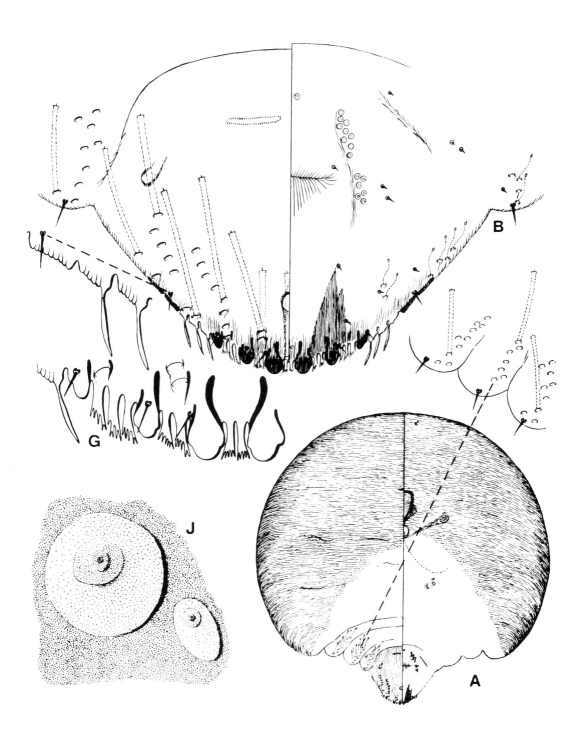

Fig. 16. *Aonidiella orientalis* (Newstead).

mesal angle. Abdominal segments 1 to 3 with a submarginal row or cluster of dorsal macroducts present on each side. Thoracic tubercles minute. Perivulvar pores present in 4 or 5 groups, total of 19-32 present; prevulvar scleroses and apophyses absent.

Material examined
PAPUA NEW GUINEA. C.P.: 1980. On *Carica papaya*.

Comments
　　A. orientalis is a tropical and subtropical species with a wide distribution and polyphagous habits, attacking almost any host except conifers; Balachowsky (1956) implies it can be a pest. It is easily distinguished from other species of *Aonidiella* found in the South Pacific area as it is less heavily sclerotized and fails to become strongly reniform at maturity, and has conspicuous submarginal groups of macroducts on either side of the free abdominal segments. It is similar to *A. simplex* (Grandpré & Charmoy), but the latter lacks perivulvar pores. It does not appear to be very common in Papua New Guinea, and has not been recorded causing damage there. The species appears to be new to the South Pacific area and could possibly spread.

Genus **ASPIDIELLA** Leonardi

Aspidiella Leonardi, 1898b: 50. Type-species *Aspidiotus sacchari* Cockerell, by original designation.

　　This genus is closely related to *Aspidiotus* and its relatives in possessing 1-barred ducts and plates, but differs in having only 2 pairs of well-developed lobes, the third pair represented by short points. It also differs in having a marginal zone of ducts on the dorsal and ventral margins of the pygidium. These ducts are more or less of equal size. A few species are recognized in the genus, mainly occurring on monocotyledons. Two species, *A. hartii* (Cockerell) and *A. sacchari* (Cockerell), are tropicopolitan. Both species are known in the South Pacific region and can be separated as follows:

Key to species of *Aspidiella*

1. Median lobes with poorly developed internal scleroses at base, parallel. Plates present lateral to position of third lobes. Margin of pygidium weakly crenulate
　　... *hartii* (Cockerell)
-- Median lobes with well-developed internal scleroses at base, diverging antero-laterally. Plates absent lateral to position of third lobes. Margin of pygidium deeply crenulate
　　... *sacchari* (Cockerell)

Aspidiella hartii (Cockerell) (Fig. 17)

Aspidiotus hartii Cockerell, 1895: 7.
Aspidiotus (Aspidiella) hartii (Cockerell), Leonardi, 1898b: 62.
Aspidiella hartii (Cockerell), MacGillivray, 1921: 404; Ferris, 1938: 188; Balachowsky, 1958: 283.

Description
　　Scale of female circular, brownish with exuviae lighter. Male scale oval, same colour as female scale.
　　Adult female recognizable by the presence of 2 pairs of sclerotized lobes, each third lobe represented by a short unsclerotized point; weak parallel scleroses present at bases of median lobes. Plates fimbriate between lobes and lateral to third lobes. Dorsal and ventral ducts present

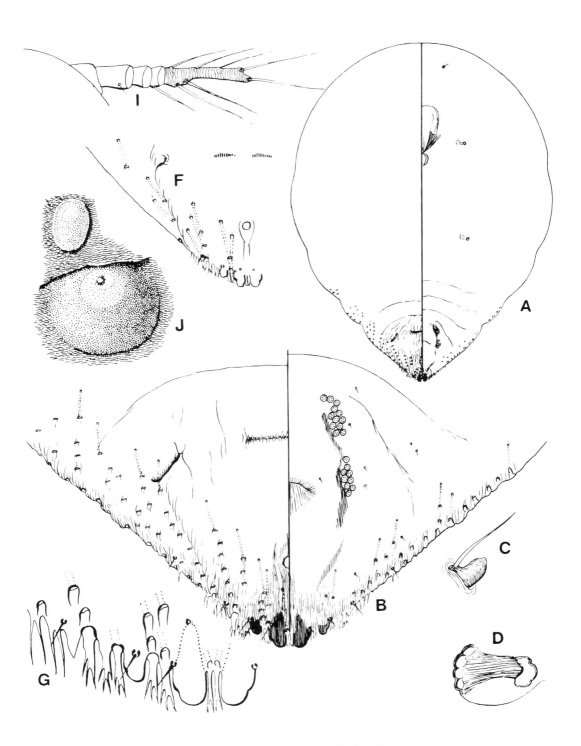

Fig. 17. *Aspidiella hartii* (Cockerell).

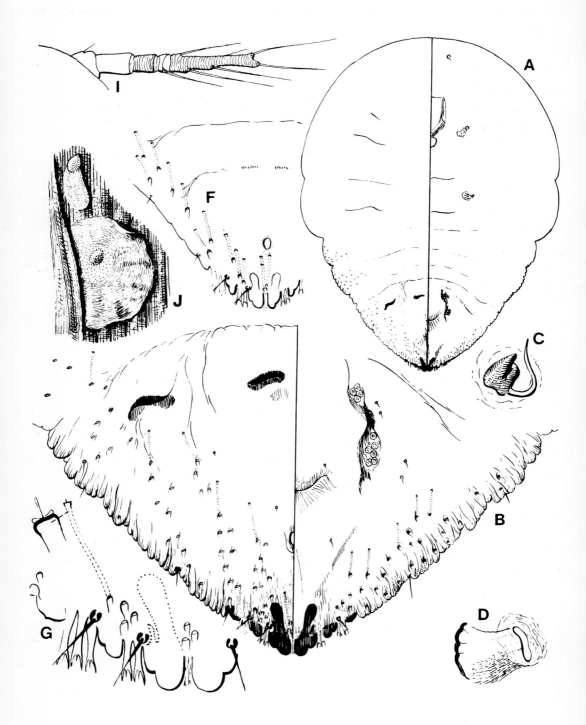

Fig. 18. *Aspidiella sacchari* (Cockerell).

in a submarginal zone as far forward as abdominal segment 2; ventral ducts slightly smaller than dorsal ducts. Perivulvar pores in 4 groups.

Material examined
FIJI. Gau, 1943; Kadavu, 1943; Lau Group, Mago, 1976; Viti Levu, 1914. On *Dioscorea* sp., *D. alata, Zingiber officinale.*
PAPUA NEW GUINEA. Morobe P.: 1976. On *Dioscorea* sp.
SOLOMON IS. G.P.: Guadalcanal, 1976. M.U.P.: San Cristobal, 1984. W.P.: New Georgia Is, Kolombangara, 1982. On *D. alata, Ipomoea batatas.*
TONGA. Tongatapu Group, Tongatapu 1973. On *D. alata.*
VANUATU. Malekula, 1983. On *D. alata.*

Comments
This tropicopolitan species is usually associated with yams of the genus *Dioscorea*, particularly with yams in storage. It is also a pest of other root crops, especially members of the family Zingiberaceae when grown in close proximity with yams; it has been intercepted at quarantine inspection on *Curcuma longa* and ginger. Another species, *A. zingiberi* Mamet, seems to be identical, but it has not been recorded from the South Pacific area.

A. hartii has been recorded from Fiji on *D. esculenta* (Lever, 1945b); from the Isle of Pines and Ouvéa in the Loyalty Is on *D. alata*, encrusting stored tubers and drying the tissues so they become fibrous (Cohic, 1958a), and from New Caledonia on *D. sativa* (Cohic, 1956).

Aspidiella sacchari (Cockerell) (Fig. 18)

Aspidiotus sacchari Cockerell, 1893b: 255.
Aspidiotus (Aspidiella) sacchari Cockerell, Leonardi, 1898b: 232.
Aspidiella sacchari (Cockerell), MacGillivray, 1921: 405; Ferris, 1938: 189; Balachowsky, 1958: 284.

Description
Female scale pale brown, circular with yellowish exuviae subcentral. Male scale same colour as that of female, elongate, exuviae terminal.

Adult females resemble *A. hartii* in many respects, but the scleroses at the bases of the median lobes are much more elongate and well defined, diverging antero-laterally. Furthermore, there are no plates beyond the third lobes, and the margin is much more crenulate.

Material examined
COOK IS. Rarotonga, 1977. On Poaceae.
FIJI. Viti Levu, 1962. On *Alocasia macrorhiza, Ischaemum* sp., *Saccharum officinarum.*
PAPUA NEW GUINEA. C.P.: 1979. On Poaceae.
SOLOMON IS. G.P.: Guadalcanal, 1978. On *Brachiaria mutica.*
WESTERN SAMOA. Savai'i, 1977. On *S. officinarum.*

Comments
This is normally a grass-infesting species, but it has been reported from Africa on *Cocos* sp. It is widely known, however, as a pest on sugarcane, and has been reported from many of the sugarcane-growing areas of the world.

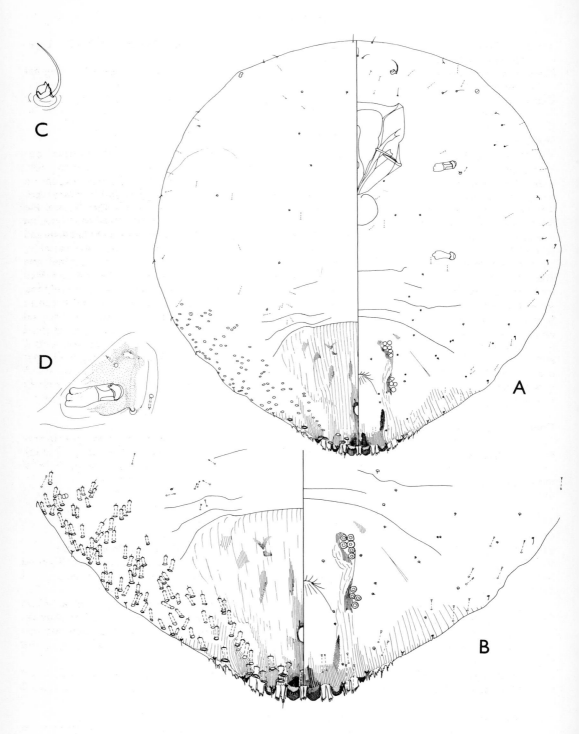

Fig. 19. *Aspidiotus cochereaui* Matile-Ferrero & Balachowsky.

Genus **ASPIDIOTUS** Bouché

Aspidiotus Bouché, 1833: 52; Ferris, 1938: 190; Balachowsky, 1956: 49; Takagi, 1969: 62. Type-species *Aspidiotus nerii* Bouché by subsequent designation.
Cryptophyllaspis Cockerell, 1897a: 14. Type-species *Aspidiotus occultus* Green by original designation.
Temnaspidiotus MacGillivray, 1921: 387. Type-species *Aspidiotus excisus* Green by original designation.

Description
Body broadly pyriform, normally membranous. Ducts 1-barred. With 3 pairs of well-developed lobes, these never bilobed; fourth lobes totally absent. Median lobes elongate or as long as wide, well sclerotized, subapically notched on outer or both edges. Basal sclerosis sometimes present on median lobe; variably sclerotized area extending from median lobe bases into pygidium. Medium lobes may be sunk in recess between second lobes, sometimes making apex of pygidium straight. Second and third lobes usually clearly notched at least on outer edges. Plates well developed, present laterally as far forward as the fifth segment; 2 present between median lobes, 3 between median and second lobe, 3 between second and third lobes; number of plates lateral to third lobe variable; fringes well developed, never clavate, becoming simpler beyond third lobe. Dorsal marginal seta present at outer basal corner of median lobe sometimes elongate; those present on second and third lobes sometimes thickened, widest at base, tapering gradually to apex. Anus fairly prominent, situated towards apex of pygidium. Dorsal macroducts variable in length, quite short and broad in type-species; marginal ducts arranged as 1 present between median lobes, 2 between median and second lobes, 2 between second and third lobes and 2-4 beyond third lobes on each side; submarginal pygidial macroducts forming intersegmental rows in poriferous furrows at least between segments 6 and 7, and 7 and 8 on each side; ducts rather scattered in some species. Dorsal submarginal bosses present on either side of segments 1 and 3, usually membranous. Perivulvar pores present in 2-5 groups (usually 4). Anterior spiracles each with an associated area of dermal granulation, a smooth tubercle and 1-3 microducts. Antennae often with sclerotized spurs, each bearing one robust seta.

Comments
Over 30 species are included in the genus at present, although some may not be congeneric with the type-species. Takagi (1969) considers the natural distribution to be Afro-Asiatic, although some species are tropicopolitan. Eight species are known from the South Pacific area, three of them introduced. Five are apparently endemic: four of these are described here as new.

Key to species of *Aspidiotus*

1	Prepygidial macroducts absent; pygidial macroducts more than 16 times as long as wide	***destructor*** Signoret
--	Prepygidial macroducts present; pygidial macroducts less than 13 times as long as wide	2
2	2 plates present between second and third lobes; prepygidial macroducts numerous, about 40 on each side of abdomen	***cochereaui*** Matile-Ferrero & Balachowsky
--	3 plates present between second and third lobes; seldom more than 20 prepygidial macroducts on either side of the abdomen	3
3	Submarginal as well as marginal prepygidial macroducts present	4
--	Only marginal prepygidial macroducts present	6
4	Third lobe rounded; 1-3 marginal pygidial macroducts present beyond third lobe	***pacificus*** sp. n.
--	Third lobe pointed; 4 marginal pygidial macroducts present beyond third lobe	5

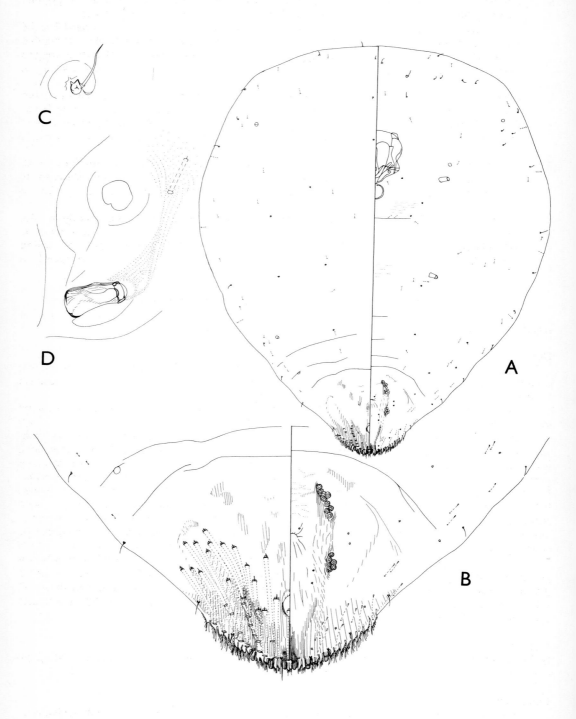

Fig. 20. *Aspidiotus destructor* Signoret. Solomon Is, taken from *Cocos nucifera*.

5	1 microduct present in the area of dermal granulation by each anterior spiracle; prepygidial and pygidial macroducts of similar size which may be 3.5-6 times as long as wide .... *nerii* Bouché	
--	3 microducts present in the area of dermal granulation by each anterior spiracle; prepygidial macroducts appreciably smaller than pygidial macroducts, which may be 7-10 times as long as wide .... *maddisoni* sp. n.	
6	Median lobes lacking basal scleroses; second lobe notches absent or very slight; median lobes strongly recessed; prosoma never reniform .... *excisus* Green	
--	Basal scleroses usually present on median lobes; second lobe with two notches; median lobes may or may not be recessed; prosoma sometimes reniform.. 7	
7	2-22 perivulvar pores present in 1-4 groups; prosoma in mature specimens reniform, with lateral lobes more or less enclosing pygidium .... *macfarlanei* sp. n.	
--	More than 23 perivulvar pores in 4-5 groups; large prosomal lobes absent 8	
8	6-7 plates and 1-3 marginal pygidial macroducts present beyond third lobe; pygidium with strongly sclerotized areas .... *pacificus* sp. n.	
--	8-11 plates and 4 marginal pygidial macroducts present beyond third lobe; pygidium lacking strongly sclerotized areas .... *musae* sp. n.	

Aspidiotus cochereaui Matile-Ferrero & Balachowsky (Fig. 19)

Aspidiotus cochereaui Matile-Ferrero & Balachowsky, 1973: 241.

Description
Scale of adult female circular, slightly convex, translucent white to fawn, with more or less central orange exuviae sometimes dusted with white wax. Male scale elongate-oval with exuviae located near one end, colouration similar to female scale.
Adult female, slide-mounted, about 0.7 mm long, subcircular; cuticle not sclerotized except for parts of pygidium.
Pygidium broad and rounded, only lightly sclerotized. Median lobes about as long as wide, slightly notched subapically on either side, with a smooth margin; separated by 0.5-1 median lobe width; lacking basal scleroses. Second lobes slightly smaller than median, with subapical notch present on outer edge. Third lobes reduced, pointed, with a subapical notch on outer margin. Plates between lobes well developed, as long as lobes, with short fringe; only 2 plates present between second and third lobes on each side; 3-5 plates present beyond third lobe, reduced, with sparse fringing. Anus wide, oval, slightly wider than median lobe, situated around posterior quarter of pygidium. Dorsal macroducts short, up to 6 times as long as wide, only marginal or submarginal, present as far forward as segment 1; more than 45 present on each side.
Venter of pygidium with 4 groups of perivulvar pores present, 2-8 in each antero-lateral group, 1-6 in each postero-lateral group, 12-22 in total. Microducts few, submarginal; about 2 present by each spiracle and a few on prosomal margin. Spiracles lightly sclerotized; sometimes 1-2 small tubercles present near each anterior spiracle. Antennae each bearing 3 lightly sclerotized spurs and 1 fairly large seta.

Material examined
NEW CALEDONIA. New Caledonia, 1966. On *Dracophyllum* sp., *D. ramosum*, *Nothofagus codonandra*.

Comments
Two paratype specimens, and additional material on *D. ramosum*, were kindly made available by D. Matile-Ferrero of the Muséum National d'Histoire Naturelle, Paris. Apart from a slightly wider

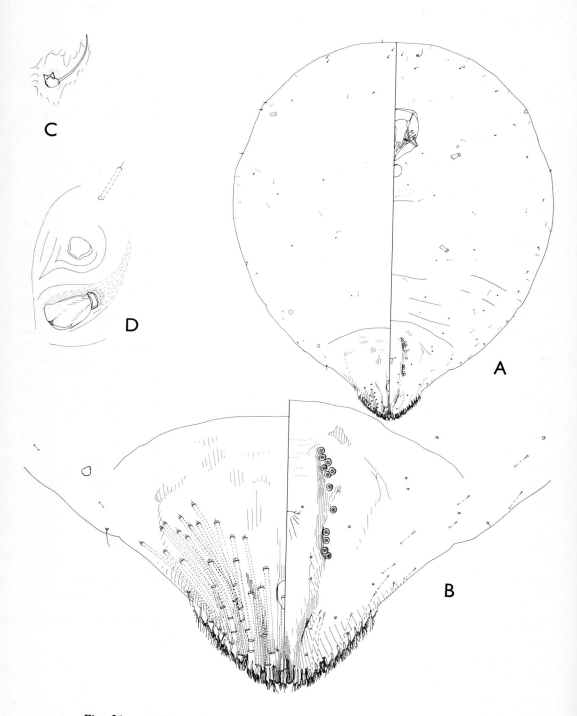

Fig. 21. *Aspidiotus destructor* Signoret. Fiji, taken from *Carica papaya*.

range in numbers of perivulvar pores, and an occasional specimen with 2 median marginal macroducts, the material fits the description well. This species seems to be confined to New Caledonia.

Aspidiotus destructor Signoret (Figs 20, 21, 22, 23)

Aspidiotus destructor Signoret, 1869b: 120; Ferris, 1941b: 51; Takagi, 1969: 65.
Aspidiotus transparens Green 1890: 22. Lectotype female, Guyana (BMNH), here designated [examined].
Aspidiotus cocotis Newstead 1893: 186. Lectotype female, Sri Lanka (BMNH), here designated [examined].
Temnaspidiotus destructor (Signoret) Borchsenius 1966: 270.

Description
Scale of adult female oval to circular, fairly flat, very thin and almost transparent; pale yellow exuviae more or less central. Male scale similar colour to that of female but smaller and more oval.
Adult female 0.6-1.1 mm long, ovoid to broadly pyriform, membranous except for parts of pygidium.
Pygidium moderately produced to fairly acute, quite lightly sclerotized; apex sometimes recessed between second lobes. Median lobes variable in size, about twice as long as wide, with tips turned very slightly inwards; notched subapically either side; apex slightly slanting outwards, sometimes serrate; separated by two fifths to four fifths lobe width; lacking dense basal scleroses. Second lobes often narrower than, and sometimes surpassing, median lobes; third lobes similar size or smaller than second lobes; second and third lobes strongly notched on outer margin and constricted basally. Dorsal seta present on outer basal corner of median lobe slender, normally more than 1.4 times lobe length. Plates extending beyond lobe apices, deeply and finely fringed; 6-10 (usually 8) plates present beyond third lobe on each side, outer 2 or 3 slightly simpler. Anus ellipsoid to oval, 1-2 times wider and longer than median lobe; situated at about posterior quarter to third of pygidium. Dorsal macroducts long, 17-45 times as long as wide, forming 3 poriferous furrows on each side between segment 7 and 8, 6 and 7 and on the fifth segment; pygidial macroducts not opening anterior to anus. Short submarginal microducts present, sparse, more numerous near eye and submarginal abdominal tubercles.
Venter of pygidium with 4-5 groups of perivulvar pores present; occasionally 1-3 in median group, 6-12 in each antero-lateral group, 3-8 in each postero-lateral group, total of 15-34. Marginal and submarginal microducts few, most numerous on pygidium and abdominal segments; 1 present near each anterior spiracle, sometimes several near each posterior spiracle. Spiracles moderately sclerotized. Antennae quite small, each with 3 short spurs and 1 fairly large seta.

Material examined
AMERICAN SAMOA. Tutuila, 1984. On *Artocarpus altilis*, *Cocos nucifera*.
FIJI. Lau Group, Namuka, 1945; Taveuni, 1931; Vanua Levu, 1976; Viti Levu, 1914; Wakaya, 1946; Yasawa Group, Viwa, 1976. On *Annona reticulata*, *Carica papaya*, *Cocos nucifera*, *Elaeis guineensis*, *Eugenia* sp., *E. malaccensis*, *Euphorbia* sp., *Musa* sp., *Persea americana*, *Piper* sp., *P. macgillivrayi*, *P. methysticum*, *Platanocephalus morindaefolius*, *Psidium guajava*, *Theobroma cacao*, *Vigna unguiculata*, *Xanthosoma sagittifolium*.
FRENCH POLYNESIA. Marquesas Is, Tahuata, 1930; Society Is, Moorea, 1925, Tahiti, 1897; Tuamotu Is, Hikueru, 1925. On *Carica papaya*, *Cocos nucifera*, *Plumeria rubra*.
IRIAN JAYA. Biak, 1959. On ?
PAPUA NEW GUINEA. E.N.B.P.: 1960. E.S.P.: 1954. Madang P.: 1979. Morobe P.: 1959. W.N.B.P.: 1956. On *C. nucifera*, *Hibiscus* sp.
SOLOMON IS. W.P.: New Georgia Is, Gizo, 1984, Rendova, 1934. On *C. nucifera*.
VANUATU. Efate, 1978; Espiritu Santo, 1983; Malekula, 1983. On *Artocarpus altilis*, *Breynia disticha*, *C. nucifera*, *Eucalyptus deglupta*, *Musa* sp., *Piper* sp., *P. methysticum*.

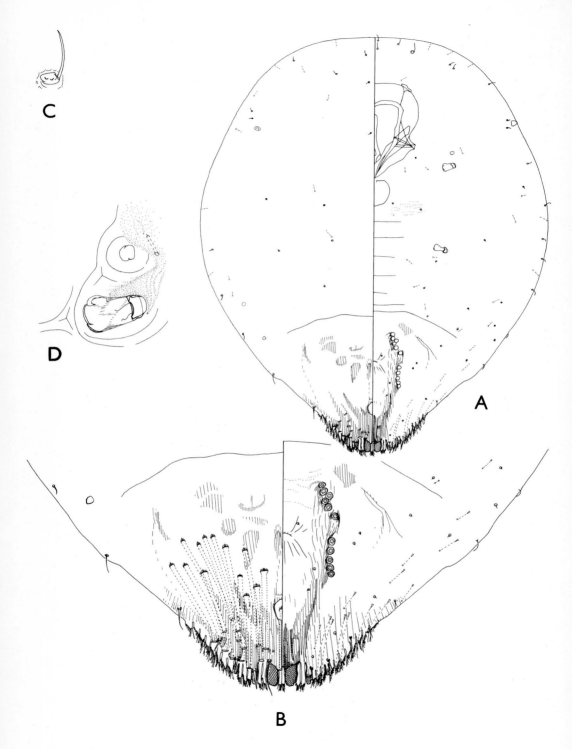

Fig. 22. *Aspidiotus destructor* Signoret. Fiji, taken from *Persea americana*.

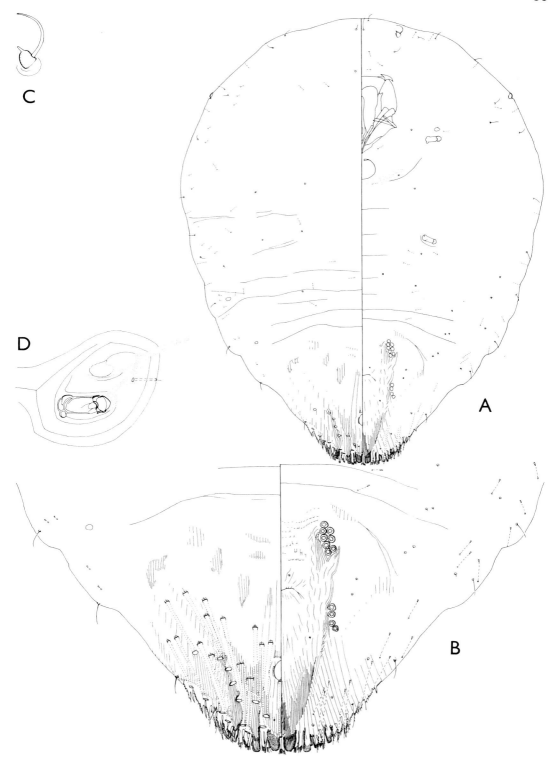

Fig. 23. *Aspidiotus destructor* Signoret. Western Samoa, taken from *Citrus maxima*.

WESTERN SAMOA. Savai'i, 1975; Upolu, 1975. On *Cinnamomum zeylanicum, Citrus grandis, C. maxima, Cocos nucifera, Cycas* sp.

Comments
    A. destructor, commonly known as the transparent scale, is a notorious pest of coconut and other crops, and is recorded from tropical and subtropical regions worldwide. Considerable variation occurs in the relative sizes of median and second lobes, macroduct numbers and lengths, and length of the dorsal seta on the outer basal corner of the median lobe; some samples contain every degree of variation, however, implying there is only one species involved.
    Green's original slide of *A. transparens,* labelled 'type', from Sri Lanka, Jallawakelle, on ? (BMNH) contains 7 specimens; 1 is clearly marked as lectotype, and the other 6 are here designated paralectotypes. The morphology of this material falls within the range of variation observed in *A. destructor,* as does that of the 16 adult specimens on the slide of *A. cocotis* Newstead marked 'co-type', from Guyana, Demarara (=Georgetown), on cocoa palm leaves, Sept. 1892, Newstead coll. (BMNH). One of the latter specimens is clearly marked as lectotype, and the remaining 15 are here designated paralectotypes.
    S. Takagi of the Entomological Institute, Faculty of Agriculture, Hokaido University, Saporo, Japan, kindly loaned authentic material of *A. watanabei* Takagi; the range of variation of *A. destructor* from the South Pacific area encompasses that of *A. watanabei* (Fig. 22), so there is no way of deciding whether the latter species is represented. In the absence of biological information, it is impossible to decide whether *A. destructor* ssp. *rigidus* Reyne occurs in the area, since it is morphologically indistinguishable from *A. destructor* (Reyne, 1948).
    In addition to the distribution given above, and to the remarks in the introduction, Risbec (1942) and Borchsenius (1966) mention *A. destructor* as being present in New Caledonia; these records were based on a misidentification of *A. nerii* (Cohic, 1958a). The species was first recorded there as a new introduction, damaging coconuts, by Brun & Chazeau (1984). *A. destructor* has also been recorded from American Samoa damaging coconuts (Ikin, 1984); from Easter I. (Charlín, 1973); from Irian Jaya (Reyne, 1961); from Vanuatu (Rao *et al.,* 1971); and from Wallis Is, probably introduced shortly after establishment of an air link with Fiji (Cohic, 1959). Maddison (1976) describes *A. destructor* as a pest on *Dioscorea* spp., *Persea americana* and *Spondias dulcis* throughout the tropical South Pacific area, and on *Zingiber officinale* in Fiji. Introduced parasites and predators play an important part in the control of this pest (Rao *et al.,* 1971; Macfarlane, 1986). The literature for the South Pacific area is too lengthy to list here, but the following hosts are given, not included above:-   *Albizia lebbek, Aleurites moluccana, A. triloba, Allemanda hendersoni, Alpinia nutans, Annona cherimolia, A. muricata, A. reticulata, A. squamosa,* Arecaceae, *Artocarpus incisa, Averrhoa carambola, Barringtonia* spp., *B. asiatica, Brassica chinensis, B. napus, B. oleracea, Calophyllum inophyllum, Canna indica, Capsicum* sp., *C. annuum, C. frutescens, C. minimum, Cassia* sp., *C. nodosa, C. occidentalis, C. tora, Ceiba pentandra, Colocasia esculenta, Crotalaria mucronata, C. saltiana, Cucumis sativus,* Cucurbitaceae, *Dillenia biflora, Dioscorea nummularia, E. pulcherrima, Ficus* spp., *Heliconia ?bahai, Hevea brasiliensis, Inocarpus fagifer, Jasminum* sp., *J. officinale, J. sambac, Lantana camara, Laportea photiniphylla, Litsea vitiensis, Lycopersicon esculentum, Macaranga seemannii, Mangifera indica, Musa* spp., *M. paradisiaca, M. sapientum,* Orchidaceae, *Passiflora quadrangularis, Phoenix* sp., *Physalis lanceolata, P. peruviana, Piper puberulum, Plumeria acutifolia, Psidium guajava, Raphanus sativus, Saccharum officinarum,* Solanaceae and *Solanum melongena.*

Aspidiotus excisus Green (Fig. 24)

Aspidiotus excisus Green, 1896a: 53; Ferris, 1941b: 53; Takagi, 1969: 70. Lectotype female, Sri Lanka (BMNH), here designated [examined].
Temnaspidiotus excisus (Green), MacGillivray, 1921: 403; Borchsenius, 1966: 271.

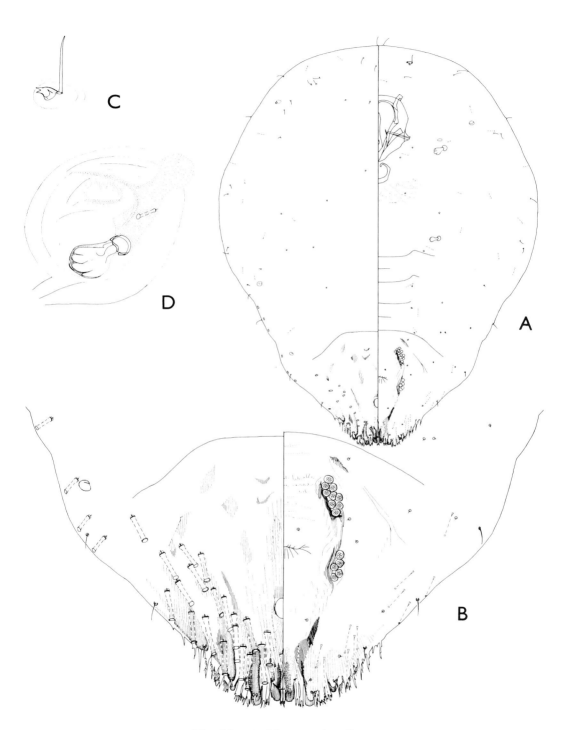

Fig. 24. *Aspidiotus excisus* Green.

Description
Scale of adult female approximately circular, often with irregular margin, convex, translucent, fawn, with approximately central slightly yellow exuviae. Male scale elongate-oval with exuviae near one end, colouration as for female scale.

Adult female, slide-mounted, about 0.75 mm long, pyriform; prosoma membranous.

Pygidium broad, some parts quite strongly sclerotized; apex deeply recessed between second lobes. Outer margin of median lobe about twice as long as that of second lobe. Median lobes quite large, about twice as long as wide, notched subapically on either side, each with a finely serrate apex; separated by about one third lobe width; lacking basal scleroses. Second lobes slightly smaller than median, yet surpassing median lobes due to apical recess; rounded; subapical notches absent or very slight if present. Third lobes one quarter to one third size of second, rounded. Plates extending beyond lobe apices, variously fringed; 6-8 plates present beyond third lobe, outer ones simple. Anus oval, wider and longer than median lobe, situated around posterior third to quarter of pygidium. Dorsal macroducts short, 4-8 times as long as wide; sometimes forming 3-4 poriferous furrows on each side, 1 between segments 7 and 8, and 6 and 7, and 2 on fifth segment; submarginal macroducts sometimes extending to segment 3 on each side. Segments 2 and 3 each with 3-5 marginal macroducts present on either side.

Venter of pygidium with 4 (rarely 5) groups of perivulvar pores present, 0-3 in median group, 4-15 present in each antero-lateral group, 2-9 in each postero-lateral group, total of 27-39. Microducts few, most numerous on pygidium; 1-2 present by each spiracle. Spiracles moderately sclerotized. Antennae each with sclerotized single or double spur and 1 robust seta.

Material examined
PAPUA NEW GUINEA. C.P.: 1978. E.S.P.: 1959. On *Carica papaya*, *Citrus* sp., *C. aurantifolia*, *Euphorbia* sp.

Comments
A. excisus was described from Sri Lanka, Punduloya, on *Cyanotis pilosa* (BMNH). Green's original slide labelled 'type' contains 7 specimens, one of which is clearly marked as lectotype, and the other 6 are here designated paralectotypes. The original material seems to represent one end of a range of variation, of which the Pacific material represents the other extreme. The Pacific material tends to have smaller median lobes, a smaller anus further from them, and fewer dorsal macroducts less randomly arranged than in the type material. Considerable variation occurs within samples, suggesting there is only one species involved.

The apical recess in the pygidium is a feature shared in varying degrees with *A. maddisoni*, *A. macfarlanei*, *A. musae*, *A. pacificus*, and some specimens of *A. destructor*. The records of *A. excisus* from Fiji (Jepson, 1915, Green, 1915, Hinckley, 1965) on *Musa* sp. fruit and *Epipremnum pinnatum* could well have been specimens of *A. destructor* with recessed median lobes; the latter species occurs on many hosts in Fiji, including *Musa*. No material of *A. excisus* from Fiji was seen in this study, and we have no evidence that it does in fact occur there.

Aspidiotus macfarlanei sp. n. (Fig. 25)

Description
Scale of adult female circular, fawn, with slightly yellower subcentral exuviae. Male scale oval with subcentral exuviae, same colour as female scale.

Slide-mounted female 0.5-1 mm long; newly emerged specimens pyriform, membranous except for parts of pygidium; prosoma expanding and sclerotizing with maturity, becoming reniform with postero-lateral lobes more or less enclosing the pygidium; postsoma often retracted into prosoma. Ultimately even the pygidium becomes heavily sclerotised.

Pygidium quite broad and rounded, with apical recess between second lobes. Outer margin of median lobes sometimes longer than that of second lobes; median lobes surpassed by second, due to apical recess. Median lobes twice as long as wide, notched subapically on either side, apex

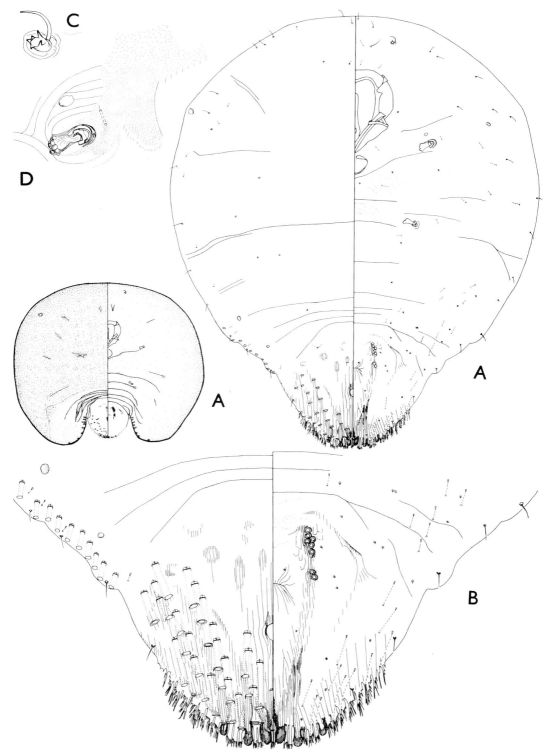

Fig. 25. *Aspidiotus macfarlanei* sp. n.

somewhat serrate; separated by about one third lobe width; small narrow basal sclerosis often present at inner basal corner of lobe, not very distinct. Second lobes usually slightly smaller than median, sometimes notched part way down each side, tip rounded. Third lobes half size of second lobes, sometimes with notch on outer edge, tip rounded. Plates extending well beyond most lobes, deeply fringed, particularly lateral to third lobes; 8-10 plates present beyond third lobe, the outer 2 or 3 shorter and simpler than the others. Anus a wide ellipsoid, about as wide as, and shorter than, median lobe; situated at posterior two fifths of pygidium. Dorsal pygidial macroducts 4.5-9 times as long as wide; single median marginal macroduct reaching one third to one half way to anus; 1 marginal macroduct present between median and second lobes on each side, 2 between second and third, 2 between third lobe and dorsal marginal seta of fourth segment, and 1-2 present between dorsal marginal setae of fourth and fifth segments. Submarginal macroducts forming 3 poriferous furrows on each side; 2-4 macroducts present between segments 7 and 8, 6-8 between segments 6 and 7, and 5-7 present on the fifth segment. Segments 2 and 3 each with 3-4, and segment 1 with 2 or 3 marginal macroducts 3 to 4 times as long as wide present on each side, which become heavily sclerotized in mature specimens. Very few short submarginal microducts present on prosoma, usually near eye and submarginal abdominal tubercles.

Venter of pygidium with 1 to 4 groups of perivulvar pores present, 1-9 in each antero-lateral group, 0-3 in each postero-lateral group, total of 2-22. Microducts few, submarginal or submedian, mostly located on abdominal segments; 1 present near each spiracle. Spiracles moderately sclerotized. Antennae each with several sclerotized spines and 1 robust seta. Pair of setae between antennae unusually long and robust for this genus.

Material examined
Holotype female. **SOLOMON IS.** G.P.: Guadalcanal, Kukum, on *Cocos nucifera*, 1.ii.1956 (*E.S. Brown*) (BMNH).
Paratypes female. **SOLOMON IS.** Same data as holotype, 3 (BMNH). Same locality and host as holotype, 6.vii.1955 (*E.S. Brown*), 24 (BMNH), 2 (DSIR), 2 (USNM), 2 (ARSDC). W.P.: Shortland Is, Maleai, on *Carica papaya*, 16.xi.1984 (*R. Macfarlane*), 1 (BMNH).

Comments
This species resembles *A. nerii* in the width, length and distribution of macroducts, and *A. excisus* in the presence of a pygidial recess; it is therefore placed in *Aspidiotus* in spite of the expansion and sclerotization of the prosoma, and the location of the anus so near the middle of the pygidium. The latter characteristics are more or less shared by *A. macfarlanei*, *A. maddisoni* and *A. musae*, suggesting these new species may originate in this area: *A. macfarlanei* is the only species of the three which becomes reniform at maturity, however. The illustration (Fig. 25) shows the general aspects of the adult female both before and after expansion of the prosomal lobes.

One paratype sample contained *A. macfarlanei* in a mixed colony with *Aonidiella eremocitri* and *Hemibelesia palmae* (Cockerell) on the upper surface of coconut fronds; the scales of the former two species are very similar, that of *A. macfarlanei* being rather paler.

Aspidiotus maddisoni sp. n. (Fig. 26)

Description
Scale of adult female dirty white with brown exuviae. Male scale not seen.
Slide-mounted female about 1 mm long, elongate-pyriform; maximum width at mesothorax about half body length. Body membranous, becoming lightly sclerotized at maturity.
Pygidium quite well produced; median lobes twice as long as wide, notched subapically on either side; separated by almost one lobe width; basal sclerosis present at inner basal corner of lobe, not very distinct. Second lobes slightly smaller than median lobes, small notch present part way down either side, tip rounded or slightly flattened. Third lobes about half size of second lobes, pointed, with or without slight notch present on outer margin. Plates extending well beyond tips of lobes, deeply and finely fringed between lobes; 7-8 plates present beyond third lobe,

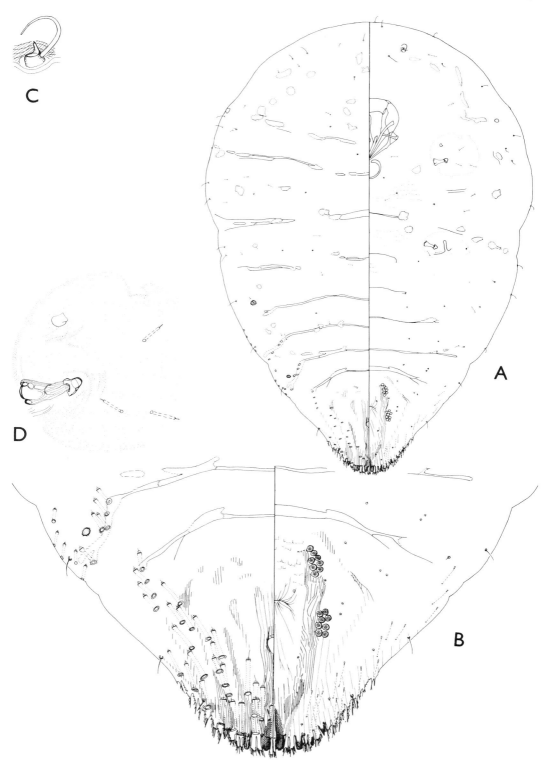

Fig. 26. *Aspidiotus maddisoni* sp. n.

shorter and simpler with increasing distance from lobe. Anus elongate-oval, nearly twice as wide and at least as long as median lobe; situated at posterior two fifths of pygidium. Dorsal pygidial macroducts 4-10 times as long as wide; median marginal macroduct reaching two fifths way to anus; 1 marginal macroduct present between second and third lobes on each side, 2 between third lobe and marginal seta of fourth segment, and 2 between dorsal marginal setae of fourth and fifth segments. Submarginal pygidial macroducts forming 3 slightly irregular poriferous furrows on each side; 2-3 macroducts present between segments 7 and 8, 3-6 between segments 6 and 7, and 4-5 present on fifth segment; 3-5 more forming a short row forward to anterior edge of pygidium. Short prepygidial macroducts present; 4-5 marginal ducts on segment 3 on each side, and 4 on segment 2 on each side. Short rows of submarginal macroducts present, 4-6 between segments 3 and 4, 4 between segments 2 and 3, and 1-2 between segments 1 and 2 on each side. Most macroduct orifices having sclerotized rims. A few short dorsal submarginal microducts and very few longer submedian microducts present.

Venter of pygidium with 4-5 groups of perivulvar pores present, 7-8 in each antero-lateral group, 7-8 in each postero-lateral group, 0-1 in median group, total of 31. Microducts present submarginally, most numerous on pygidium; 2-3 present in area of fine dermal granulation by each spiracle. Spiracles quite heavily sclerotized. Antennae each with one sclerotized spine and 1 robust seta. Pair of setae between antennal bases unusually long and robust for this genus.

Material examined
Holotype female. **WESTERN SAMOA.** Upolu, Utumapu, on *Asplenium nidus*, 7.i.1977 (*P.A. Maddison*) (BMNH).
Paratype female. **WESTERN SAMOA.** Savai'i, Vaipouli Coll., on leaflets of unidentified fern, 11.ii.1977 (*P.A. Maddison*). 1 (DSIR).

Comments
A. maddisoni differs from the type-species by having more than 1 microduct in the area of dermal granulations by the anterior spiracle, and pygidial macroducts more than 7 times as long as wide. The location of the anus near the middle of the pygidium, and the tendency for the prosoma to become slightly sclerotized, suggest *A. maddisoni*, *A. macfarlanei* and *A. musae* are related; but *A. maddisoni* differs from these species by possessing submarginal prepygidial macroducts. The sclerotization of macroduct orifices is more characteristic of the genus *Oceanaspidiotus*, but the macroduct size and distribution are typical of *Aspidiotus*.

Aspidiotus musae sp. n. (Fig. 27)

Description
Slide-mounted female 0.8 mm long, pyriform; prosoma moderately sclerotized, abdomen less so.
Pygidium well produced, apex recessed between second lobes. Outer margin of median lobe slightly shorter than that of second lobe; median lobes obviously surpassed by second lobes, due to apical recess. Median lobes almost twice as long as wide, sharply notched subapically on either side, with smooth apex; separated by 0.6 times lobe width; narrow basal sclerosis present at inner basal corner of lobe. Second lobes slightly larger than median, notched part way down each side, tip rounded. Third lobes three quarters size of second lobes, notched on outer edge, tip rounded. Plates extending just beyond lobe apices, deeply and finely fringed between the lobes; 8-11 plates present lateral to third lobe, progressively simpler and shorter with distance from the lobe. Anus ovoid, about twice as wide as and slightly longer than median lobe, situated at posterior two fifths of pygidium. Dorsal pygidial macroducts 7-11 times as long as wide; single median marginal macroduct reaching two fifths to anus; 1 marginal macroduct present between median and second lobe, 2 between second and third lobes on each side, 2 between third lobe and dorsal marginal seta of fourth segment, and 2 between dorsal marginal setae of fourth and fifth segments. Submarginal macroducts forming 3 poriferous furrows on each side; 2 macroducts present between segments 7 and 8, 6 between segments 6 and 7, and 4 on the fifth segment. Segment 3 with 3-4, and segment 2 with 1-2

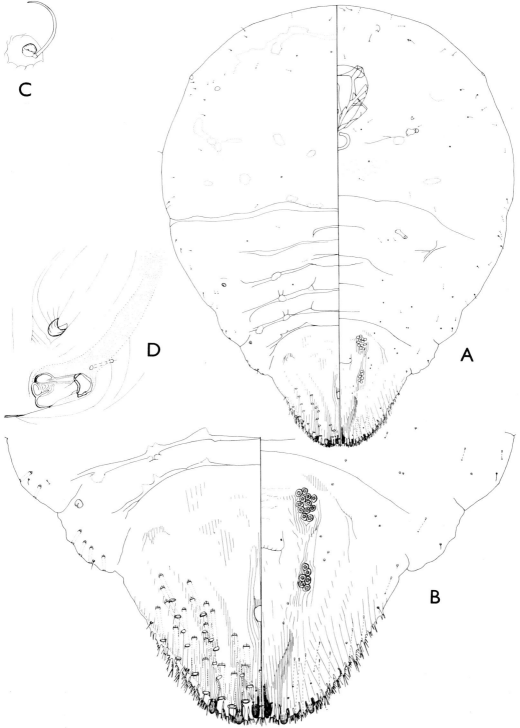

Fig. 27. *Aspidiotus musae* sp. n.

marginal macroducts about 5 times as long as wide on each side. A few short dorsal submarginal macroducts present on prosoma and abdomen, especially near eye and submarginal abdominal tubercles.
　　Venter of pygidium with 4 groups of perivulvar pores present, 9-12 in each antero-lateral group, 5-6 in each postero-lateral group, total of 32. Microducts few, submarginal or marginal, most numerous on pygidium; 1 present by each spiracle. Spiracles moderately sclerotized; tubercle by each anterior spiracle more thickly sclerotized. Antennae small, simple, each bearing 1 robust seta.

Material examined
Holotype female. **PAPUA NEW GUINEA.** Morobe P.: Lasanga Is, on *Musa* sp., 18.ix.1979 (*J.H. Martin*) (BMNH).

Comments
　　This species is described from a single specimen because the host is economically important. It differs from *A. excisus* in the presence of distinct notches on either side of the second lobes, a narrower, more produced pygidium, and a tendency for the prosoma to become sclerotized. This sclerotization, and the proximity of the anus to the middle of the pygidium, suggest *A. musae*, *A. maddisoni* and *A. macfarlanei* are related.

Aspidiotus nerii Bouché (Fig. 28)

Aspidiotus nerii Bouché, 1833: 52; Borchsenius, 1966: 261.
Aspidiotus hederae Signoret, 1869a: 856; Ferris, 1938: 192; 1941b: 54; Balachowsky, 1956: 70.

Description
　　Scale of adult female circular, quite flat, pale grey to white, with more or less central yellow exuviae, sometimes coated with white. Male scale smaller, more elongate; exuviae often subcentral, colouration as for female scale.
　　Adult female, slide-mounted, 0.7-1.2 mm long, broadly ovoid to almost circular, membranous except for parts of pygidium.
　　Pygidium very short and wide, central area of dorsum sclerotized. Median lobes usually slightly longer than wide, subapically notched on either side, apex smooth; separated by half to two thirds lobe width; basal scleroses well developed, wide, sometimes longer than median lobe. Second lobes much smaller than median lobes, outer margin notched; third lobes even smaller than second lobes, more or less pointed, sometimes with notches. In individuals with very small lobes, long dorsal marginal setae may almost reach the tips of the plates. Plates longer than lobes, especially near third lobe; finely fringed; 5-8 plates present lateral to third lobe on each side, often tapering, simpler and shorter with increasing distance from lobe. Anus oval, almost as wide as median lobe, and longer; situated near posterior third of pygidium. Dorsal macroducts short, less than 6 times as long as wide; submarginal ducts sometimes forming 3 poriferous furrows on each side, and extending as far forward as abdominal segment 2. Marginal prepygidial macroducts present, 3-6 on each of segments 2 and 3 on each side, sometimes a few on segment 1.
　　Venter with 4-5 groups of perivulvar pores present, 0-6 in median group, 3-15 in each antero-lateral group, 4-11 in each postero-lateral group, total of 24-54. Microducts few, mostly submarginal, most present on pygidium and segments 2 and 3; 1 present in area of fine dermal granulation by anterior spiracle. Spiracles quite strongly sclerotized. Antennae quite small, each with 3 small sclerotized spurs and 1 robust seta.

Material examined
LORD HOWE I. 1923. On *Diospyros* sp., *Tmesipteris tannensis*.
NEW CALEDONIA. New Caledonia, 1899, Ouen, 1928. On *Citrus* sp., *Cocos nucifera*.
NORFOLK I. 1983. On *Asplenium nidus*.

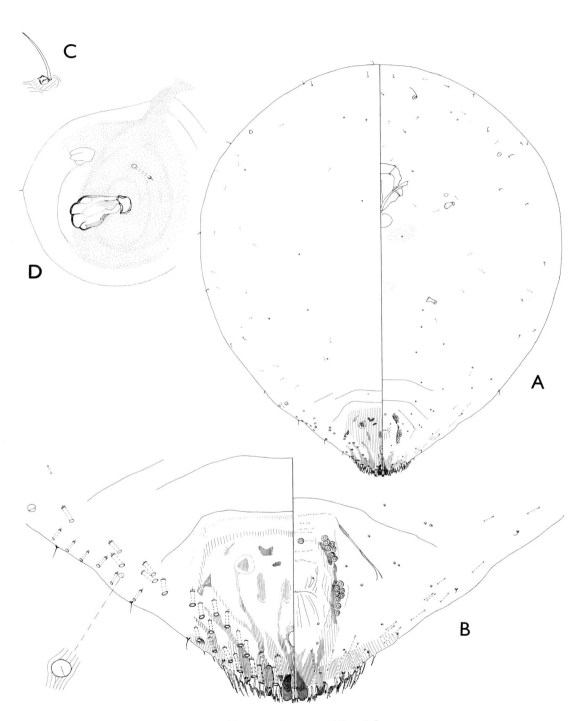

Fig. 28. *Aspidiotus nerii* Bouché.

Comments

A. nerii is a cosmopolitan species found in many tropical and subtropical regions; it has a limited distribution within the South Pacific area. It occurs on many hosts, but seldom on conifers. Its morphology is variable, especially the length of the second and third lobes and their associated marginal setae. The presence of basal scleroses on the median lobes distinguishes it from *A. cochereaui*. It is assumed that, since none of the material was collected from *Pittosporum*, *A. paranerii* Gerson, a morphologically indistinguishable but parthenogenetic species found on *Pittosporum undulatum*, does not occur in the South Pacific area.

In addition to the material listed above, *A. nerii* has been recorded from Easter I. on *Melia azederach* and *Phoenix* sp. (Charlín, 1973), and from New Caledonia on *Ananas sativus*, *Musa sapientum*, *Pandanus* sp., *Plumeria acutifolia*, and *Vitis vinifera* (Cohic, 1956 and 1958a). This species has been intercepted at quarantine in New Zealand on *Citrus* from other parts of the Pacific region, and is probably more widespread than the distribution given above.

Aspidiotus pacificus sp. n. (Fig. 29)

Description

Appearance of scale not known. Adult female, slide-mounted, 0.8-1.2 mm long, broadly pyriform to almost circular; prosoma membranous.

Pygidium quite well produced, fairly broad, parts quite strongly sclerotized; apex obviously recessed between second lobes. Outer margin of median lobe slightly longer than that of second lobe. Median lobes twice as long as wide, notched subapically on either side, with a finely serrate apex; separated by about half lobe width; well-developed moderate-sized basal scleroses present, sometimes slightly clavate. Second lobes usually larger than median lobes, rounded, usually with a very slight notch present on each side. Third lobes one quarter to one half as large as second lobes, rounded, outer margin usually notched. Plates extending to or beyond lobe apices, variously fringed; 6-7 plates present lateral to third lobe, shorter and simpler with increased distance from lobe. Anus broad, oval to circular, about 1.3 times as wide as, and shorter than, median lobe; situated around posterior third to two fifths of pygidium. Dorsal macroducts short, 3-7 times as long as wide; single median macroduct present between median and second lobes, 2 between second and third lobes on either side, 1-2 between third lobe and marginal seta of fourth segment, 0-1 between dorsal marginal setae of fourth and fifth segments. Submarginal pygidial macroducts forming 4 rather irregular poriferous furrows on each side, with macroducts less widely spaced towards their anterior ends; 1-3 macroducts present between segments 7 and 8; 7-11 between segments 6 and 7, with the most anterior ducts sometimes almost submedian; fifth segment with 6-8 ducts present in main furrow, and 2-6 ducts in a second outer row or furrow, normally reaching to anterior edge of pygidium. Segment 3 with 1-6, and segment 2 with 2-4 marginal macroducts present on each side; occasionally 1-3 submarginal macroducts present on each side of segment 3. All macroduct orifices with rims at least slightly sclerotized. Microducts sparsely scattered over dorsal surface.

Venter with 4-5 groups of perivulvar pores present, 0-2 in median group, 4-11 in each antero-lateral group, 4-9 in each postero-lateral group, total of 24-37. Microducts marginal and submarginal, most numerous on pygidium; 1 present by each anterior spiracle. Spiracles moderately sclerotized. Antennae each with 2 sclerotized spurs and 1 robust seta.

Material examined

Holotype female. **SOLOMON IS.** Guadalcanal: Lungga, on *Cocos nucifera*, 18.vii.1956 (*E.S. Brown*) (BMNH).
Paratypes female. **SOLOMON IS.** Same data as holotype. 3 (BMNH), 1 (DSIR).
Non-type material. **AMERICAN SAMOA.** Pago Pago, on *C. nucifera*, 15.xii.1943 (*J. Michel*), intercepted at San Pedro, California. **'SOUTH SEAS.'** On *C. nucifera*, 13.vii.1944 (*J. Michel*), intercepted at San Pedro, California.

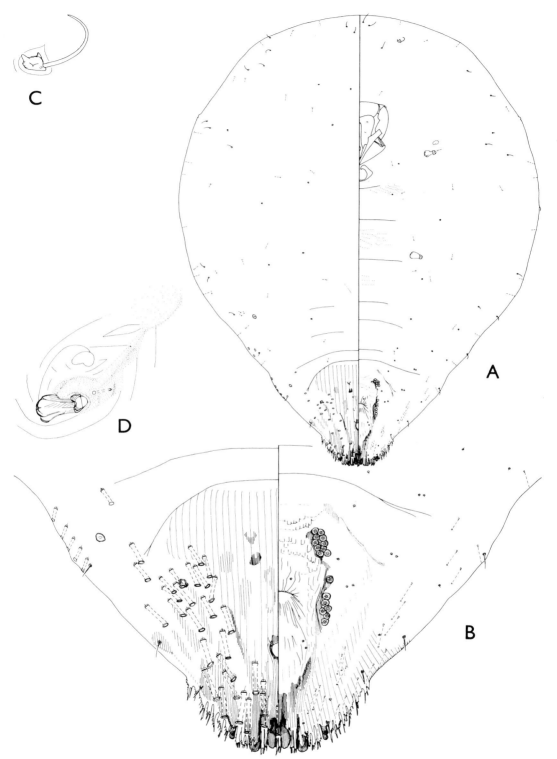

Fig. 29. *Aspidiotus pacificus* sp. n.

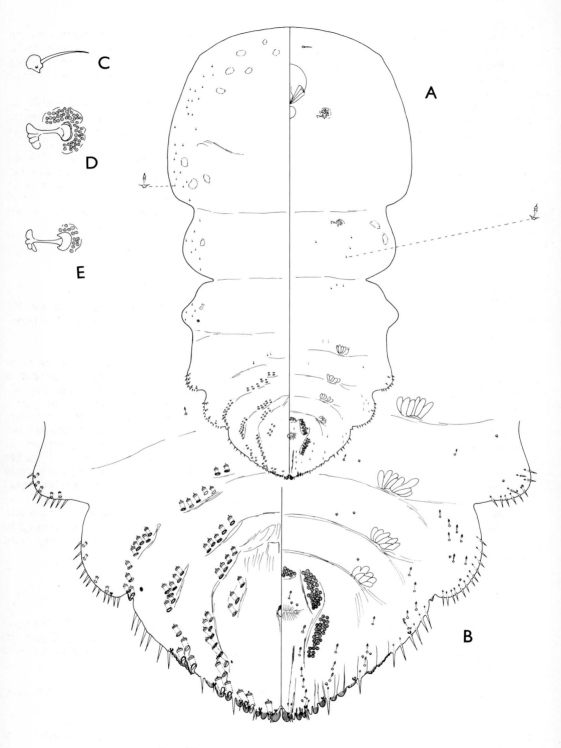

Fig. 30. *Aulacaspis madiunensis* (Zehntner).

Comments

This species is close to *A. excisus*, from which it differs by having basal scleroses on the median lobes, notched second and third lobes, an anal opening shorter than the median lobe, and a preference for *Cocos nucifera* as its host. It may be distinguished from *A. musae* by its membranous prosoma, more heavily sclerotized pygidium, and the presence of pygidial macroducts further forward than the anus.

A. pacificus shows a tendency for enlargement of the prosoma like the genus *Oceanaspidiotus*; but the dorsal macroducts are wider and usually lack sclerotization of the rims, the pygidial recess is rather well developed, and the plates beyond the third lobe are fringed, so the species is assigned to *Aspidiotus*.

The material from American Samoa was kindly sent on loan from the collections of the USNM by D.R. Miller. It varies slightly from the type material, tending to have more numerous pygidial and marginal prepygidial macroducts, and fewer macroducts in the fourth poriferous furrow; some specimens have a very enlarged prosoma with a constriction between the second and third abdominal segments. These differences probably represent variation between populations and different stages of maturity.

Genus **AULACASPIS** Cockerell

Aulacaspis Cockerell, 1893a: 180; Scott, 1952: 33; Balachowsky, 1954: 240; Takagi, 1970: 82.
Type-species *Aspidiotus rosae* Bouché by subsequent designation (see Morrison & Morrison, 1966).

This genus usually includes species which, like the type-species, are swollen in the prosoma. The median lobes are often divergent and recessed into the pygidium, forming a notch, and the dorsal 2-barred ducts are usually in submarginal and submedian rows, the submedian rows often stepped. Numerous species have been assigned to the genus, and sometimes the differences between the species are extremely small. The genus as a whole is in great need of revision, but Takagi (1985) has given extra guidance as to which species should be included in the genus. These include species without setae between the median lobes and without marginal gland spines and ducts anterior to segment 2. The new definition includes certain species without a swollen prosoma, but immediately separates the genus from *Chionaspis*, which has marginal gland spines and ducts anterior to segment 2 but without setae between the median lobes, and *Pseudaulacaspis* which has setae between the median lobes. Takagi's concepts are followed here pending further research on the whole group.

In the tropical South Pacific area some species have been introduced, but endemic species may yet be found in Papua New Guinea and the Solomon Is. *A. tubercularis* Newstead, common on mango in southern Asia, has still not been reported from the area but could easily be introduced. At present 6 species are known from the tropical South Pacific region.

Key to species of *Aulacaspis*

1	Prosoma rounded without lateral tubercles. Pore prominences on pygidium poorly developed	*tegalensis* (Zehntner)
--	Prosoma quadrate or rounded, with lateral tubercles. Pore prominences on pygidium well developed	2
2	Head rounded, with pronounced lateral tubercles	*vitis* (Green)
--	Head sub-quadrate, with poorly developed lateral tubercles	3
3	Dorsal ducts present on segment 1 of abdomen	*martini* sp. n.
--	Dorsal ducts absent from segment 1 of abdomen	4

| 4 | Submarginal ducts present on segment 2 of abdomen .... *sumatrensis* Green |
| -- | Submarginal ducts absent from segment 2 of abdomen .... 5 |

| 5 | Submedian macroducts present on segment 2 .... *rosarum* Borchsenius |
| -- | Submedian macroducts absent from segment 2 .... *madiunensis* (Zehntner) |

Aulacaspis madiunensis (Zehntner) (Fig. 30)

Chionaspis madiunensis Zehntner, 1898: 1085.
Aulacaspis madiunensis (Zehntner), Takahashi, 1940: 26; Scott, 1952: 38; Munting, 1977: 4.

Description
    Scale of adult female white, subcircular, the exuviae marginal. Male scale with median ridge, same colour as female scale.
    Adult female with swollen prosoma and lateral tubercles; prepygidial segments with well-developed lateral lobes. In mature specimens, head and thorax more heavily sclerotized than pygidium.
    Pygidium with 4 pairs of lobes, all rounded, the median pair slightly divergent, yoked at base. Pore prominences well developed and approaching the lobes in size. Gland spines usually in pairs between median and second lobes, between second and third lobes, singly on segments 5 and 6, and numerous around edges of segments 2 to 4. Dorsal ducts in submedian and submarginal rows on segments 3 to 5, and in a submarginal group of 1-4 (usually 3) on segment 6. A few smaller ducts on margins of segments 2 and 3. Minute ducts present anteriorly on margins to head. Lateral bosses present on segments 1, 4 and 6.
    Ventral surface with numerous perivulvar pores in 5 groups. Microducts in rows on pygidium and on margins of free abdominal segments. Minute ducts present near margins of thorax. Antennae each with a single seta. Anterior and posterior spiracles with groups of disc pores.

Material examined
TUVALU. Funafuti, 1976. On *Pandanus* sp.

Comments
    The record on *Pandanus* is unusual, but the single specimen available agrees well with material herein illustrated from Madiun, Java, the type locality. This species is often found on grasses, and is a sugarcane pest. Although it has not been found on sugarcane in the South Pacific area, the record shows that it could be a potential pest. It is known from southern Asia, Australia, East Africa and South Africa. The record by Reyne (1961) from New Guinea probably refers to another species.
    On sugarcane it could be confused with *A. tegalensis*, but the latter species has a rounded prosoma, whereas in *A. madiunensis* the prosoma is angled, with lateral tubercles.

Aulacaspis martini sp. n. (Fig. 31)

Description
    Appearance of scale not known. Adult female, on slide, about 1.2 mm long, with swollen prosoma and lateral tubercles; mesothorax and abdomen narrower, the prepygidial segments with lateral lobes well developed.
    Pygidium with median lobes parallel or only slightly diverging, longer than wide, separated by a space equal to the width of 1 lobe; narrowly yoked well anterior to base of lobes. Second and third lobes similar size and shape. Gland spines arranged singly on pygidium, shorter and more numerous on segments 2 and 3. Marginal macroducts 2-3 times longer than wide. Dorsal ducts shorter, in submarginal rows of 2-3 on segments 3 to 5, present on or absent from segments 1 and 2,

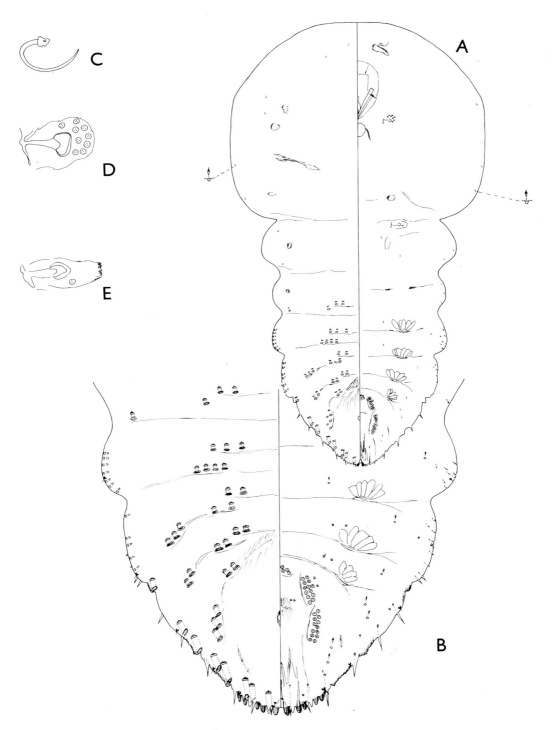

Fig. 31. *Aulacaspis martini* sp. n.

and in submedian rows on segments 1 to 5, those on the anterior segments stepped; absent from segment 6. Smaller ducts present on margins of segments 2 and 3. Minute ducts sparse, present around head, prothorax and mesothorax.

Ventral surface with perivulvar pores in 5 elongate groups. Microducts sparse on submargins of pygidium and in median areas of free abdominal segments. Antennae each with a single curved seta. Anterior spiracles each with 0-2 pores.

Material examined
Holotype female. **PAPUA NEW GUINEA.** Morobe P., Buso, on *Rhizophora* sp., 28.x.1979 (*J.H. Martin*) (BMNH).
Paratypes female. **PAPUA NEW GUINEA.** Same data as holotype. 2 (BMNH). Morobe P., Lasanga I., on *Rhizophora ?kandelia*, 21.x.1979 (*J.H. Martin*), 5 (BMNH).

Comments
Among the species from the South Pacific area, *A. martini* is easily identifiable in possessing dorsal ducts on the first abdominal segment. Some of the well-known species, such as *A. thoracica* (Robinson), *A. crawii* (Cockerell) and *A. greeni* Takahashi, which have ducts on the first abdominal segment, also have submedian ducts on segment 6. These are lacking in *A. martini*.

Aulacaspis rosarum Borchsenius (Fig. 32)

Aulacaspis rosarum Borchsenius, 1958: 165; Chen, 1983: 52.

Description
Adult female, slide-mounted, with swollen prosoma, the sides almost parallel; head rounded with lateral tubercles; mesothorax and abdomen narrower, subparallel before a rounded pygidium.

Pygidium with well-developed median lobes, divergent, strongly zygotic at base, recessed into pygidium forming a deep notch at apex; inner edge serrate. Second lobes reaching same level apically, bilobed, rounded. Third lobes similar shape to second lobes. Gland spines arranged singly on pygidium, smaller and more numerous on segments 2 and 3. Marginal macroducts 2-3 times longer than wide. Dorsal ducts shorter, in a submedian stepped row on each of segments 2 and 3, and in 1 row on each of segments 4 to 6. Submarginal rows present on segments 3 to 5. Smaller marginal ducts present on segments 2 and 3. Minute ducts, as illustrated, around margins to thorax and in median area of head. A marginal boss present on each side of segment 3.

Ventral surface with perivulvar pores numerous in 5 groups. Microducts in rows on pygidium and around abdominal margins, becoming minute as on dorsum on mesothorax. Antennae each with a single long seta. Anterior spiracles each with a compact group of numerous disc pores, posterior spiracles with fewer pores.

Material examined
COOK IS. Rarotonga, 1977. On *Rosa indica*.
FIJI. Viti Levu, 1941. On *Rosa* sp.
FRENCH POLYNESIA. Gambier Is, Mangareva, 1966. On *Rosa* sp.
PAPUA NEW GUINEA. C.P.: 1957. E.H.P.: 1960. On *Rosa* sp., *Rubus occidentalis*.
TONGA. Tongatapu Group, Tongatapu, 1975. On *Rosa indica*.
VANUATU. Tanna, 1925. On *Rosa* sp.

Comments
This species has become well established in the area on the family Rosaceae, particularly on *Rosa* spp. It was described originally from Chendu in China, and specimens from the original material have been kindly made available for comparison by Dr E.M. Danzig, Academy of Sciences, Leningrad. In possessing rows of submedian ducts on segment 2, it differs from *A. rosae* (Bouché), a species common on roses.

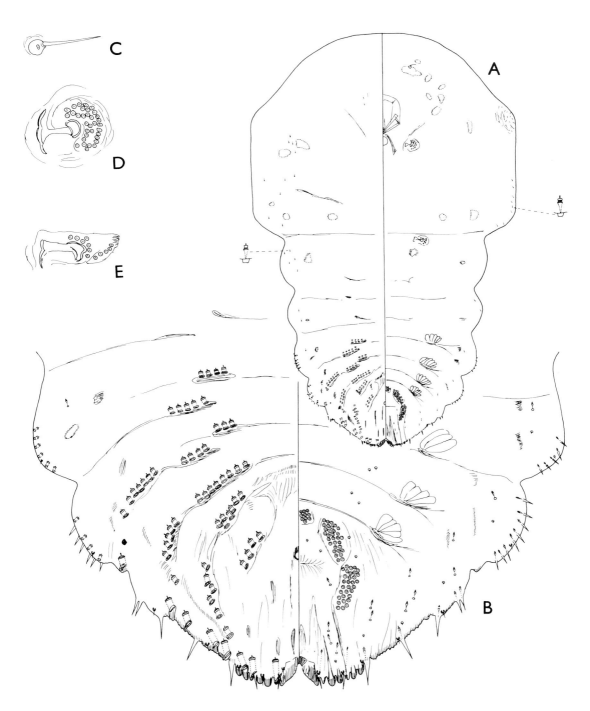

Fig. 32. *Aulacaspis rosarum* Borchsenius.

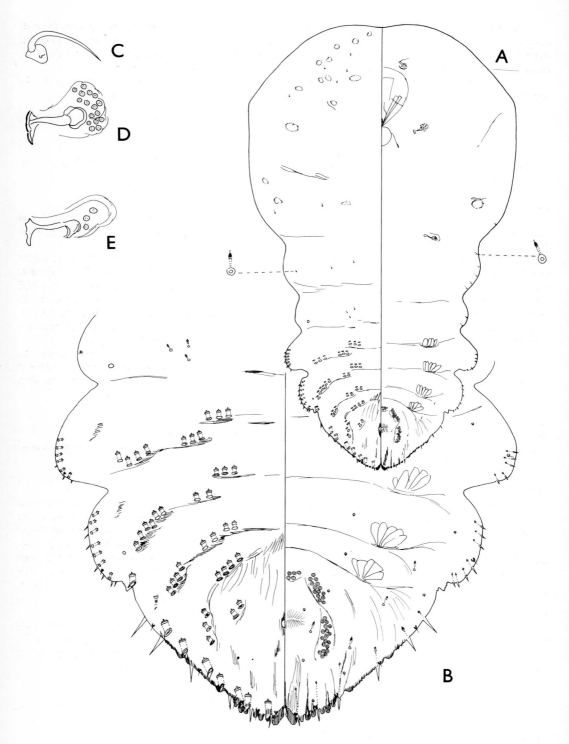

Fig. 33. *Aulacaspis sumatrensis* Green.

The species recorded by Dumbleton (1954) as *A. rosae* from New Caledonia is now known to be *A. rosarum*. It may have been misidentified in other parts of the world as *A. rosae*, but at present, apart from the Pacific records, *A. rosarum* is known only from South East Asia. All records of *A. rosae* or *Diaspis rosae* from the South Pacific area are probably erroneous.

A. sumatrensis is very similar to *A. rosarum*, but has a row of submarginal ducts on segment 2, and the ventral microducts on the pygidium are much less numerous than in the latter species.

Aulacaspis sumatrensis Green (Fig. 33)

Aulacaspis sumatrensis Green, 1930: 292. Lectotype female, Indonesia, Sumatra (BMNH), here designated [examined].

Description
An elongate species with lateral lobes on a wide subquadrate prosoma; other prepygidial segments with lateral lobes moderately developed, the widest at abdominal segment 2; pygidium rounded.

Pygidium with median lobes forming a narrow notch at apex, then diverging, with inner edges serrate and longer than outer edges, rounded apically. Second and third lobes smaller, bilobed, rounded. Gland spines arranged singly on pygidium, slightly longer than lobes, more numerous and smaller on segments 2 and 3. Marginal macroducts longer than wide. Dorsal ducts shorter, in submarginal and submedian rows on segments 2-5, and 1 or 2 submedially on each side of segment 6. Smaller ducts on margins of segments 2 and 3. A few submedian microducts usually present on anterior segments and on head. Small submarginal bosses situated on segments 1 and 3.

Ventral surface with perivulvar pores in 5 groups. Microducts sparse, present lateral to vulva, submarginally as far forward as segment 4, and on margins of metathorax and abdominal segment 1. Antennae each with a single seta. Anterior and posterior spiracles each with groups of disc pores. Labium with lateral scleroses.

Material examined
NEW CALEDONIA. New Caledonia, 1980. On *Cocos nucifera*, Arecaceae.
VANUATU. Ambrym, 1925; Efate, 1963; Epi, 1925; Espiritu Santo, 1956; Malekula, 1923; Tanna, 1978. On *Cocos nucifera*.

Comments
Although this species was described from Sumatra on *Mangifera indica*, and has not been reported since, the South Pacific specimens on coconut seem to agree with the type series in every respect. In the type material there are fewer microducts in the submedian areas forward from abdominal segment 2, but otherwise all the material at hand seems to represent the same species. *A. javanensis* Newstead is similar, but possesses more dorsal ducts on abdominal segment 6 and anteriorly. *A. sumatrensis* is also similar to *A. tubercularis* Newstead, but this species lacks the submarginal and submedian ducts on segment 2. Risbec (1937) and Dumbleton (1954) recorded *A. sumatrensis* as *A. cinnamomi* on citrus from Vanuatu, but this host record probably refers to *Cocos nucifera*.

A. cinnamomi Newstead was recorded by Simmonds (1925), Risbec (1937) and Dumbleton (1954) from Vanuatu, but these records were shown by Williams & Butcher (1987) to refer to *A. sumatrensis*. *A. cinnamomi* is a synonym of *A. tubercularis*, but the record under the latter name by Borchsenius (1966) from New Caledonia probably refers to the above material, which he studied.

Aulacaspis tegalensis (Zehntner) (Fig. 34)

Chionaspis tegalensis Zehntner, 1898: 1091.
Aulacaspis major Rutherford, 1914: 259. Syn. n.

Aulacaspis tegalensis (Zehntner), Ferris, 1921: 213; Scott, 1952: 40.
Aulacaspis rutherfordi Morrison, 1924: 232. Syn. n.
Miscanthaspis tegalensis (Zehntner), Borchsenius, 1966: 135.

Description
	Scale of female subcircular, white, exuviae at edge, yellow, convex. Male scale same colour as female scale, smaller.
	Adult female with prosoma swollen and rounded; abdomen gradually tapering, almost pyriform; head region and pygidium becoming sclerotized.
	Pygidium with 3 pairs of prominent lobes. Median lobes set close together but forming a deep notch at apex. Second and third lobes rounded. Pore prominences poorly developed. Gland spines in pairs on fifth and posterior segments; numerous on edges of segments 2-4. Dorsal ducts fairly numerous in submarginal and submedian rows on segments 3-5; submedian group present on segment 6. Short microducts present in submarginal band around free abdominal segments and thorax, and in submedian areas of abdominal segments 1 and 2. A small boss present on each side of segment 3. Anus situated towards centre of pygidium.
	Ventral surface with numerous perivulvar pores in 5 groups. Microducts present in rows on pygidium and in small marginal groups forward to head. Antennae each with a single seta. Anterior and posterior spiracles with numerous disc pores.

Material examined
PAPUA NEW GUINEA. C.P.: 1984. On *Saccharum officinarum*, 'pitpit' (Monocotyledon).

Comments
	Rutherford (1915) described *Aulacaspis major* from New Guinea (presumably Papua) on sugarcane. Because this became a junior homonym of *A. major* (Cockerell), Morrison (1924) proposed the new name *A. rutherfordi*, for Rutherford's species. A careful assessment of Rutherford's description gives every indication that the species was actually *A. tegalensis*, and Rutherford's name is here sunk as a synonym, but no original material has been traced for comparison. Although the species has not been found again until recently in the island of New Guinea, the present record adds support to this action.
	The species infests grasses and is often injurious to sugarcane. It was described originally from Java, and is now known throughout southern Asia, the Malagasian area and East Africa. Williams, J.R. (1970) discussed its importance in Mauritius, and Greathead (1975) discussed its ecology in East Africa. It is a potential pest of sugarcane throughout the tropical Pacific area.

Aulacaspis vitis (Green) (Fig. 35)

Chionaspis vitis Green, 1896b: 3; Chou, 1986: 484. Lectotype female, Sri Lanka (BMNH), here designated [examined].
Phenacaspis vitis (Green), Takahashi, 1942: 33; Ferris, 1955: 53.
Aulacaspis vitis (Green), Takagi, 1985: 50.

Description
	Adult female with prosoma swollen, rounded, but with pronounced lateral tubercles; widest at mesothorax, body tapering to a narrow abdomen with lateral lobes of the free abdominal segments moderately developed.
	Pygidium elongate with base of median lobes recessed into apex, space narrow at base, then diverging, with the inner edges serrate and longer than outer edges. Second and third lobes well developed, rounded. Pore prominences similar size to lobes. Gland spines single on margins forward to segment 4; anteriorly much smaller and more numerous on segments 2 and 3. Marginal ducts 2 to 3 times longer than wide. Anterior dorsal ducts shorter, present in submarginal rows on segments 3-5, each group numbering 2-5, and in submedian groups of 1-5 on the same segments, and singly on segment

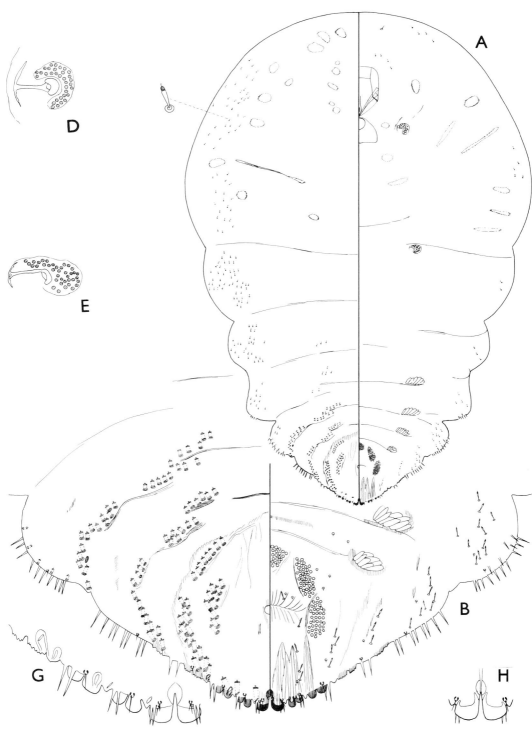

Fig. 34. *Aulacaspis tegalensis* (Zehntner).

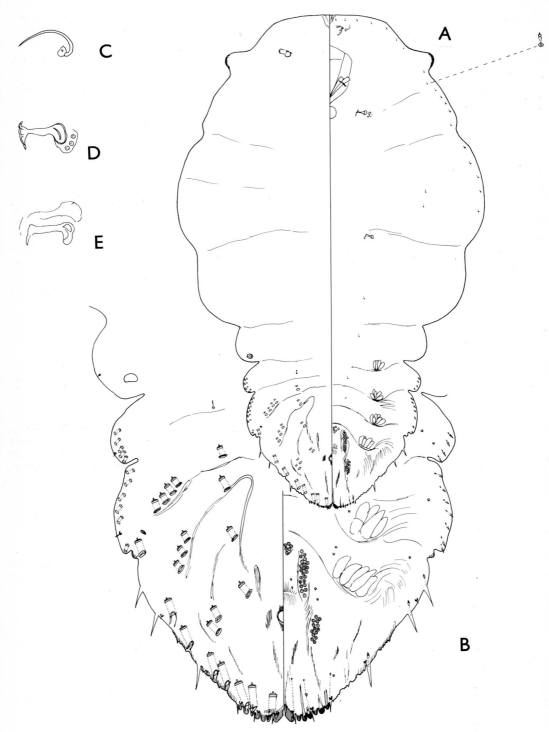

Fig. 35. *Aulacaspis vitis* (Green).

6. Smaller ducts present on margins of segment 2 and 3, and an occasional microduct in median areas of segment 2. Intersegmental line medially between segments 4 and 5 curling posteriorly. Anus situated near centre of pygidium. Submarginal bosses present on segments 1 and 3.

Ventral surface with perivulvar pores in 5 elongate groups. Minute microducts sparse, present in median areas of free abdominal segments, and around thoracic and head margins. Antennae each with a single seta. Anterior spiracles each with a small group of disc pores; posterior spiracles without pores.

Material examined
PAPUA NEW GUINEA. Morobe P.: 1957. On *Durio zibethinus*.

Comments

Specimens from Papua New Guinea, on which the accompanying illustration is based, agree in every respect with specimens from Java, also on *D. zibethinus*. These have single or occasionally 2 submedian ducts on each side; but in other specimens from southern Asia, including original material from Sri Lanka, collected on *Vitis* sp., the submedian groups are more numerous. Pending further research on all material, the material from Papua New Guinea is here regarded as *A. vitis*.

In having the head rounded with prominent lateral tubercles, the species is not typical of the genus, but in lacking pores and gland spines on segment 1 it apparently follows Takagi's concepts.

Green's original 'type' slide contains five specimens. It is labelled *Chionaspis vitis* Green, Ceylon, Punduloya, from *Vitis* sp. May 97. The lectotype selected is clearly marked, and the other four specimens are here designated paralectotypes.

Genus **CARULASPIS** MacGillivray

Carulaspis MacGillivray, 1921: 305; Ferris, 1937a: 11; Balachowsky, 1954: 202; Boratynski, 1957: 247. Type-species *Diaspis juniperi* Bouché, by original designation.

This genus contains a few species found on conifers only. It is closely related to *Diaspis*, discussed herein, but differs mainly in possessing a pair of short gland spines between the median lobes. The genus is regarded as Palaearctic in origin. Two species are included here from Pacific records only. Unlike some other *Carulaspis* species, both lack a macroduct between the median lobes. They can be separated as follows:

Key to species of *Carulaspis*

1 Duct tubercles present on thorax and first abdominal segment .... *giffardi* (Adachi & Fullaway)
-- Duct tubercles absent from thorax and first abdominal segment .... *minima* (Targioni)

Carulaspis giffardi (Adachi & Fullaway)

Pseudoparlatoria giffardi Adachi & Fullaway, 1953: 87.
Carulaspis giffardi (Adachi & Fullaway), Borchsenius, 1966: 160.

This species was described from Hawaii on *Araucaria excelsa* (now *A. heterophylla*) and *A. cunninghamii*. It has since been recorded from New Caledonia by Cohic (1956, 1958a) on *A. columnaris* (now *A. cookii*), and by Brun & Chazeau (1980) on *Araucaria* sp.

Carulaspis minima (Targioni)

Diaspis minima Targioni, 1868: 736.
Carulaspis minima (Targioni), Borchsenius, 1949: 226; Balachowsky, 1954: 210; Boratynski, 1957: 249.

Apart from Europe, where this species was described, other records are from North Africa, North and South America and Hawaii, all on Coniferae. The only record from the South Pacific area is by Charlín (1973) from Easter I. on *Cupressus macrocarpa*.

Genus **CHIONASPIS** Signoret

Chionaspis Signoret, 1869a: 844; Ferris, 1937a: 13; Balachowsky, 1954: 317; Takagi, 1985: 5. Type-species *Coccus salicis* Linnaeus.

The species discussed here are probably not related to each other, but are included in this genus following the concepts presented by Takagi (1985). *Chionaspis* is normally a holarctic genus, but it extends into certain montane tropical areas of southern Asia. Principal characters are: median lobes larger than the lobules of the second and third lobes, joined by a basal zygosis; dorsal ducts 2-barred, present marginally as far as abdominal segment 1, and sometimes on thorax. Gland spines usually present at least as far forward as abdominal segment 1.

All the following species have most of these characters. They do not belong to *Pseudaulacaspis* because they lack setae between the median lobes; they are excluded from *Aulacaspis* in possessing dorsal ducts and (except for one species) gland spines anterior to the second abdominal segment. The species are described in *Chionaspis* for the time being and it should be possible to identify them from the following key:

Key to species of *Chionaspis*

1	Median lobes parallel, the space between very narrow	2
--	Median lobes divergent, recessed, forming a notch at apex of pygidium	3
2	Body broadly oval, widest at thorax, gland spines present as far forward as prothorax .... *rhaphidophorae* sp. n.	
--	Body fusiform, widest at first abdominal segment. Gland spines present as far forward as fifth abdominal segment .... *keravatana* sp. n.	
3	Median lobes rounded, each without serrations or notches on inner margin .... *pandanicola* sp. n.	
--	Median lobes each with at least 1 notch on inner margin, usually with a few serrations	4
4	Submedian groups of ducts absent. Gland spines on free abdominal segments with more than 1 microduct .... *comys* sp. n.	
--	Submedian groups of ducts present. Gland spines on free abdominal segments with a single microduct	5
5	Submedian ducts present on segments 4 and 5 only. With 2 ducts between each median and second lobe .... *freycinetiae* sp. n.	
--	Submedian ducts present on segments 3-6. With 1 duct between each median and second lobe .... *broughae* sp. n.	

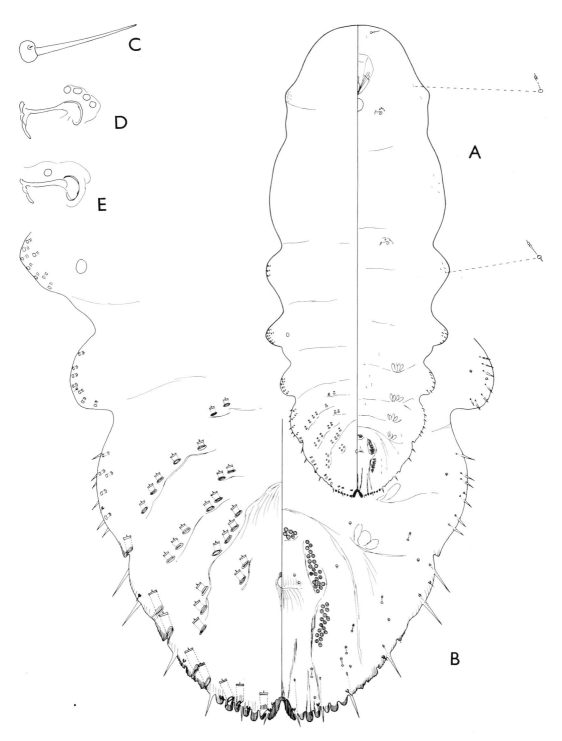

Fig. 36. *Chionaspis broughae* sp. n.

Chionaspis broughae sp. n. (Fig. 36)

Description
Female scale elongate. Male scale not known.
Adult female on slide about 1.6 mm long, membranous except for pygidium; elongate, narrow; prosoma with lateral tubercles, pygidium rounded. Lateral lobes of prepygidial segments well developed.
Pygidium with well-developed median lobes divergent, joined at base; inner edges serrated, much longer than outer edges, recessed into apex of pygidium, forming a deep notch. Second lobes well developed, bilobed, longer than wide, rounded, the inner lobule longer than the median lobes. Third lobes similar in shape to second lobes. Pore prominences well developed, almost as long as lobes. Gland spines arranged singly; narrow between the lobes, becoming longer and wider anteriorly to segment 4; smaller and more numerous on margins of segments 1 and 2. Marginal macroducts 2-3 times longer than wide. Dorsal ducts present in submarginal rows on segments 3-5 and in submedian groups on segments 3-6, the groups on segments 3 and 4 in stepped rows, and the groups on segment 6 represented by 1-3 on each side. Smaller ducts present on margins of mesothorax and abdominal segments 1-3. A submarginal boss situated on segment 1.
Ventral surface with perivulvar pores in 5 elongate groups. Microducts present around submargins of pygidium, in median areas of free abdominal segments and on thoracic margins; those on metathorax often with minute spines at orifices. Antennae each with a single long seta. Anterior spiracles each with a group of 3-5 disc pores; posterior spiracles each with a single pore.

Material examined
Holotype female. **PAPUA NEW GUINEA.** M.B.P., Trobriand Is, Kiriwina, on stems of dicotyledon, 27.x.1985 (*E.J. Brough*) (BMNH).
Paratypes female. **PAPUA NEW GUINEA.** Same data as holotype. 3 (BMNH).

Comments
If it were not for the gland spines on the first abdominal segment, and the dorsal ducts present as far forward as the mesothorax, this species would be a normal component of *Aulacaspis*. It does, however, seem to be related to *C. linderae* (Takahashi), described from Japan, in the general arrangement of the dorsal ducts and the shape of the median lobes. It differs from *C. linderae* in possessing three pairs of well-developed lobes instead of two, and in the development of the prosomal tubercles.

Chionaspis comys sp. n. (Fig. 37)

Description
Scale of adult female elongate, whitish, smooth, with yellow-brown exuviae. Male scale smaller than female scale, white, carinated.
Adult female elongate-oval, about 0.7 mm long, widest at about first abdominal segment. Lateral lobes of free abdominal segments moderately developed.
Pygidium with well-developed median lobes joined at base, divergent, with inner edges dentate and much longer than outer edges, forming a deep notch at apex. Second lobes bilobed, rounded, smaller, but apices reaching same level as those of median lobes. Third lobes represented by points. Gland spines on pygidium normal, each with 1 microduct, arranged singly; but single gland spines on margins of segments 1-4 blunt, each with 2-4 microducts. Marginal macroducts about twice as long as wide. Dorsal ducts shorter, few, represented by 1 or 2 submarginally on segments 3 to 5. Smaller ducts present on margins as far forward as mesothorax.
Ventral surface with perivulvar pores in 5 elongate groups. Microducts few, on submargins of pygidium, median areas of anterior abdominal segments, and near posterior spiracles. Small ducts, similar to those on dorsum, on margins of mesothorax. Antennae each with a single long seta.

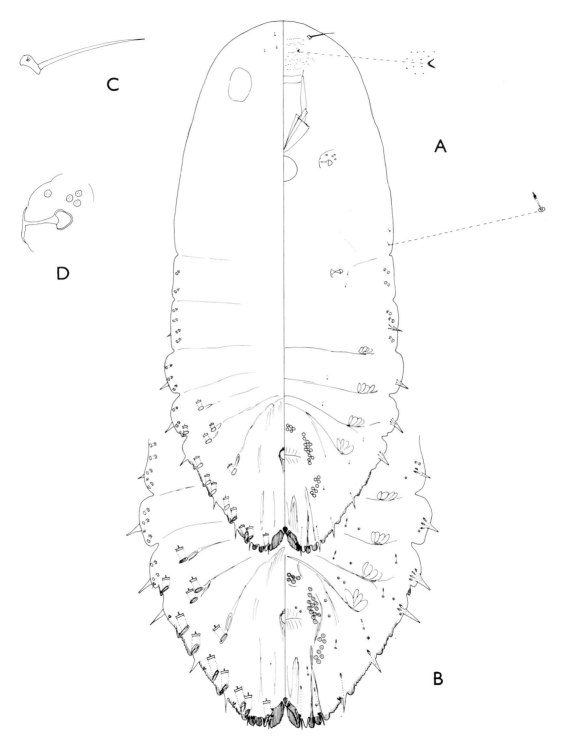

Fig. 37. *Chionaspis comys* sp. n.

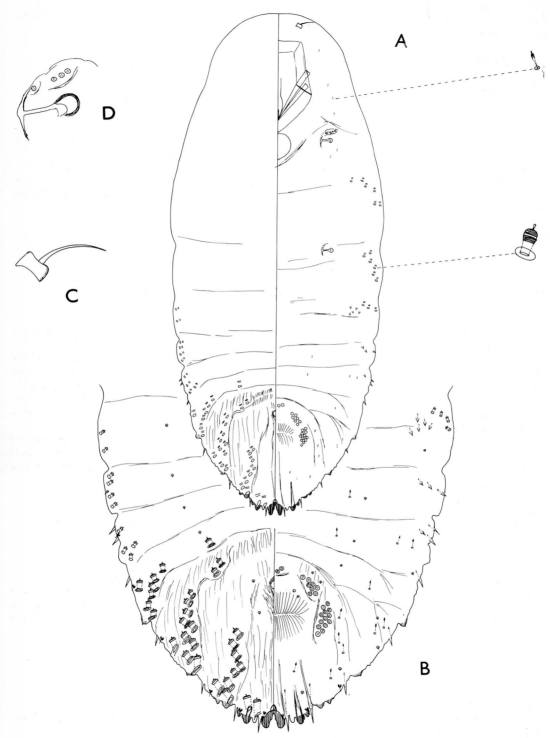

Fig. 38. *Chionaspis freycinetiae* sp. n.

Anterior spiracles each with a small group of disc pores. Minute spicules and a pair of crescentic sclerotized areas present anterior to clypeolabral shield.

Material examined
Holotype female. **IRIAN JAYA.** Sentani, nr Jayapura, on leaves of undetermined tree, 24. vi. 1959 (*T. Maa*) (BPBM).
Paratype female. **IRIAN JAYA.** Same data as holotype. 5 (BPBM), 7 (BMNH), 2 (DSIR), 2 (USNM).

Comments
Where to place this species is not yet clear. It bears a superficial resemblance to the species described as *C. eugeniae* var. *litzeae* Green from India, currently assigned to *Aulacaspis* by Takagi (1985), in lacking submedian rows of ducts. *A. litzeae* has setae between the median lobes, and should be referred to *Pseudaulacaspis*. The new species should probably be included in a new genus possessing unusual gland spines, each with 2-4 microducts, on the free abdominal segments, but is left in *Chionaspis* for the present, for reasons given in the comments to the genus.

Chionaspis freycinetiae sp. n. (Fig. 38)

Description
A small fusiform species; adult female about 0.6 mm long, head and pygidium rounded, lateral lobes of free abdominal segments only moderately developed.

Pygidium with well-developed prominent median lobes, inner margins each with 1 or 2 notches, longer than outer margins, joined at base and recessed into pygidium forming a notch. Second lobes bilobed, well developed, each with inner lobules smaller than median lobes and outer lobules smaller than inner lobules. Third lobes represented by serrations on the margin. Gland spines arranged singly on pygidium; smaller and more numerous gland spines present as far forward as segment 1. Marginal macroducts about twice as long as wide. Two present between each median and second lobe. Dorsal ducts arranged in submarginal rows on segments 4 to 6, and 1 or 2 present in submedian areas of segments 4 and 5. Smaller ducts present on margins of segments 1 to 3.

Ventral surface with perivulvar pores in 5 groups. Microducts sparse on pygidium, present on free abdominal segments and lateral to clypeolabral shield. Small ducts, similar to the small ducts on dorsum, present on margins of mesothorax, metathorax and abdominal segment 1. Antennae each with a single seta. Anterior spiracles each with a group of 4-8 disc pores.

Material examined
Holotype female. **FIJI.** Sawani, on *Freycinetia* sp., 18.iii.1957 (*B.A. O'Connor*) (BMNH).
Paratypes female. **FIJI.** Same data as holotype. 2 (BMNH).

Comments
There may be some reason for including this species in *Shansiaspis* Tang, in possessing two ducts between the median and second lobes, but species of *Shansiaspis* possess many more dorsal ducts including a submedian group on segment 6. *C. freycinetiae*, however, seems to be so closely related to *C. pandanicola*, herein described, that it is left in *Chionaspis* for the present. It is difficult to compare this species with any other thus far described in *Chionaspis*.

Chionaspis keravatana sp. n. (Fig. 39)

Description
Adult female, on slide, broadly fusiform, about 0.65 mm long, membranous except for pygidium; free abdominal segments with lateral lobes moderately developed.

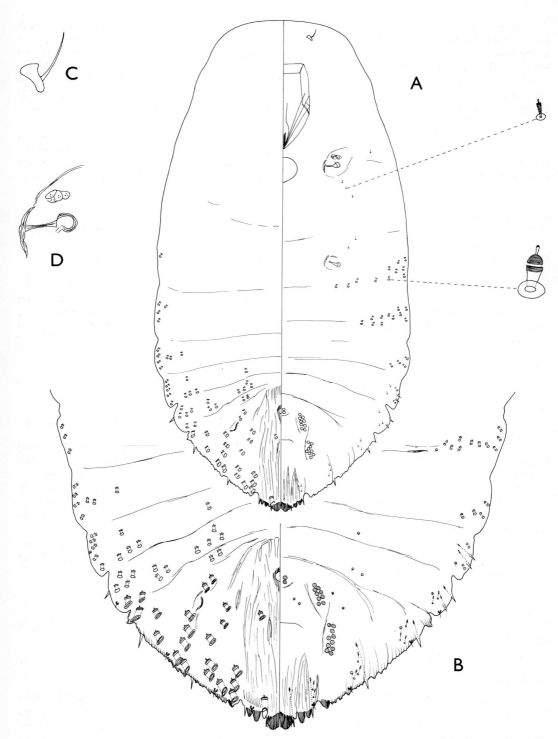

Fig. 39. *Chionaspis keravatana* sp. n.

Pygidium with median lobes prominent, yoked at base, the space between extremely narrow, each lobe notched at either side, rounded apically. Second lobes bilobed, each lobule about same length as median lobes but narrower. A deep notch present on margin at each side just lateral to second lobes. Third lobes represented by points. Gland spines arranged singly on pygidium, in pairs on segments 3 and 4. Dorsal ducts on pygidium all about same size, but becoming smaller anteriorly to mesothorax; in more or less submarginal and submedian rows on segments 3 to 6, and usually a single submedian duct present on segment 7. Marginal ducts becoming fewer anteriorly. Anus situated near base of pygidium.

Ventral surface with perivulvar pores in 5 groups. Microducts present around margins of pygidium, and submarginally between the anterior and posterior spiracles. Ducts, similar to the smaller ducts on dorsum, present in small groups on margins of segments 2 and 3, and in larger groups on mesothorax and segment 1 of abdomen. Antennae each with 1 long seta. Anterior spiracles each with 4-8 disc pores.

Material examined
Holotype female. **PAPUA NEW GUINEA.** E.N.B.P., Keravat, on the bark of *Eucalyptus deglupta*, vii. 1976 (*T. Porpiu*) (BMNH).
Paratypes female. **PAPUA NEW GUINEA.** Same data as holotype. 9 (BMNH), 2 (DSIR), 2 (USNM).

Comments
It is difficult to relate this species to others in the genus. *C. lumbiniana* Takagi described from Nepal has similar median lobes, separated by only a narrow space, but has gland spines much further forward. In the South Pacific area *C. keravatana* should be identifiable from the key. It may be found in other areas where there is reafforestation with *Eucalyptus deglupta*.

Chionaspis pandanicola sp. n. (Fig. 40)

Description
Adult female on slide fusiform, about 0.85 mm long; head often straight; pygidium rounded, prepygidial segments with lateral lobes moderately developed.

Pygidium with median lobes divergent, slightly recessed into apex, rounded, strongly zygotic at base. Second lobes bilobed, each with inner lobule rounded, outer lobule smaller. Third lobes represented by serrations on margin. Gland spines arranged singly on fifth and posterior segments, smaller and more numerous on segments 2 to 4, and usually present singly on first abdominal segment. Dorsal ducts all about same size on pygidium, the marginal macroducts slightly longer; arranged in submarginal rows on segments 3 to 5, and in submedian groups of 1 or 2 on segments 4 and 5. Smaller ducts present on margins of mesothorax and segments 1 to 3 of abdomen.

Ventral surface with perivulvar pores in 5 elongate groups. Microducts sparse, present on submargins of pygidium, submedian areas of a few abdominal segments, and on head. Ducts, similar to the small ducts on dorsum, situated in groups around margins of thorax and abdominal segment 1. Antennae each with a single long seta. Anterior spiracles each with 9-15 pores.

Material examined
Holotype female. **FIJI.** Viti Levu, Nadi-Sigatoka road, on *Pandanus* sp., 28.ix.1955 (*B.A. O'Connor*) (BMNH).
Paratypes female. **FIJI.** Same data as holotype. 4 (BMNH).

Comments
Among the Pacific species this is closely related to *C. freycinetiae*, but has fewer dorsal ducts in a different arrangement. Superficially it resembles *C. castanopsidis* Takagi, described from Nepal, in having a similar arrangement of dorsal ducts; but the median lobes of this species are serrated, whereas those of *C. pandanicola* are smooth.

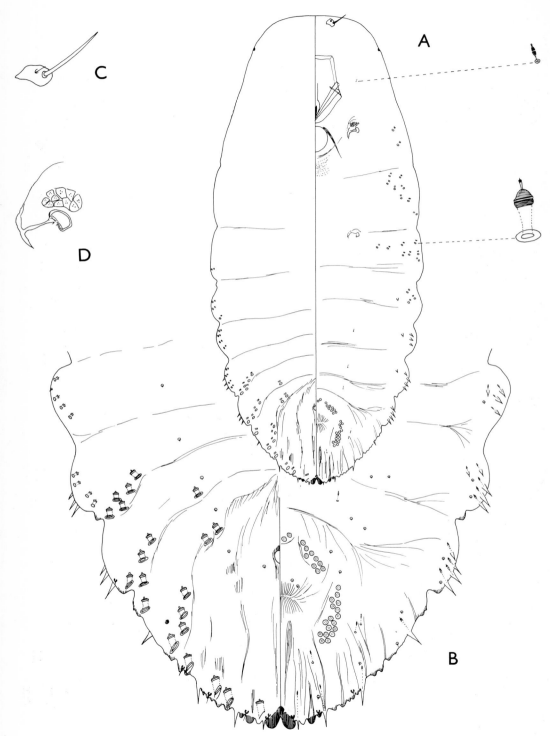

Fig. 40. *Chionaspis pandanicola* sp. n.

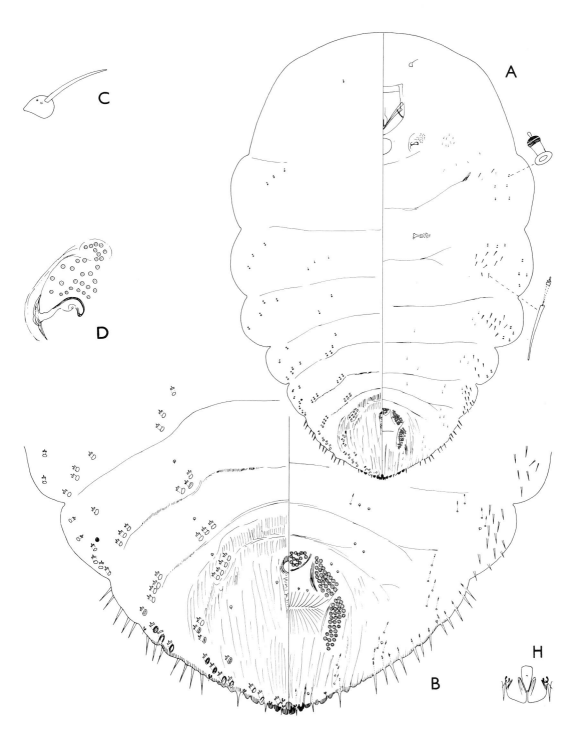

Fig. 41. *Chionaspis rhaphidophorae* sp. n.

Chionaspis rhaphidophorae sp. n. (Fig. 41)

Description
A broadly oval, almost turbinate species, membranous except for pygidium; adult female about 1.55 mm long, widest at mesothorax; head and pygidium rounded, lateral lobes of prepygidial segments well developed.

Pygidium with median lobes small in comparison to body size, rounded to quadrate, the space between very narrow, but united at base by a well-developed basal zygosis. Second lobes bilobed, each lobule about same size as a median lobe but more pointed. Third lobes represented by wide projections from margin. Gland spines single between the lobes and lateral to positions of third lobes; more numerous on anterior segments of pygidium and forming groups on the latero-ventral margins of the free abdominal segments and thorax. On the prothorax and mesothorax they are replaced entirely by gland spines that are extremely narrow but elongate, blunt, and sclerotized, there being also 1 or 2 among the normal gland spines on abdominal segment 1. Marginal ducts on pygidium in normal positions. Dorsal ducts on pygidium all about same size, forming submarginal and submedian rows posteriorly to segment 5; but anteriorly they become smaller to metathorax, and on mesothorax they are represented by a small marginal group.

Ventral surface of pygidium with numerous perivulvar pores in 5 groups. Microducts arranged in submarginal groups on pygidial segments and as far forward as first abdominal segment; others present in median areas of free abdominal segments and in a group lateral to each anterior spiracle. Small ducts, similar to the small ducts on dorsum, present on margins of thoracic and first abdominal segments. Antennae each with a single seta. Anterior spiracles each with a large group of small disc pores; posterior spiracles each with a smaller group.

Material examined
Holotype female. **FIJI.** Viti Levu, Naivicula, on *Rhaphidophora* sp., 4.xii.1957 (*B.A. O'Connor*) (BMNH).
Paratypes female. **FIJI.** Same data as holotype. 3 (BMNH).

Comments
When compared with the species from the South Pacific area, this species has a superficial resemblance to two turbinate species of *Pseudaulacaspis* discussed herein. It does not possess setae between the median lobes, however, and is therefore excluded from *Pseudaulacaspis*. Among the Holarctic species of *Chionaspis* that have median lobes almost fused throughout their length, or close together, are *C. lumbiniana* Takagi, *C. caryae* Cooley, and *C. ortholobis* Comstock, but these species are fusiform, whereas *C. rhaphidophorae* is broadly oval or even turbinate. The extremely slender thoracic gland spines, which are sclerotized and blunt, are characters that separate this species from any other in the Pacific area.

Genus **CHRYSOMPHALUS** Ashmead

Chrysomphalus Ashmead, 1880: 267; Ferris, 1938: 198; McKenzie, 1939: 51; Balachowsky 1956: 82.
Type-species *Chrysomphalus ficus* Ashmead = *Coccus aonidum* Linnaeus, by monotypy.

Description
Adult female usually pyriform, with the head and thorax membranous, and pygidium sclerotized but small in comparison to body. Lobes in 3 pairs, well developed, fourth pair represented by serrations on margin. Plates fimbriate, but often some clavate, absent from lateral to position of fourth lobes. Dorsal ducts 1-barred, long and slender, in lines or furrows. Marginal paraphyses well developed, usually longer than the lobes, arising from inner and outer basal angles of the lobes, also with 1 between each second and third lobe. Anus situated near apex of pygidium. Perivulvar pores present in 4 or 5 groups.

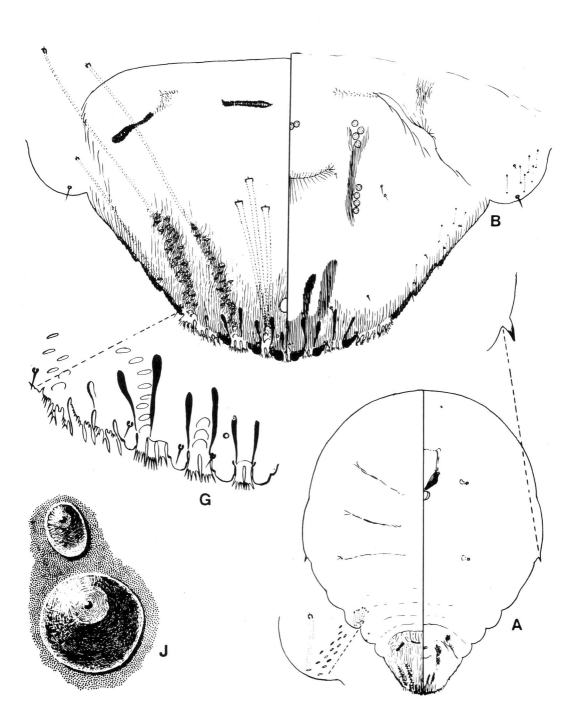

Fig. 42. *Chrysomphalus aonidum* (Linnaeus).

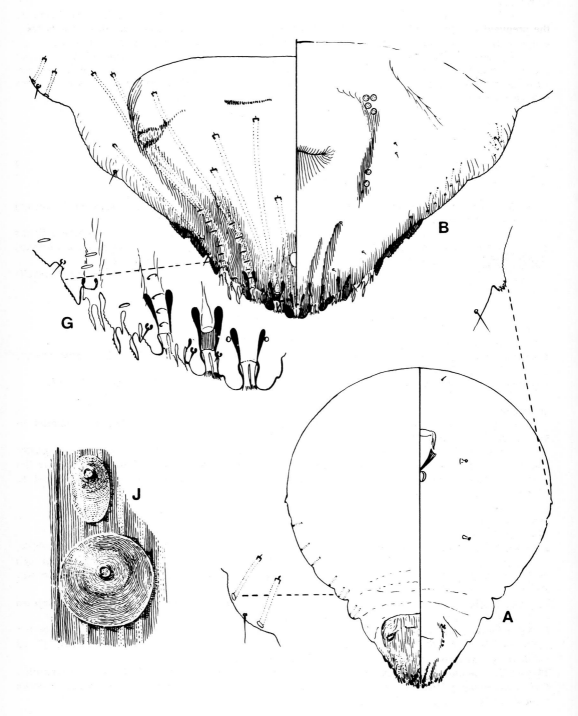

Fig. 43. *Chrysomphalus dictyospermi* (Morgan).

Comments

This genus is closely related to *Aonidiella*, but differs in never having the body reniform or the prepygidial lobes curving posteriorly to pygidium; furthermore, the paraphyses in *Chrysomphalus* are as long as or longer than the lobes, whereas in *Aonidiella* they are usually much shorter.

The genus contains a few destructive and polyphagous species; *C. aonidum* and *C. dictyospermi* are tropicopolitan species, often found in large numbers on many economic plants including *Citrus*. At present, four species are known from the South Pacific area. All are probably introduced, even though *C. pinnulifer* (Maskell) was described from Fiji as early as 1891.

Key to species of *Chrysomphalus*

1	Plates just lateral to third lobes fringed	2
--	Plates just lateral to third lobes clavate	3
2	With a submarginal cluster of ducts on second abdominal segment .... ***aonidum*** (Linnaeus)	
--	Without a submarginal cluster of ducts on second abdominal segment .... ***propsimus*** Banks	
3	Ducts in second and third furrows few, in single rows .... ***dictyospermi*** (Morgan)	
--	Ducts in second and third furrows more numerous, in double to triple rows .... ***pinnulifer*** (Maskell)	

Chrysomphalus aonidum (Linnaeus) (Fig. 42)

Coccus aonidum Linnaeus, 1758: 455.
Chrysomphalus ficus Ashmead, 1880: 267; Ferris, 1938: 201, McKenzie, 1939: 59; Balachowsky, 1956: 88; Takagi, 1969: 86.
Chrysomphalus aonidum (Linnaeus), Cockerell, 1899a: 273; Morrison & Morrison, 1966: 37.

Description

Female scale round, usually dark purple; exuviae paler, subcentral. Male scale similar to that of female but smaller and oval.

Adult female easily recognizable by the rather wide pygidium with the posterior end gently rounded, not pointed; by the fairly numerous ducts in the second and third furrows; and by the cluster of small submarginal ducts on segment 2 of abdomen, these being the most conspicuous feature of the species.

Material examined
FIJI. Viti Levu, 1976. On *Ludwigia octovalvis*.
KIRIBATI. Abaiang, 1971; Abemama, 1976; Onotoa, 1971; Tarawa, 1963. On *Artocarpus altilis*, *Bougainvillea* sp., *Citrus* sp., *C. aurantifolia*, *C. limon*, *Cocos nucifera*, *Euphorbia* sp., *Morinda citrifolia*, *Musa* sp., *M. sapientum*, *Nerium oleander*, *Pandanus* sp., *P. odoratissimus*, *Plumeria rubra*, *Premna* sp., *Ricinus communis*.
NEW CALEDONIA. New Caledonia, 1925. On *Agathis moorei*, Arecaceae, *Cocos nucifera*, *Fagraea* sp., *Nerium oleander*, *Nothofagus aequilateralis*, *Pandanus* sp.
PAPUA NEW GUINEA. C.P.: 1959. E.N.B.P.: 1968. E.S.P.: 1968. M.B.P.: 1985. Morobe P.: 1959. N.I.P. 1983. N.S.P. 1983. W.N.B.P. 1956. On *Citrus* sp., *C. limon*, *C. paradisi*, *Cocos nucifera*, *Melaleuca* sp., *Musa* sp., *Pinus caribaea*, *Vanilla* sp.
TUVALU. Funafuti, 1926; Nukufetau, 1976; Nukulaelae, 1976; Vaitupu, 1976. On *Artocarpus altilis*, *Calophyllum inophyllum*, *C. nucifera*, *Cycas* sp., *Gardenia* sp., *Heliconia* sp., *Musa* sp., *Nerium oleander*, *Pandanus* sp., *Plumeria* sp., *P. rubra*.

Comments
This species is known as the Florida red scale and is common in the South Pacific area. Although its host-plant range is wide, it is particularly common on citrus wherever the scale has been introduced.

In addition to the above material, *C. aonidum* has been recorded from American Samoa on *Citrus* sp. and *Cocos nucifera* by Dumbleton (1954); from French Polynesia, Tahiti, on *Psidium guajava* (Doane & Hadden, 1909); from New Caledonia, where Cohic (1950a) found it was most prolific in the dry season, spreading from the leaves of *Citrus sinensis* to the twigs and branches, and often killing the tree; other hosts listed from New Caledonia by Cohic (1956, 1958a) and Brun & Chazeau (1980, 1984) are *Agathis lanceolata, Aleurites moluccana, Annona muricata, A. reticulata, A. squamosa, Aspidistra* sp., *Barringtonia asiatica, B. speciosa, Bauhinia variegata, Calophyllum inophyllum, Citrus* sp., *C. aurantium, C. grandis, C. limon, C. paradisi, C. reticulata, C. sinensis, Dodonaea viscosa, Eriobotrya japonica, Eugenia* sp., *E. cumini, Ficus* sp., *F. pumila, Gardenia* sp., *Gerbera* sp., *G. jamesoni, Gladiolus* sp., *Jasminum* sp., *J. sambac, Latania commersonii, Laurus nobilis, Leucopogon* sp., *Melaleuca quinquenervia, M. leucadendron, Musa sapientum, Rosa* sp., *Sansevieria* sp., *Santalum austrocaledonicum* and *Tamarindus indica*. Szent-Ivany & Stevens (1966) mention it from Papua New Guinea, Morobe P., severely damaging seedlings of *Pinus* sp.; Reddy (1970) records it from Western Samoa on *Cocos nucifera*, and Maddison (1976) mentions it as a major pest of citrus there.

Chrysomphalus dictyospermi (Morgan) (Fig. 43)

Aspidiotus dictyospermi Morgan, 1889: 352.
Chrysomphalus dictyospermi (Morgan), Leonardi, 1899: 218; Ferris, 1938: 200; McKenzie, 1939: 57; Balachowsky 1956: 86.

Description
Female scale round, light brown or yellowish, rather delicate, exuviae more or less central. Male scale similar to female scale but oval.

Adult female on slide tending to have a pointed pygidium because the lobes are oblique and point postero-medially. Important diagnostic characters are 3 pairs of elongate lobes, the fourth pair each represented by an acute point; plates between the first 3 pairs of lobes fringed, but between third and fourth lobes they are clavate with serrate edges. The dorsal ducts are few, those in the second and third furrows present in single rows except for an occasional duct out of line.

Material examined
COOK IS. Aitutaki, 1975; Atiu, 1977; Mangaia, 1977; Mauke, 1954; Rakahanga, 1977; Rarotonga, 1973. On *Artocarpus heterophyllus, Asparagus plumosus, Citrus maxima, C. sinensis, Cocos nucifera, Dianella intermedia, Malus sylvestris, Mangifera indica*.
FIJI. Taveuni, 1977; Vanua Levu, 1974; Viti Levu, 1922. On *Alpinia nutans, Anacardium occidentale*, Arecaceae, *Artocarpus altilis, A. incisa, Barringtonia* sp., *B. racemosa, Calathea* sp., *Calophyllum inophyllum, Camellia sinensis, Epipremnum pinnatum, Macadamia tetraphylla, Mangifera indica, Musa* sp., *Myristica* sp., *Phaeomeria speciosa, Pinus caribaea, Psidium guajava, Rosa indica, Roystonea regia, Spondias dulcis, Terminalia catappa, Vanilla fragrans*.
FRENCH POLYNESIA. Society Is, Tahiti, 1934. On ?
IRIAN JAYA. Biak, 1959; Tenwer, 1959. On *Cocos nucifera*.
KIRIBATI. Butaritari, 1976; Marakei, 1976; Nonouti, 1976. On *Artocarpus altilis, Calophyllum inophyllum, Ficus tinctoria, Xanthosoma* sp.
NIUE. 1975. On *Citrus aurantifolia, Schefflera* sp.
PAPUA NEW GUINEA. C.P.: 1959. E.S.P.: 1958. M.B.P.: Trobriand Is, Kiriwina, 1985. Morobe P.: 1959. N.I.P.: 1983. N.S.P.: 1983. S.H.P.: 1983. W.H.P.: 1984. On *Carica papaya, Casuarina* sp., *Citrus* sp., *Cocos nucifera, Ficus* sp., 'Laulau tree', *Schuurmansia henningsii, Vanilla* sp.

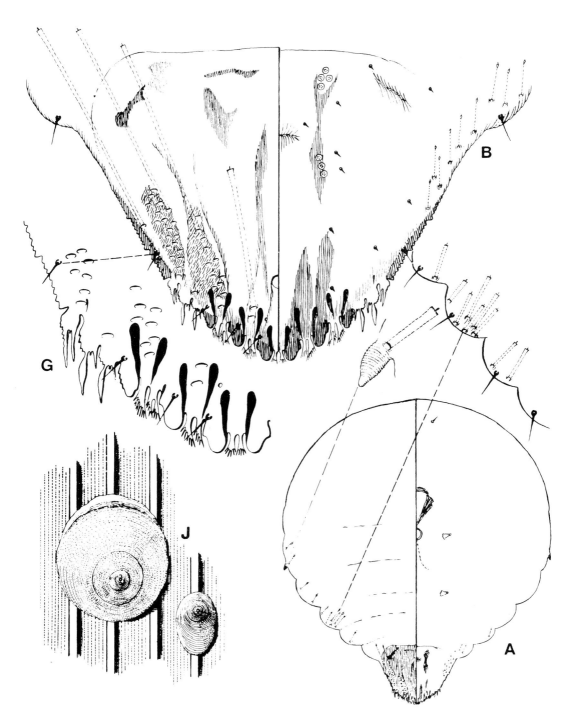

Fig. 44. *Chrysomphalus pinnulifer* (Maskell).

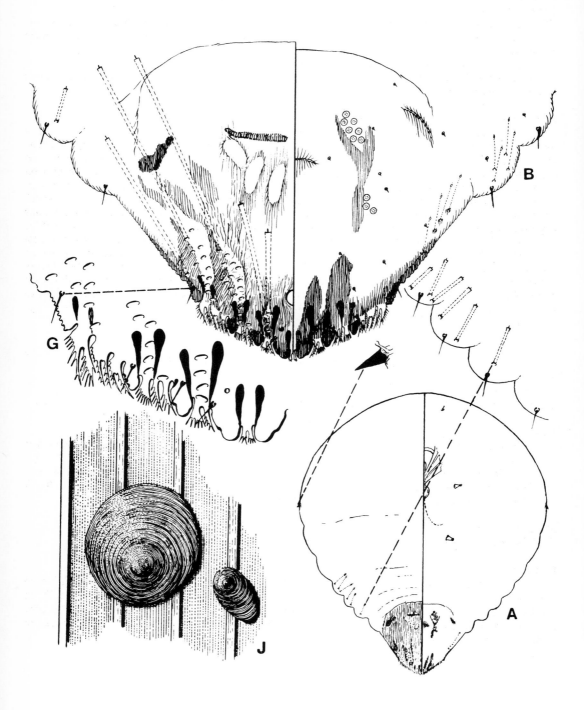

Fig. 45. *Chrysomphalus propsimus* Banks.

SOLOMON IS. G.P.: Guadalcanal, 1954. T.P.: Reef Is, 1985. W.P.: New Georgia Is, Gizo, 1984. On *Areca* sp., *Cocos nucifera*.
TONGA. Tongatapu Group, 'Eua, 1977, Tongatapu, 1973; Vava'u Group, Vava'u, 1974. On *Citrus limon, C. maxima, Cycas* sp., *Mangifera indica, Manihot esculenta, Meryta macrophylla, Persea americana*.
TUVALU. Funafuti, 1976. On *Artocarpus altilis, Musa* sp.
WESTERN SAMOA. Savai'i, 1975; Upolu, 1975. On *A. heterophyllus, Elaeis guineensis, Mangifera indica, P. americana, Swietenia macrophylla*.

Comments

This tropicopolitan species has a wide distribution in the South Pacific area on numerous plant species, usually on the leaves. Interceptions at quarantine stations from the area, on citrus and *Howeia forsteriana*, suggest that it has an even wider distribution in the Pacific region. In addition to the material listed above, *C. dictyospermi* has been recorded from Cook Is as a pest on *Citrus aurantifolia* (Maddison, 1976); from Easter I. on *C. aurantium* by Charlín (1973); from Fiji on *Cocos nucifera* (Simmonds, 1925; Greenwood, 1929), *Areca catechu, Macaranga* sp., *Veitchia joannis* (Hinckley, 1965), *Musa sapientum* (Veitch & Greenwood, 1921) and *Zingiber* sp. (Lever, 1945a); it has been recorded from French Oceania on *Cocos nucifera* (Dumbleton, 1954), Solanaceae (Cohic, 1955), *Citrus grandis, Lagerstroemia flos-reginae* and *Solanum melongena* (Reboul, 1976), and is a major pest of citrus (Maddison, 1976); from Tahiti, Cohic (1955) mentions it on *Eugenia malaccensis, Mangifera indica*, and damaging *Citrus paradisi* and *C. sinensis*, covering the tree, turning the leaves chlorotic, and drying and killing the branches. Cohic (1958a) recorded it from New Caledonia on *Spermolepis gummifera*. In Niue it is a pest on *Citrus sinensis* fruit (Maddison, 1976) and in Tonga it is a major pest of citrus (Maddison, 1976) and occurs on *Cocos nucifera* and *Plumeria* sp. (O'Connor, 1949).

Chrysomphalus pinnulifer (Maskell) (Fig. 44)

Diaspis pinnulifera Maskell, 1891: 4.
Chrysomphalus pinnulifer (Maskell), McKenzie 1939: 61.

Description

Female scale reddish-brown, circular, thin, exuviae central. Male scale oval, similar colour to female scale.

Adult female on slide resembling *C. dictyospermi*, but differing mainly in the arrangement and number of the dorsal pygidial ducts. In *C. dictyospermi* they are few, in single rows, but in *C. pinnulifer* they are more numerous in double to triple rows.

Material examined
PAPUA NEW GUINEA. E.H.P.: 1959. W.H.P.: 1984. On *Citrus* sp.

Comments

This species can only be confused with *C. dictyospermi*, but in life the scale is reddish-brown whereas the scale of *C. dictyospermi* is paler. Although *C. pinnulifer* was described originally in 1891 from Fiji on an unknown plant, apparently it has not been recorded from there since. It is now known to be a tropicopolitan species extending into temperate areas and is found on the leaves of numerous plant species.

Chrysomphalus propsimus Banks (Fig. 45)

Chrysomphalus propsimus Banks, 1906a: 230; McKenzie, 1939: 62.

Description
Scale of female described as dark chocolate brown, round, exuviae subcentral. Male scale similar colour to that of female, but smaller and oval.
Adult female on slide similar to that of *C. aonidum* but lacking the clusters of ducts on segment 2 of abdomen. A few ducts, however, present on prepygidial margins of abdomen, but not in clusters.

Material examined
KIRIBATI. Tarawa, 1976. On *Pandanus odoratissimus*.
TUVALU. Nanumea, 1977. On *Cocos nucifera*.

Comments
Original specimens were described from the Philippines as occurring on the leaves of *Cocos nucifera*, and the species has been recorded from Sumatra on *Calamus spectabilis*. Identifications of the above Pacific specimens are based on the description and illustration by McKenzie (1939), and on a specimen from Sumatra on *Calamus spectabilis*, presumably from the same material on which the record was based. The species may have a much wider distribution.

Genus CLAVASPIS MacGillivray

Clavaspis MacGillivray, 1921: 391; Ferris, 1938: 202; Balachowsky, 1956: 90. Type-species *Aspidiotus subsimilis* var. *anonae* Houser = *Aspidiotus herculeanus* Cockerell & Hadden.

Description
This genus is easily recognizable in the Pacific region by the presence of median lobes only, each with a well-developed clavate paraphysis arising laterally from the base. Smaller paraphyses are present laterally. The plates resemble gland spines, and the dorsal ducts are submarginal and 1-barred.

Comments
About 20 species have been described in the genus, all with paraphyses varying considerably in size, but none with the large paraphyses as large as those in the type-species. The genus occurs in various parts of the world, but only *C. herculeana* is known to be tropicopolitan, occurring on a wide range of plants.

Clavaspis herculeana (Cockerell & Hadden) (Fig. 46)

Aspidiotus herculeanus Cockerell & Hadden *in* Doane & Hadden, 1909: 298.
Clavaspis herculeana (Cockerell & Hadden), MacGillivray, 1921: 441; Ferris, 1938: 206; Balachowsky, 1956: 94.

Description
Female scale whitish or brownish, circular, the exuviae central; but, according to Ferris (1938), the scales are so mingled with the epidermal layers of the plant, that it is difficult to give a good description.
Adult female on the slide almost pyriform, pygidium acute. Other characters as in the description of the genus.

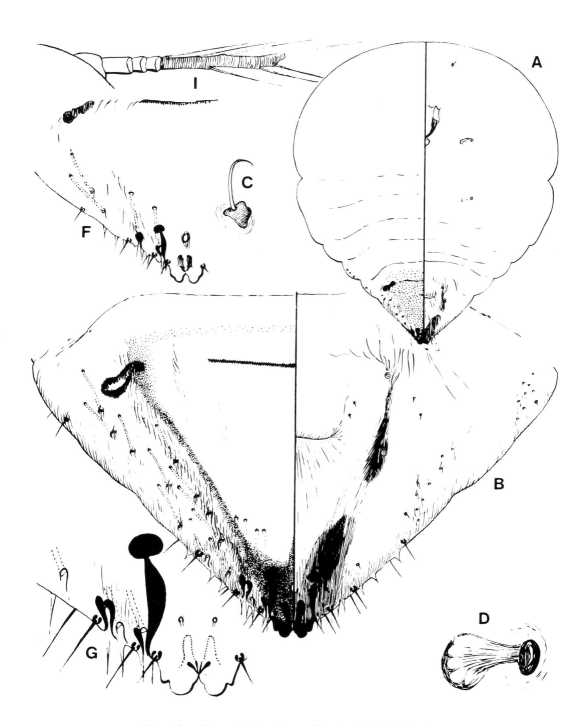

Fig. 46. *Clavaspis herculeana* (Cockerell & Hadden).

Material examined
COOK IS. Rarotonga, 1977. On *Rosa indica*.
FIJI. Taveuni, 1977; Viti Levu, 1942. On *Gossypium* sp., 'Leguminosae', *Mimosa pudica*, *Pithecellobium saman*.
TONGA. Ha'apai Group, Foa, 1977. On *Myristica hypargyraea*.
WESTERN SAMOA. Savai'i, 1975. On *Mangifera indica*.

Comments
This species was described originally from the Society Islands, presumably Tahiti, on an unknown plant, but it was probably introduced there. No material has been seen from there for this work. In addition to the material examined, *C. herculeana* has been recorded from Fiji on *Pithecellobium saman* (Hinckley, 1965); and from French Polynesia on *Erythrina indica* by Reboul (1976). The species probably has a wider distribution in the Pacific area. It is not very conspicuous on plants, however, and may escape detection.

Genus **DIASPIS** Costa

Diaspis Costa, 1835: 19. Type-species *Diaspis calyptroides* Costa = *Aspidiotus echinocacti* Bouché, by subsequent designation.

Description
Body turbinate, membranous except for pygidium. Ducts 2-barred. Median lobes, although sometimes close together, never yoked or zygotic. Second and third lobes, bilobed, usually well developed. Often a pronounced spur in position of each fifth lobe. Marginal pygidial macroducts usually arranged as 6 on each side. Submarginal ducts the same size or smaller, in varying numbers. Dorsal ducts usually smaller, in definite submarginal and submedian groups. Perivulvar pores present in 5 groups. Gland spines small, never between median lobes, but a pair of setae always present in this position.

Comments
Over 50 species are included in the genus at present, although some may not be congeneric with the type-species. The distribution is mainly tropicopolitan, but the genus extends into some temperate areas. Four species are known from the South Pacific region. Three have been introduced, and the other, described here, is apparently endemic to New Caledonia. It is included in *Diaspis* for the present, but it has distinctive characters.

Key to species of *Diaspis*

1	Submedian groups of ducts on pygidium, present and conspicuous	2
--	Submedian groups of ducts on pygidium absent	3
2	Dorsal ducts present on median areas of head. Ventral duct tubercles and spicules present on thorax	***casuarinae*** sp. n.
--	Dorsal ducts absent from median areas of head. Ventral duct tubercles and spicules absent from thorax	***echinocacti*** (Bouché)
3	Submarginal macroducts numbering only 2 on each side	***boisduvalii*** Signoret
--	Submarginal macroducts numbering at least 6 on each side	***bromeliae*** (Kerner)

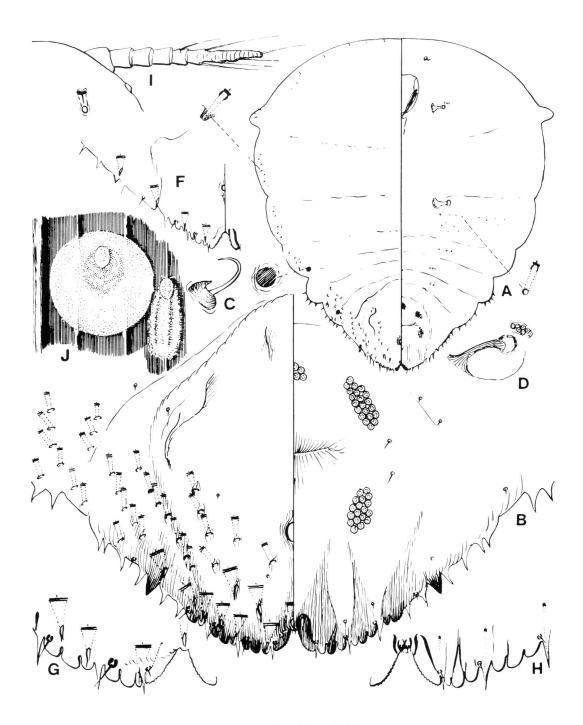

Fig. 47. *Diaspis boisduvalii* Signoret.

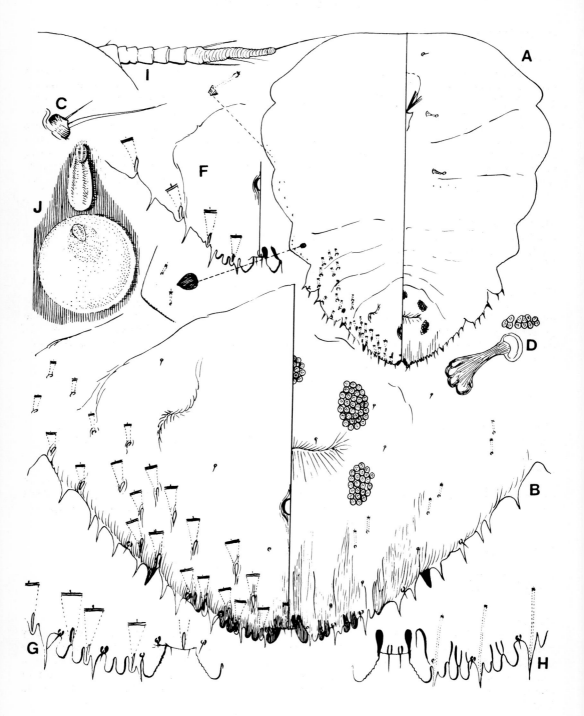

Fig. 48. *Diaspis bromeliae* (Kerner).

Diaspis boisduvalii Signoret (Fig. 47)

Diaspis boisduvalii Signoret, 1869b: 432; Ferris, 1937a: 32; Balachowsky, 1954: 178.

Description
   The prosoma of this species usually has lateral lobes. Main pygidial characters are the 6 macroducts on each side and 1 macroduct between median lobes. Median lobes forming a deep notch at apex; the inner margins of each lobe diverging, serrate, and much longer than the outer margins. Second and third lobes rounded, bilobed. Lateral spur well developed at position of each sixth macroduct. A pair of macroducts present at each side above second and third lobes. Dorsal ducts smaller than marginal ducts, forming submarginal groups forward to about abdominal segment 2. Duct tubercles present anteriorly on margins of dorsum as far forward as mesothorax. Submarginal bosses present on first and third abdominal segments. Perivulvar pores present in 5 compact groups. Antennae each with a single seta. Anterior spiracles each with 6-10 disc pores. Anus situated about one third length of pygidium from apex.

Material examined
COOK IS. Atiu, 1975. On *Ananas comosus*, [?]*Citrus sinensis*.
FIJI. Viti Levu, 1984. On *A. comosus*.
SOLOMON IS. G.P.: Guadalcanal, 1984. On *A. comosus*.

Comments
   This is a tropicopolitan species found on numerous plant species, although it seems to prefer monocotyledons. It is often found on palms and is common on pineapple, when it can be confused with *D. bromeliae*. The record on citrus above is unusual, but there is no way of checking the host. *D. boisduvalii* has been recorded from Tahiti on *Rosa* sp. (Doane & Hadden, 1909), and has been intercepted at quarantine in New Zealand from the area.

Diaspis bromeliae (Kerner) (Fig. 48)

Coccus bromeliae Kerner, 1778: 20.
Diaspis bromeliae (Kerner), Signoret, 1869b: 434; Ferris, 1937a: 33; Balachowsky, 1954: 182.

Description
   This species is morphologically similar to *D. boisduvalii*, described above, but differs in easily distinguishable characters. Whereas in *D. boisduvalii* the large submarginal macroducts number 2 on each side just anterior to the second and third lobes, in *D. bromeliae* they number 6 or more.

Material examined
COOK IS. Aitutaki, 1975; Atiu, 1973; Mangaia, 1960; Rarotonga, 1954. On *A. comosus*.
FRENCH POLYNESIA. Society Is, Tahiti, 1969. On *A. comosus*.
WESTERN SAMOA. Upolu, 1975. On *A. comosus*.

Comments
   This tropicopolitan species is found on a wide variety of plant species; it is generally known, however, as a pest of pineapples, although its pest status is not fully understood. It is usually found wherever pineapples are grown. Veitch & Greenwood (1921) record it from Fiji on *Ananas sativus*, and Lever (1940) and several subsequent authors mention that it is a pest of pineapples there; it has also been intercepted at quarantine in Hawaii on produce from the Pacific islands.

Diaspis casuarinae sp. n. (Fig. 49)

Description
   Scale of adult female appearing elongate, but sides 'clasping' the branchlets; about 1.2 mm long, white, smooth, convex, with yellow-brown exuviae terminal. Male scale elongate, about 0.75 mm long, ridged, white, but more transparent than female scale.
   Adult female, in common with most species of the genus, turbinate, about 1.3 mm long, 0.75 mm wide, membranous except for pygidium.
   Pygidium with median lobes divergent, forming an apical notch; appearing yoked because of heavy sclerotization, but separated by a minute space, with a pair of setae between. Second lobes bilobed, inner and outer lobules rounded. Third lobes each represented by 2 rounded projections. Fourth lobes not recognizable except as marginal swellings. Gland spines single on either side of each segment. Marginal macroducts present as 1 between median lobes, 1 between median and second lobes and numerous around margins to segment 4. Submarginally they are slightly smaller and form groups anteriorly to abdominal segment 1. A single submedian duct usually present on segment 6 and other submedian ducts present as far forward as segment 1. A few submedian microducts also present on metathorax. Mid-head region with single or groups of 2-4 small ducts; orifices of ducts within each group uniting, but not coalescing, within a common base. A single submarginal boss present on abdominal segment 1. Anus situated about two thirds length of pygidium from apex.
   Ventral surface with perivulvar pores present in 5 groups, varying considerably in numbers from 43-82. Microducts sparse, present on submargins of pygidium and in median areas of free abdominal segments. Duct tubercles pointed, present in submedian areas of prothorax, mesothorax and metathorax; small groups also present laterally on abdominal segments 1 to 3. Small ducts present in groups lateral to the duct tubercles on mesothorax, metathorax and first abdominal segment. Antennae each with 1 long seta. Disc pores present, 6-11 by each anterior spiracle and 2-3 by each posterior spiracle; each pore with 3 loculi. Spicules conspicuous but few, in median areas of mesothorax and metathorax.

Material examined
Holotype female. **NEW CALEDONIA**. 'Plume Farm', on *Casuarina* sp., 31.v.1928 (*T.D.A. Cockerell*) (BMNH).
Paratypes female. **NEW CALEDONIA**. Same data as holotype, 3 (BMNH). Nouméa, on *Casuarina collina* (iron wood), 22. viii. 1940 (*F.X. Williams*), 4 (BPBM), 3 (BMNH); La Coulée, on *Casuarina* sp., 23.i.1963 (*N.L.H. Krauss*), 8 (BPBM), 7 (BMNH).

Comments
   This is an unusual species in having groups of tubular ducts in the mid-head region. It bears some relationship to *D. chilensis* Cockerell, another species with large numbers of ducts, but this species has much larger pygidial lobes. Normally the scale in *Diaspis* is round with the exuviae subcentral or submarginal, but in *D. casuarinae* the exuviae are terminal.

Diaspis echinocacti (Bouché) (Fig. 50)

Aspidiotus echinocacti Bouché, 1833: 53.
Diaspis echinocacti (Bouché), Fernald, 1903: 229; Ferris, 1937a: 36; Balachowsky, 1954: 189.

Description
   This species differs from all the others in the Pacific area in possessing a prominent pair of median lobes which are rounded and, although they may diverge, they do not form a deep notch at the apex of pygidium. The dorsal ducts are numerous on segments 5 and 6, where they are particularly conspicuous in compact submedian groups.

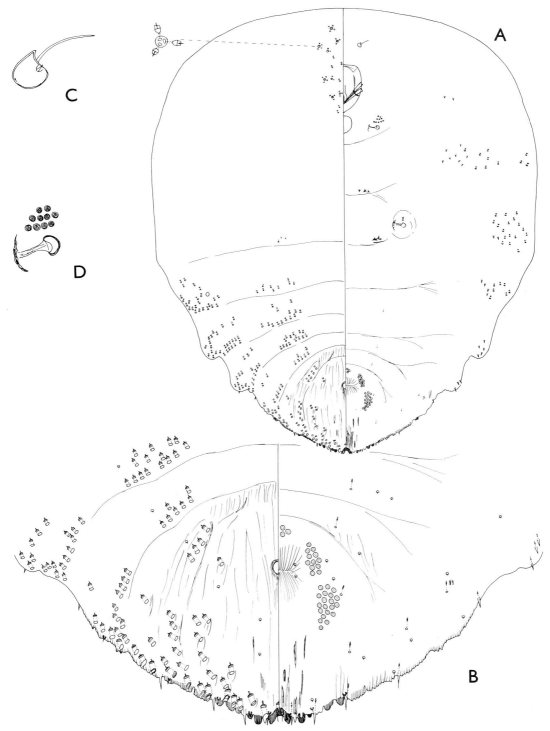

Fig. 49. *Diaspis casuarinae* sp. n.

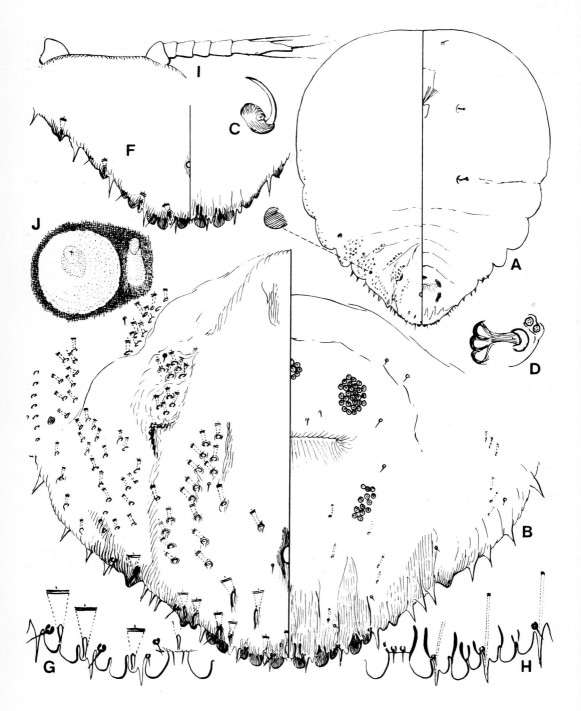

Fig. 50. *Diaspis echinocacti* (Bouché).

Material examined
PAPUA NEW GUINEA. C.P.: 1962. On ornamentals.

Comments

There should be no difficulty in recognizing this species using the characters given in the key. It is often found wherever species of Cactaceae are grown, and the record above, on ornamentals, is almost sure to refer to a cactus. *D. echinocacti* has been recorded as a pest of *Opuntia* sp. in French Polynesia by Reboul (1976), and it is probably widespread throughout the Pacific area.

Genus **DUPLASPIDIOTUS** MacGillivray

Duplaspidiotus MacGillivray, 1921: 394; Ferris, 1938: 226; Balachowsky 1958: 257. Type-species *Pseudaonidia clavigera* Cockerell, by original designation.

Description

A genus with 1-barred ducts and with plates usually fringed. Body usually heavily sclerotized, turbinate or pyriform, but with a deep constriction between the prothorax and the rest of the body. It resembles *Pseudaonidia* in possessing a pygidium with much of the dorsum sclerotized and areolated. These are good characters for recognizing both genera, but *Duplaspidiotus* differs in possessing well-developed paraphyses arising from the outer basal angles of the median lobes, the inner end of each paraphysis usually circular and detached from the stem.

Duplaspidiotus claviger (Cockerell) (Fig. 51)

Pseudaonidia clavigera Cockerell, 1901: 226.
Duplaspidiotus claviger (Cockerell), MacGillivray, 1921: 453; Ferris, 1937b: 70; Balachowsky, 1958: 258.

Material examined
COOK IS. Rarotonga, 1975. On *Citrus sinensis*, *Gardenia* sp.
FIJI. Viti Levu, 1955. On *Dodonaea viscosa*.
NIUE. 1977. On *Macadamia tetraphylla*.
PAPUA NEW GUINEA. C.P.: 1959. On *Ficus* sp.
WESTERN SAMOA. Savai'i, 1975. On *Citrus limon*.

Comments

Of about 20 species described in *Duplaspidiotus*, this is the only one established in the South Pacific area. Hinckley (1965) records it on *Gardenia* sp. from Fiji, in addition to the host given above. It is known from southern Asia, South Africa, the Malagasian area and Hawaii on numerous species of plants, and should be recognizable from the key and generic description.

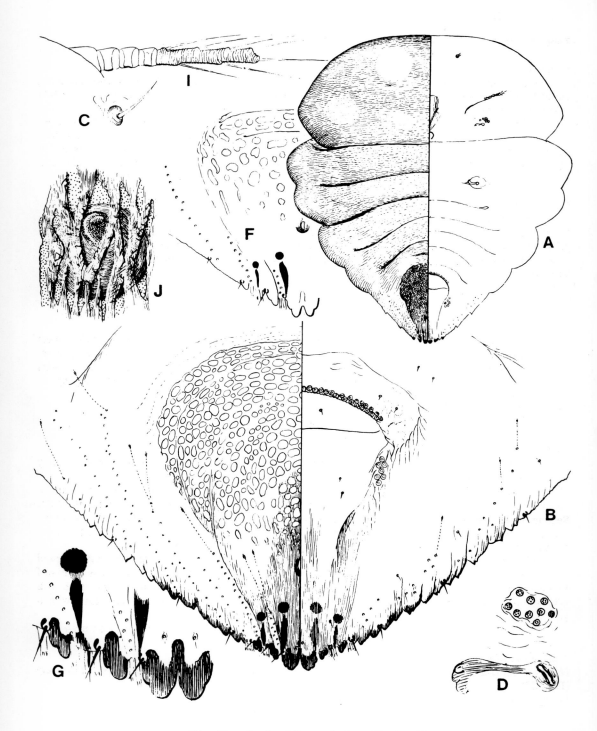

Fig. 51. *Duplaspidiotus claviger* (Cockerell).

Genus **FIJIFIORINIA** gen. n.

Type-species *Fijifiorinia oconnori* sp. n.

Description
Pupillarial scale of female black, resembling many black Aleyrodoidea; fusiform, but appearing wider because of a covering of white wax extending past margins; exuviae brown, terminal. Male scale possibly white, but represented only by pieces.

Adult female elongate-oval, membranous except for pygidium. Median lobes prominent, without setae between, yoked at base by an elongate process, fused except towards apex; distal edge notched. Second and third lobes present or absent. Marginal ducts 2-barred. Gland spines present on pygidium. Perivulvar pores present in 5 groups. Microducts on pygidium with minute spines at orifices.

Comments
This genus bears a superficial resemblance to *Pinnaspis* in possessing median lobes that are fused for most of their length. It is pupillarial, however, and its true affinities are with the genus *Fiorinia*. In *Fiorinia* the lobes are separated and there are always two setae between at the base. There appear to be two distinct species in the material at hand, and these can be separated as follows:

Key to species of *Fijifiorinia*

1   Perivulvar pores in 3 elongate groups. Marginal ducts numbering 4 on each side. Microducts present on venter, thorax, free abdominal segments and pygidium .... ***astronidii*** sp. n.
--   Perivulvar pores in 3 compact groups. Marginal ducts numbering 3 on each side. Microducts present on venter confined to the pygidium .... ***oconnori*** sp. n.

Fijifiorinia astronidii sp. n. (Fig. 52)

Description
Adult female on slide elongate-oval, about 1.0 mm long, head and pygidium rounded, body membranous except for pygidium.

Pygidium with prominent median lobes separated only by a narrow space nearly full length of lobes, each lobe pointed, with 3 notches on oblique outer edge; yoke at base of lobes elongate. Second lobes either absent or represented by sclerotized points. Gland spines well developed, arranged singly, there being 6 on each side. Marginal ducts slender, numbering 4 on each side. Preanal scars present singly and lateral scars present in pairs.

Ventral surface with 5 elongate groups of perivulvar pores present. Microducts on submarginal area of pygidium with spines at orifice; other microducts without spines, present in median areas of free abdominal segments and thorax and laterally on prothorax. Antennae near head margin, each with a single seta. Anterior spiracles each with a group of disc pores; posterior spiracles each with a smaller group.

Second instar with lobes represented by median pair. Marginal ducts numbering 4 on each side.

Material examined
Holotype female. **FIJI.** Viti Levu, Lomaivuna, *Astronidium* sp., 11.viii.1963 (*B.A. O'Connor*) (BMNH).
Paratypes female. **FIJI.** Same data as holotype. 5 (BMNH).

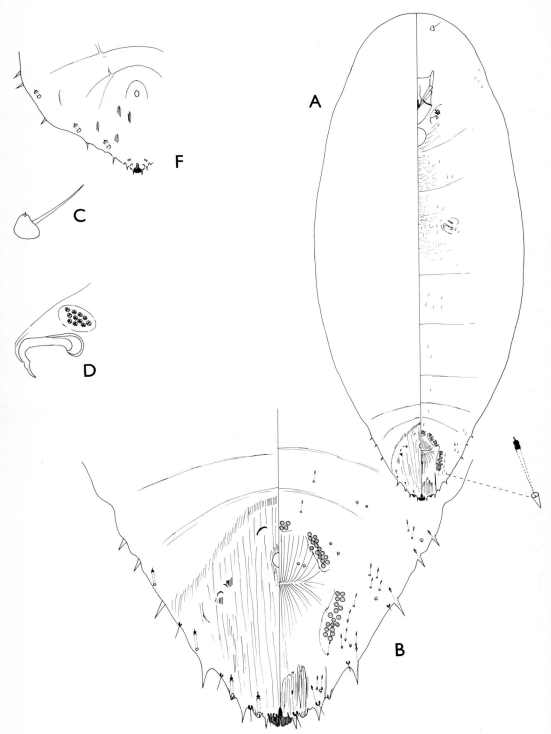

Fig. 52. *Fijifiorinia astronidii* sp. n.

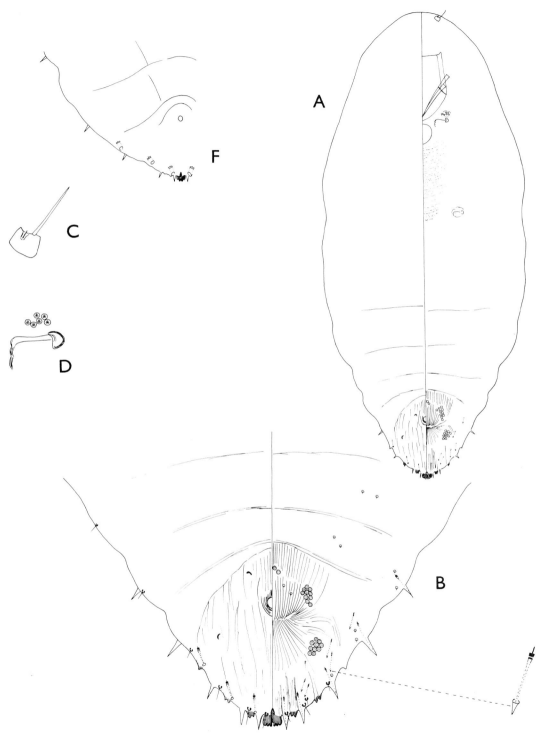

Fig. 53. *Fijifiorinia oconnori* sp. n.

Fijifiorinia oconnori sp. n. (Fig. 53)

Description
Adult female, slide-mounted, elongate-oval, about 0.7 mm long, head and pygidium rounded. Only pygidium sclerotized.

Pygidium with median lobes prominent, contiguous almost to apices, where they are separated by a short cleft, each lobe subrectangular, parallel, with about 3 notches distally; yoke at base of lobes elongate. Second lobes present as notched sclerotized projections, smaller than median lobes; position of third lobes represented by short dentations on the margin. Gland spines robust, arranged singly on segmental margins, usually 5 on each side. Marginal ducts slender, 3 on each side. Anus situated near base of pygidium. Preanal and lateral crescentic scars present.

Ventral surface with 5 compact groups of perivulvar pores present. Microducts present submarginally on pygidium, each with a minute spine at orifice. Antennae near head margin, each with prominent base and a single seta. Anterior spiracles each with a small group of disc pores.

Second instar with median lobes only present and with 3 pairs of marginal ducts.

Material examined
Holotype female. **FIJI.** Viti Levu, Sawani, on *Myristica macrantha*, 2.vii.1956 (*B.A. O'Connor*) (BMNH).
Paratypes female. **FIJI.** Same data as holotype. 5 (BMNH), 1 (DSIR), 3 (USNM); Koronovia, on *Erythrina lithosperma*, 11.i.1957 (*B.A. O'Connor*), 4 (BMNH).

Comments
The record listed by Hinckley (1965) as *Fiorinia* sp. nr *geigerae* (Maskell) refers to this species.

Genus **FIORINIA** Targioni

Fiorinia Targioni, 1868: 735. Type-species *Fiorinia pellucida* Targioni = *Diaspis fioriniae* Targioni, by monotypy.

Description
A pupillarial genus, usually elongate, often with parallel sides, but some species wide. Median lobes zygotic, often forming a deep notch at apex of pygidium, or prominent; always with a pair of setae between. Second lobes usually well developed, but sometimes reduced or absent. Gland spines, if present, arranged singly. Duct tubercles, if present, on ventral margins. Marginal ducts large or slender, 2-barred. Submarginal ducts absent from pygidium but occasionally present in submedian area at pygidial base. Perivulvar pores often present in 5 or fewer groups, occasionally absent entirely. Antennae usually close together near head margin, often conical; each with a single seta. Some species with a variably-shaped interantennal process present. Anterior spiracles each with disc pores containing 3 loculi. Second instar with pygidium often similar to that of adult female.

Comments
This genus is in great need of revision. It is probably of southern or eastern Asian origin and some species have become established throughout the world. Although the characters are fairly easy to study, the species are pupillarial and adult females are difficult to remove intact from the second instars. The genus may be widespread in the west Pacific region and although species may often be intercepted at quarantine inspection, their pest status is not yet known. Pacific species discussed here exclude *Fiorinia neocaledonica* Lindinger described from New Caledonia on *Baeckea pinifolia*, by Lindinger (1911). A syntype female has kindly been forwarded for study by H. Strümpel, Zoologisches Institut und Zoologisches Museum, Hamburg, but this is immature.

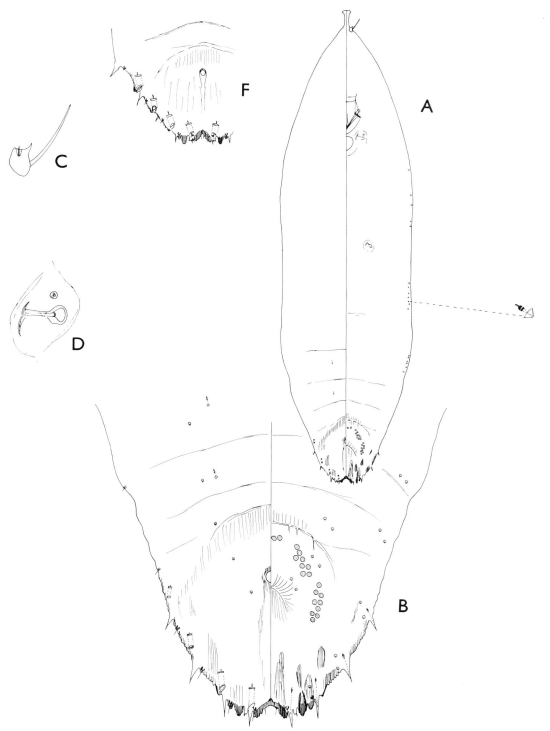

Fig. 54. *Fiorinia biakana* sp. n.

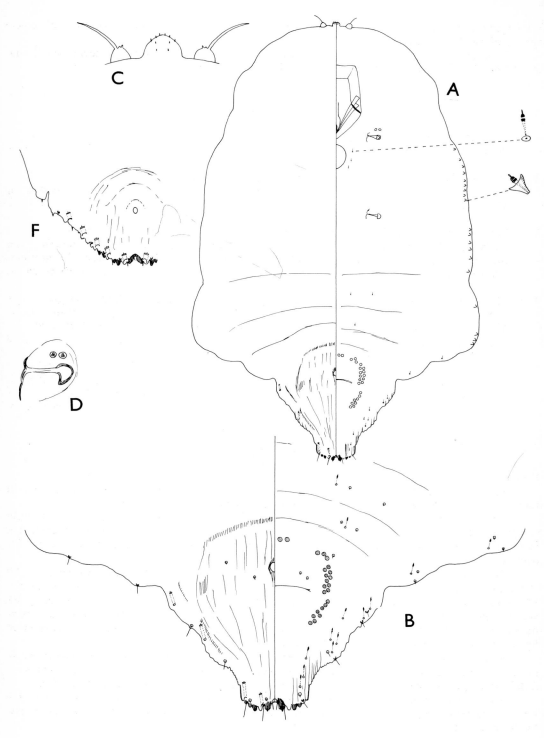

Fig. 55. *Fiorinia coronata* sp. n.

Lindinger's illustration of the adult female suggests that the species is not congeneric with other species of *Fiorinia*, and needs further study.

F. theae Green has been recorded from Jayapura, Irian Jaya on *Citrus* sp., by Reyne (1961) and by Thomas (1962a, 1962b), but no specimens have been available for this work.

Key to species of *Fiorinia*

1 Prominent interantennal process present 2
-- Prominent interantennal process absent 4

2 Median lobes separated by a space equal to width about 3 times width of 1 lobe
 .... *biakana* sp. n.
-- Median lobes separated by a space equal to width of 1 lobe or narrower 3

3 Interantennal process elongate and spatulate. Body narrow with almost parallel sides.
 Marginal ducts numbering 5 pairs .... *proboscidaria* Green
-- Interantennal process rounded with spicules, almost crown-like. Body wide, narrowing abruptly
 to triangular pygidium. Marginal ducts numbering 4 pairs .... *coronata* sp. n.

4 Median lobes prominent, rounded, not forming deep notch at apex of pygidium. Marginal ducts
 slender .... *fijiensis* sp. n.
-- Median lobes divergent, forming a deep notch at apex of pygidium. Marginal ducts wide
 5

5 With 5 groups of perivulvar pores. Marginal macroducts numbering 3 or 4 on each side
 .... *fioriniae* (Targioni)
-- With 2 groups of perivulvar pores. Marginal macroducts numbering 6 on each side
 .... *reducta* sp. n.

Fiorina biakana sp. n. (Fig. 54)

Description

Scale of female about 2.0 mm long, elongate with a median dorsal ridge, pale brown, covered in translucent white wax, exuviae terminal, same colour. Male scale not observed.

Adult female, slide-mounted, elongate, about 1.65 mm long; sides parallel, curving posteriorly to a truncated pygidium and anteriorly to a pointed head with a prominent interantennal process that is elongate; but usually with straight apex with lateral projections. Body membranous except for pygidium.

Pygidium with median lobes widely separated, yoked at base, divergent; apex of each lobe with about 4 notches, both lobes forming a deep notch at apex of pygidium. Second lobes each with inner lobule narrower than median lobes, outer lobule about half width of inner lobule. Gland spines single, represented by 4 on each side, well developed and longer than lobes. Marginal ducts numbering 4 on each side, each fairly wide but becoming smaller anteriorly. Microducts represented by one or two in median areas of free abdominal segments.

Ventral surface with perivulvar pores present in 5 elongate groups. Duct tubercles present in marginal rows, separated into groups on each of mesothorax, metathorax and abdominal segment 2. Antennae prominent, each with a single seta. Anterior spiracles each with a single disc pore.

Second instar with robust median lobes forming notch at apex of pygidium. Second lobes with inner lobes projecting. Gland spines single, short on pygidium, but becoming wider anteriorly. Marginal macroducts wide, numbering 4 on each side.

Material examined
Holotype female. **IRIAN JAYA.** Biak, on leaves of tree, 22.v.1959 (BPBM).
Paratypes female. **IRIAN JAYA.** Same data as holotype. 5 females (BPBM), 4 (BMNH), 1 (USNM).

Comments
This species has close affinities with *F. proboscidaria* Green, but differs in having the median lobes wider apart, the marginal ducts wider and the ventral marginal duct tubercles separated into segmental groups instead of in a continuous row.

Fiorinia coronata sp. n. (Fig. 55)

Description
Female pupillarial; scale pale brown, elongate, flat, exuviae terminal. Male scales apparently white. Found on the leaves.
Adult female, on slide, about 0.65 mm long, broadly sub-rectangular, widest at about second abdominal segment, then narrowing abruptly to a triangular-shaped pygidium; body membranous except for lightly sclerotized pygidium; interantennal process rounded with short to long spicules, sometimes reduced in size with only one or two spicules.
Pygidium with median lobes forming a notch at apex, separated by a space equal to width of 1 lobe; narrowly yoked at base, each lobe pointed at apex with one or two distal notches. Second lobes each with rounded inner lobule narrower than, but at about the same level as median lobes; outer lobule smaller and shorter than inner lobule. Gland spines apparently absent. Long setae present between median and second lobes and lateral to second lobes. Marginal ducts narrow, each about 6 times longer than wide, numbering 4 on each side.
Ventral surface with perivulvar pores present in 5 elongate groups. Microducts present around submargins of pygidium, forward to about abdominal segment 3; others present in median areas of free abdominal segments, and lateral to labium. Duct tubercles in single marginal row from mesothorax to abdominal segment 2, but interrupted. Antennae set close together at anterior edge of head. Anterior spiracles each with 2 disc pores.

Material examined
Holotype female. **SOLOMON IS.** Guadalcanal, on *Cocos nucifera*, vi. 1954 (*E.S. Brown*) (BMNH).
Paratypes female. **SOLOMON IS.** Same data as holotype. 9 (BMNH), 2 (DSIR), 2 (USNM).
Non-type material. **IRIAN JAYA.** Biak, on *Pandanus* sp., 26. v. 1959. **PAPUA NEW GUINEA.** W.N.B.P., Iboki, Linga, on *Nypa fruticans*.

Comments
This species is probably related to *F. euryae* Kuwana, described from Japan and known from Vietnam. It differs in possessing only 4 marginal pygidial ducts, whereas in *F. euryae* they number 6-9. Present records are only from Arecaceae and Pandanaceae. There is considerable variation in the development of the interantennal process. It is well developed, resembling a crown in many specimens from Irian Jaya and less so in the Solomon Is, and sometimes it is reduced to a small prominence: however, this variation exists in the whole range of material at hand.

Fiorinia fijiensis sp. n. (Fig. 56)

Description
Female scale reddish brown, hemispherical. Male scale white, flat, almost same length as female scale.
Adult female, slide-mounted, about 0.9 mm long, elongate-oval, sides often parallel; pygidium triangular; head rounded, without interantennal process; body membranous, except for sclerotized pygidium.

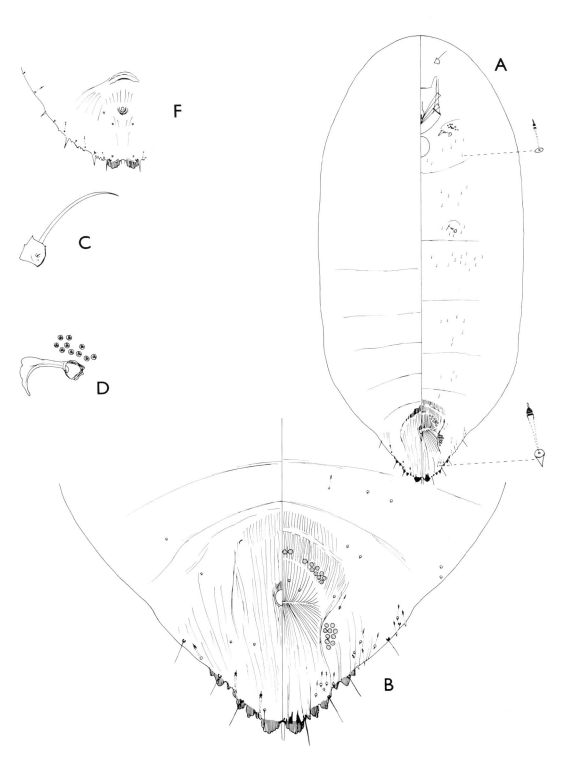

Fig. 56. *Fiorinia fijiensis* sp. n.

Pygidium with median lobes prominent, each rounded and with many notches, separated by a narrow space, yoked at base, with a pair of setae between, these longer than lobes; ventral paraphyses well developed. Second lobes with inner and outer lobules almost same size, but narrower and smaller than median lobes. Third lobes represented by sclerotized projections. Gland spines absent. Pygidial marginal setae conspicuously long. Marginal ducts slender, numbering 3 on each side. Anus situated near centre of pygidium.

Ventral surface with perivulvar pores present in 5 elongate groups. Microducts on pygidium each with a minute spine at orifice, these spines more prominent at margins; other microducts numerous in median areas of free abdominal and thoracic segments. Duct tubercles absent. Antennae situated between clypeolabral shield and head margin, each with a single seta. Anterior spiracles each with a group of disc pores.

Second instar with a pair of prominent rounded lobes; second lobes barely perceptible. Gland spines present, longer than lobes, arranged singly on pygidial margins. Marginal ducts slender.

Material examined
Holotype female. **FIJI.** Viti Levu, Vulgalei, on *Alpinia boia*, 31.xii.1957 (*J.S. Pillai*) (BMNH).
Paratypes female. **FIJI.** Same data as holotype. 15 (BMNH), 2 (DSIR), 2 (USNM); Suva, on *Heliconia* sp., 19.iii.1956 (*B.A. O'Connor*), 9 (BMNH).

Comments
This is a distinctive species with well-developed and rounded median lobes and should be easily recognizable in the South Pacific area. It seems to have some relationship to *F. nachiensis* Takahashi, described from Japan on *Rhododendron*, another species with well-developed median lobes that are rounded, but *F. nachiensis* has gland spines and the second lobes are not so well developed. *F. fijiensis* was listed from Fiji by Hinckley (1965) as *Fiorinia* sp. on *Alpinia boia* and *Heliconia* ?*bihai*. It will probably be found on a wide range of plant species.

Fiorinia fioriniae (Targioni) (Fig. 57)

Diaspis fioriniae Targioni, 1867: 14.
Fiorinia pellucida Targioni, 1868: 735.
Fiorinia fioriniae (Targioni), Cockerell, 1893c: 39; Ferris, 1937a: 55; Balachowsky, 1954: 303;
    Takagi, 1970: 59.

Description
Female scale about 1.5 mm long, elongate, yellow-brown, flat, exuviae terminal. Male scale whitish, smaller than female scale.

Adult female on microscope slides elongate, sides subparallel, head rounded, without interantennal process; pygidium triangular.

Pygidium with well-developed median lobes, forming a distinct notch at apex, strongly divergent and zygotic. Second lobes prominent, with inner and outer lobules. Gland spines arranged singly on margins. Marginal macroducts large, numbering 3 or 4 on each side. Dorsal ducts smaller than marginal ducts, represented by a few in submedian area at base of pygidium.

Ventral surface with perivulvar pores present in 5 elongate groups. A few duct tubercles present on margins of mesothorax, metathorax and abdominal segment 1. Antennae conical, each with a single seta. Anterior spiracles each with a small group of disc pores.

Material examined
COOK IS. Rarotonga, 1975. On *Camellia* sp., *Persea americana*.
FRENCH POLYNESIA. Society Is, Tahiti, 1976. On *P. americana*.
NEW CALEDONIA. New Caledonia, 1928. On *Cocos nucifera*.
VANUATU. Espiritu Santo, 1983. On *P. americana*.
WESTERN SAMOA. Savai'i, 1975; Upolu, 1975. On *Citrus reticulata*, *P. americana*.

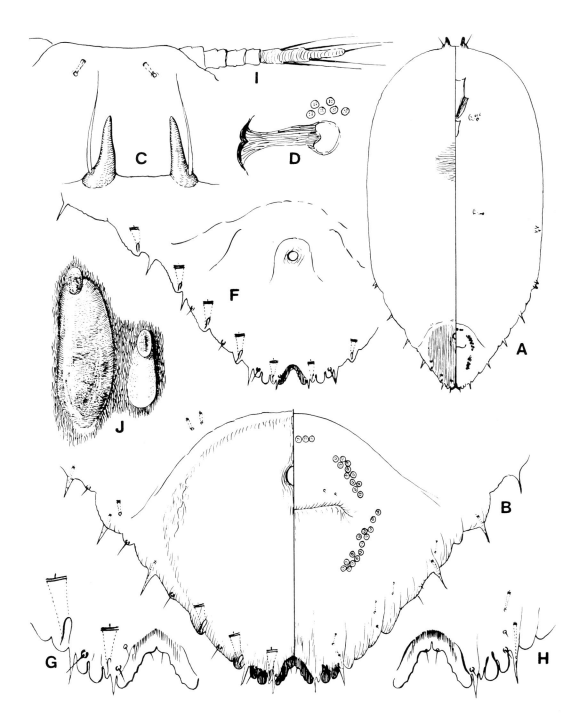

Fig. 57. *Fiorinia fioriniae* (Targioni).

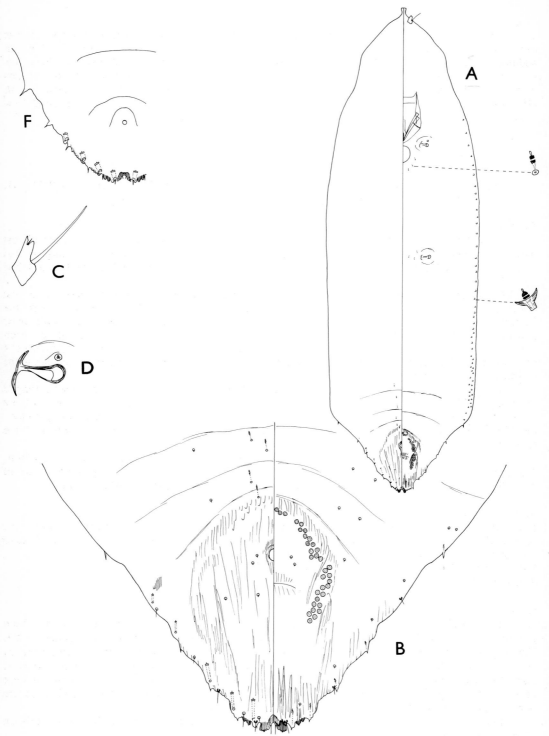

Fig. 58. *Fiorinia proboscidaria* Green.

Comments

This is normally a tropicopolitan species, but it extends into some temperate areas. It has been reported on numerous species of plants and will probably be found throughout the Pacific. In addition to the above records, Cohic (1956, 1958a) found *F. fioriniae* in New Caledonia on *Anthurium* sp. and *Santalum austrocaledonicum*, and reported it as common on *Persea americana*, causing serious defoliation; it is also present in Wallis Is on the latter host (Cohic, 1959). Szent-Ivany *et al.* (1956) mention it as a pest on *Cycas* sp. in New Guinea. This species has been intercepted at New Zealand quarantine on *Howeia forsteriana* and *P. americana* from the area.

Fiorinia proboscidaria Green (Fig. 58)

Fiorinia proboscidaria Green, 1900: 256; Takagi, 1970: 65. Lectotype female, Sri Lanka (BMNH), here designated [examined].

Description

Adult female on slides about 1.5 mm long, membranous except for pygidium; sides almost parallel, pygidium narrowing abruptly; head narrowing almost to point from lateral angles, but apically prolonged into an elongate interantennal process, the anterior end straight and lobed laterally.

Pygidium with median lobes slightly divergent, separated by a space that varies from just under to just over width of a single lobe; serrate at apices, yoked at base, not prominent; the ventral surface of each lobe with well-developed paraphyses. Second lobes short, each with inner and outer lobules. Gland spines short, rarely reaching lobes in length, arranged singly on each side of segments. Marginal pygidial setae surpassing lobes in length. Marginal ducts numbering 4 or 5 on each side, slender, becoming smaller anteriorly. Dorsal microducts present in median areas of free abdominal segments. Anus situated near base of pygidium.

Ventral surface with perivulvar pores present in 5 elongate groups. Duct tubercles arranged on margins of prothorax to about segment 3 of abdomen, in a single row on each side, except posteriorly where they form double or triple rows. Microducts present between posterior spiracles, and posterior to labium. Antennae prominent, close together at apex of head, each with a single long seta and a minute seta. Anterior spiracles each with 0-1 disc pores.

Second instar female with the median lobes divergent, forming a notch at apex of pygidium. Marginal macroducts robust, numbering 4 on each side. Gland spines arranged singly on each side of posterior abdominal segments, slender towards apex, but becoming more robust anteriorly.

Material examined
FIJI. 1935. Taveuni, 1959; Vanua Levu, 1975; Viti Levu, 1940. On *Areca catechu*, *Citrus* sp., *C. aurantifolia*, *C. aurantium*, *C. limon*, *C. maxima*, *C. paradisi*, *C. reticulata*, *C. sinensis*, *Daucus carota*, *Fortunella japonica*, *Mangifera indica*, *Rhaphidophora vitiensis*.
FRENCH POLYNESIA. Society Is, Tahiti, 1975. on *C. aurantifolia*.
TONGA. Tongatapu Group, Tongatapu, 1979. On *C. limon*.

Comments

This species was described from Sri Lanka, on *Gelonium lanceolatum*, but has since been recorded from India, Taiwan and the Ryukyu Is. It is probably widespread throughout southern Asia. The angled head, narrowing towards a long interantennal process, separates this species from *F. biakana*. Green's original slides are labelled '*Fiorinia proboscidaria*, from *Gelonium lanceolatum*, Pundaluoya, Ceylon, viii.98.' One adult female is on the slide labelled 'type', and this is here selected as lectotype. The other two specimens are on a slide labelled 'co-type', and these are here designated paralectotypes.

In addition to the above material, Hinckley (1965) records *F. proboscidaria* from Fiji on *Epipremnum pinnatum* and *Rosa* sp.

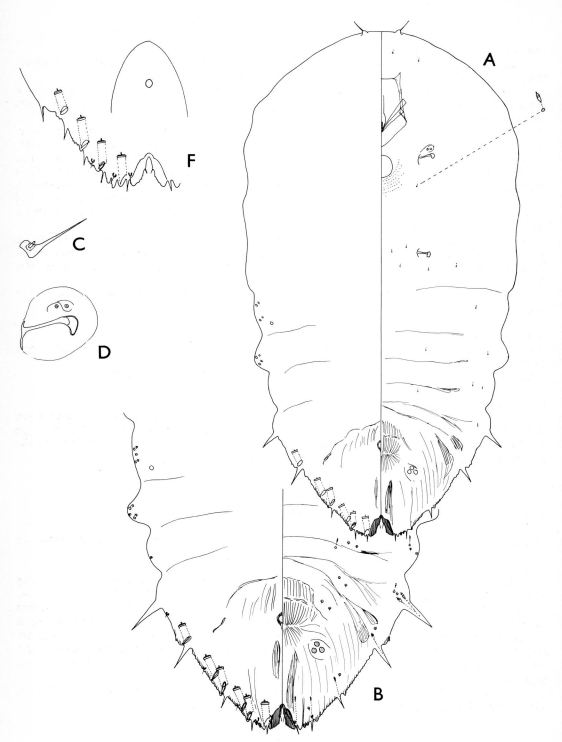

Fig. 59. *Fiorinia reducta* sp. n.

Fiorinia reducta sp. n. (Fig. 59)

Description
Scale of female pale brown and transparent laterally, elongate, flat. Male scale white, elongate, but smaller than female scale. Found on the leaves.

Adult female, slide-mounted, about 0.5 mm long, oval, widest at head or prothorax; free abdominal segments showing some lateral development; body membranous except for pygidium. Interantennal process lacking.

Pygidium with median lobes well developed, forming a deep notch at apex, the inner edges long, curving to serrate distal margins before curving laterally to short outer edges, yoked at base. Second lobes each represented by minute inner and outer lobules. Gland spines projecting from apex, narrow and single between median and second lobes and lateral to second lobes; anteriorly to fourth segment they become progressively larger, those on segment 4 often with more than one microduct, these also well developed. Marginal macroducts large, numbering 6 on each side. Anterior ducts smaller than marginal ducts, few, present on margins of segments 1 and 2, there being also a submarginal boss on each side of segment 1.

Ventral surface with only the postero-lateral groups of perivulvar pores present, each group with 3-4 pores. Microducts sparse across the prepygidial segments, and few between clypeolabral shield and head margin. Antennae set close together on head margin, each with a long pointed seta. Anterior spiracles each with 2 disc pores.

Second instar female with similar shaped median lobes to those of adult; second lobes projecting. Marginal macroducts fairly large, numbering 4 on each side.

Material examined
Holotype female. **IRIAN JAYA.** Jayapura (formerly Hollandia), on leaves of unidentified tree, 24.i.1960 (*T. Maa*) (BPBM).
Paratype females. **IRIAN JAYA.** Same data as holotype. 3 (BPBM), 4 (BMNH).

Comments
It is difficult to relate this to any other species of *Fiorinia*, although it does have some affinities with *F. fioriniae*; the latter species, however, possesses only 3 or 4 pairs of large pygidial macroducts, whereas *F. reducta* has 6 pairs. The perivulvar pores, restricted to the postero-lateral groups only, distinguish it from any other species in the genus.

Genus **FROGGATTIELLA** Leonardi

Froggattiella Leonardi, in Cockerell, 1900: 72. Type-species *Aspidiotus inusitatus* Green, by monotypy.

This genus is accepted in the predominantly grass-feeding tribe Odonaspidini, characterized by lacking plates and gland-spines, but possessing abundant short ducts on both the dorsum and the venter, so that the scale, when formed, includes a well-developed ventral scale, also incorporating the ventral exuviae. The dorsal exuviae are almost terminal. Although a single median lobe is usually developed in *Odonaspis*, it is obsolete in *Froggattiella*. Furthermore, perivulvar pores are absent in *Froggattiella* but present in *Odonaspis*. The only species known from the South Pacific area is *F. penicillata*, easily recognizable by a tuft of 6 spiniform processes at the apex of the pygidium.

The description has been kept to a minimum because Ben-Dov (?1987) has recently revised the whole tribe. Thanks are due to Dr Ben-Dov for help with this difficult group.

Froggattiella penicillata (Green) (Fig. 60)

Odonaspis penicillata Green, 1905, 346; Ferris, 1938: 164; Beardsley, 1966: 502; Takagi, 1969: 60.
Froggattiella penicillata (Green); Rutherford, 1915: 104; Borchsenius, 1966: 227.

Material examined
FIJI. Viti Levu, 1951. On bamboo stems.

Comments
   This species feeds on bamboos, and its origin is probably southern Asia, where it is fairly widely distributed. It has been introduced to USA, South America and North Africa. Beardsley (1966) recorded it from Palau in Micronesia, and Hinckley (1965) mentions its presence in Fiji on *Schizostachyum glaucifolium*.

Genus **FURCASPIS** Lindinger

Furcaspis Lindinger, 1908: 98; Ferris, 1938: 230. Type-species *Aspidiotus biformis* Cockerell, by subsequent designation.

   Important characters of this genus are discussed in the description of the following species. At present the genus contains seven species and is in great need of revision.

Furcaspis biformis (Cockerell) (Fig. 61)

Aspidiotus biformis Cockerell, 1893d: 548.
Furcaspis biformis (Cockerell), Sanders, 1909: 54; Ferris, 1938: 231.

Description
   Scale of adult female reddish brown, round, with exuviae subcentral. Male scale similar colour to female scale, but smaller and elongate. Adult female with a combination of characters affording easy recognition. There are 3 pairs of well-developed and rounded lobes; plates fringed in a peculiar form, chelate or claw-like; dorsal ducts 1-barred, but filiform; anus situated near middle of pygidium; and perivulvar pores normally present in 4 groups, but often in a loose arrangement.

Material examined
FIJI. Viti Levu, 1977. On *Oncidium* sp., Orchidaceae.
FRENCH POLYNESIA. Society Is, Tahiti, 1976. On *Miltonia regnelli*.

Comments
   The distribution of this species is mainly in tropical America, but it has been introduced to Hawaii. It seems to be confined to the plant family Orchidaceae, and is often found heavily infesting plants.
   There are no records of this species in Micronesia; but another species, *F. oceanica* Lindinger, is common there on palms, and may yet be found in the South Pacific area. It differs from *F. biformis* in lacking perivulvar pores.

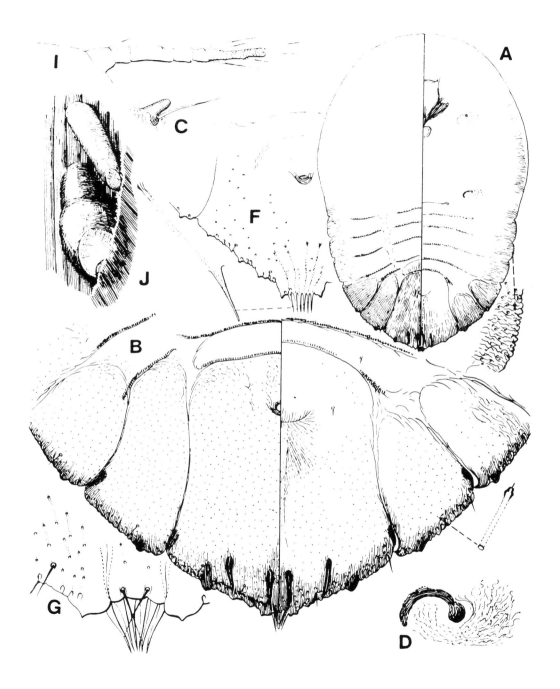

Fig. 60. *Froggattiella penicillata* (Green).

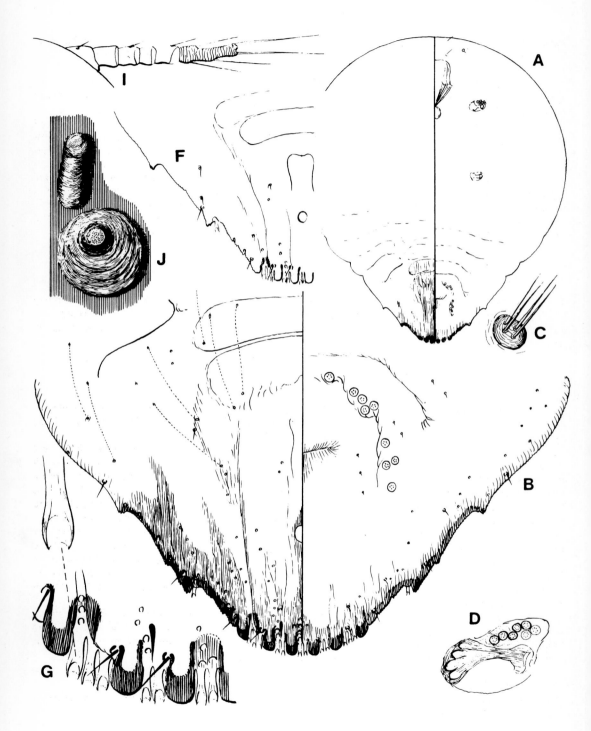

Fig. 61. *Furcaspis biformis* (Cockerell).

Genus **GENAPARLATORIA** MacGillivray

Genaparlatoria MacGillivray, 1921: 248; Ferris, 1937a: 60; McKenzie, 1945: 80; Balachowsky, 1958: 318. Type-species *Parlatoria pseudaspidiotus* Lindinger, by original designation.

This genus is closely related to *Parlatoria*; the main difference is that the body of *Parlatoria* is membranous except for the pygidium, whereas in *Genaparlatoria* the head and thorax becomes heavily sclerotized at maturity. Important characters are dorsal 2-barred ducts, those on the pygidial margin with the orifices parallel to the margin and surrounded by sclerotized semi-lunate rims. There are 3 pairs of well-developed lobes, and a fourth and sometimes a fifth pair represented by sclerotized points. The plates, each with a microduct, are deeply fringed and become sclerotized.

It has been usual to separate this genus from other parlatoriine genera by the rounded sclerotized prosoma and the lack of perivulvar pores; but a new species, herein described, has a deep constriction between the prothorax and mesothorax, and some perivulvar pores are present. The only reliable character to distinguish the genus, therefore, is the heavily sclerotized head and thorax. Both species should be easily recognizable from the accompanying illustrations.

Key to species of *Genaparlatoria*

1 Prosoma rounded. Perivulvar pores absent .... *pseudaspidiotus* (Lindinger)
-- Prosoma with a constriction between prothorax and mesothorax. Perivulvar pores present
    .... *araucariae* sp. n.

Genaparlatoria araucariae sp. n. (Fig. 62)

Description
External appearance not noted. Adult female on slide about 1.25 mm long, almost turbinate, but with a pronounced constriction between prothorax and mesothorax; head and pygidium rounded; body heavily sclerotized, more so on head and thorax.

Pygidium with 5 pairs of recognizable lobes. Median lobes longer than wide, each with 1 or 2 notches on inner margin and usually with a single notch on outer margin; separated by a space equal to the width of 1 lobe. Anterior lobes narrower than median lobes, notched on each side, becoming more pointed anteriorly. Plates, each with slender filamentous duct, more or less parallel between the first, second and third lobes, fimbriate distally; anteriorly becoming more triangular and serrated, but around prothorax, mesothorax and anterior abdominal segments, represented by points or duct tubercles. Marginal macroducts with orifices surrounded by sclerotized semi-lunate rims, the orifice of each parallel to margin; represented by 1 duct between median lobes and about 5, all the same size, on each side, anteriorly; marginal ducts progressively narrower to abdominal segment 1. Submarginal ducts about same length as marginal macroducts but narrower, on pygidium only, in approximately single rows. Anus situated near apex of pygidium. Sclerotized submarginal bosses present on segments 1 and 4.

Ventral surface with about 6 small perivulvar pores present on each side. Microducts present in submedian rows on pygidium, sparse in median areas of free abdominal segments and lateral to labium. Antennae with setae missing in available specimen. Anterior spiracles each with 4 disc pores each containing 5 loculi. Posterior spiracles without disc pores.

Material examined
Holotype female. **PAPUA NEW GUINEA**. E.H.P., Kratke Mts, Kassam, on leaves of *Araucaria* sp., 1.xi.1959 (*T. Maa*) (BPBM).

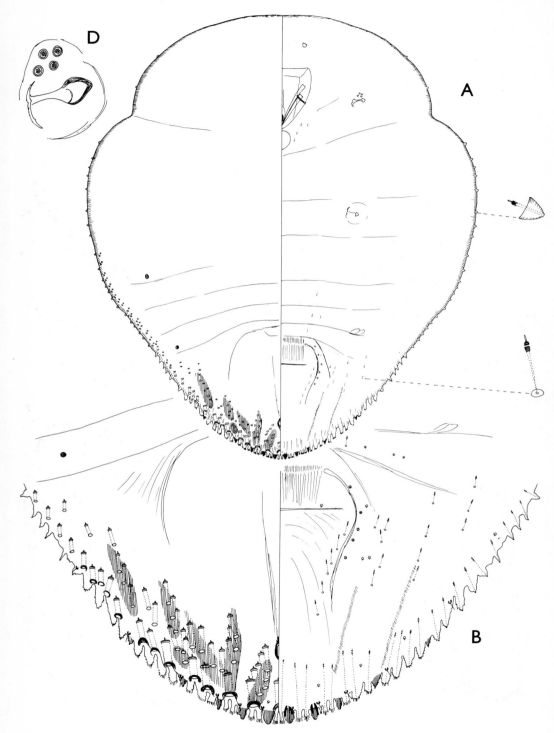

Fig. 62. *Genaparlatoria araucariae* sp. n.

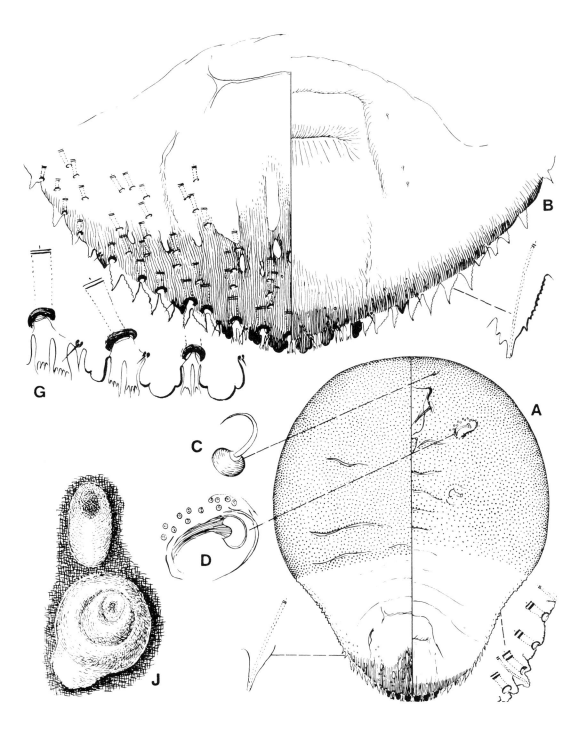

Fig. 63. *Genaparlatoria pseudaspidiotus* (Lindinger).

Comments
This species differs from the type-species, the only other known species, in the pronounced constriction between the prothorax and mesothorax, and in possessing perivulvar pores and ventral microducts. It has been described from a single available specimen in good condition, in case the species may be a pest of *Araucaria*, and because it may be mistaken at quarantine inspection for *G. pseudaspidiotus*.

Genaparlatoria pseudaspidiotus (Lindinger) (Fig. 63)

Parlatoria pseudaspidiotus Lindinger, 1905: 131.
Genaparlatoria pseudaspidiotus (Lindinger); MacGillivray, 1921: 255; Ferris, 1937a: 61; McKenzie, 1945: 81; Balachowsky, 1958: 318.

Description
Adult female recognizable by the broadly oval body and rounded sclerotized head and thorax, tapering posteriorly to a rounded but less sclerotized pygidium except for marginal areas. Lobes represented by 5 pairs, not bilobed. Plates present, each fringed, with a microduct; those anterior to fourth lobe tapering; present also as far forward as about segment 4, but anteriorly on abdomen represented by marginal points. Marginal macroducts present with orifices surrounded by semi-lunate rims. Dorsal ducts numerous, decreasing in size forward around margins to about segment 2 of abdomen. Perivulvar pores and ventral microducts absent. Anterior spiracles each with 10-12 disc pores.

Material examined
FRENCH POLYNESIA. Society Is, Tahiti, 1976. On *Vanda* sp.
PAPUA NEW GUINEA. C.P.: 1959. On *Cymbidium* sp., Orchidaceae, *Vanda* sp.
WESTERN SAMOA. Savai'i, 1977. On *Vanda* sp.

Comments
This is a tropicopolitan species, but it inhabits some temperate areas. It is usually associated with the family Orchidaceae, although records are known on *Euphorbia* sp. and *Mangifera* sp.

Genus HEMIBERLESIA Cockerell

Hemiberlesia Cockerell, *in* Leonardi, 1897: 375; Ferris, 1938: 232; Balachowsky, 1956: 104; Takagi, 1969: 76. Type-species *Aspidiotus rapax* Comstock (see Morrison & Morrison, 1966).

Important characters of this genus are 1-barred ducts; fringed plates; median lobes prominent; the second and third lobes, when present, reduced to points. Paraphyses on pygidial margin robust. The anus is large and is usually situated less than twice its diameter from the apex of the pygidium. In this respect it differs from *Abgrallaspis*, which has the anus situated further forward. Furthermore, in *Abgrallaspis* the paraphyses are much less developed, and the second and third lobes show some development and are not spiniform as in *Hemiberlesia*.

About 25 species of *Hemiberlesia* are currently recognized from different parts of the world. Three of these are cosmopolitan and have become established in the South Pacific area.

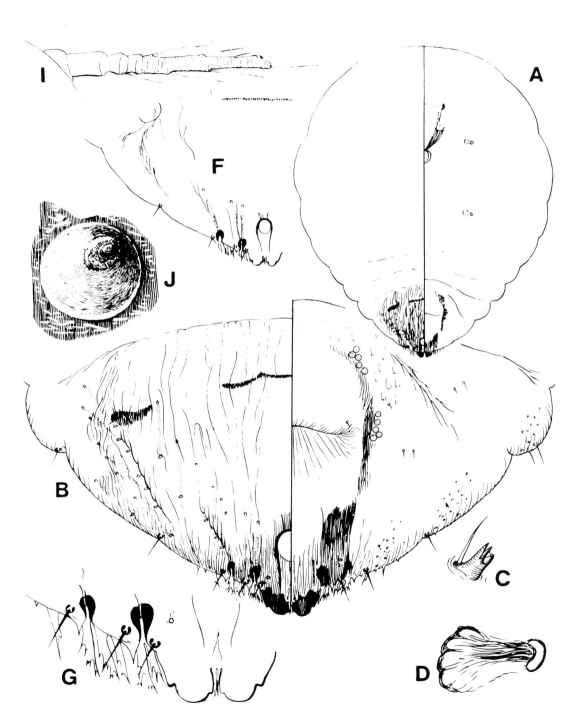

Fig. 64. *Hemiberlesia lataniae* (Signoret).

Key to species of *Hemiberlesia*

1. Perivulvar pores absent .... *rapax* (Comstock)
-- Perivulvar pores present 2

2. Median lobes close together, the space between very narrow .... *lataniae* (Signoret)
-- Median lobes separated by a space almost equal to width of one lobe ..... *palmae* (Cockerell)

Hemiberlesia lataniae (Signoret) (Fig. 64)

Aspidiotus lataniae Signoret, 1869b: 124.
Aspidiotus cydoniae Comstock, 1881: 295.
Hemiberlesia lataniae (Signoret), Cockerell, 1905a: 202; Ferris 1938: 241; Balachowsky, 1956: 108; Takagi, 1969: 78.

Description
    Female scale varying in colour, but usually whitish, with the subcentral exuviae yellow; subcircular. Male scale not known.
    Adult female, slide-mounted, with prominent median lobes appearing to converge medially, the space between narrow. Second and third lobes almost unrecognizable, but represented by membranous points. Plates narrow, reduced to only 1 or 2 lateral to third lobes. Dorsal ducts few, present in short single rows. Perivulvar pores present in 4 groups.

Material examined
COOK IS. Aitutaki, 1975; Atiu, 1977; Mangaia, 1977; Manihiki, 1969; Penrhyn, 1977; Rakahanga, 1969; Rarotonga, 1969. On *Abutilon* sp., *Artocarpus heterophyllus, Begonia* sp., *Chrysanthemum* sp., *Chrysophyllum cainito, Citrus maxima, C. reticulata, Cocos nucifera, Cordia* sp., *Eriobotrya japonica, Eugenia malaccensis, Ficus carica, Fitchia* sp., *Gomphrena globosa, Heliconia* sp., *Malus sylvestris, Manihot esculenta, Musa* sp., *M. sapientum, Nerium oleander, Pemphis acidula, Persea americana, Plumeria rubra,* 'putakava', *Vitis vinifera*.
FIJI. Kadavu, 1963; Taveuni, 1927; Viti Levu, 1944. On *Avicennia nitida, Barringtonia* sp., *B. asiatica, Citrus paradisi, Cocos nucifera, Dodonaea viscosa, Elephantopus scaber, Gladiolus* sp., *Gossypium* sp., mangrove, *Mimosa pudica, Parinari laurinum, Pithecellobium dulce, Psidium guajava, Schefflera actinophylla, Vitis vinifera*.
FRENCH POLYNESIA. Gambier Is, Makaroa, 1934; Marquesas Is, Eiao, 1929, Hivaoa, 1928. On *Abutilon graveolens, Pandanus* sp.
IRIAN JAYA. Wamena, 1960. On *Casuarina* sp.
KIRIBATI. Butaritari, 1976; Line Is, Kiritimati, 1924, Tabuaeran, 1924; Onotoa, 1971; Phoenix Is, Enderbury, 1950; Tarawa, 1976. On *Cassia* sp., *Cordia* sp., *Ficus* sp., *F. carica, Leucaena glauca, Morinda citrifolia, Rhizophora mucronata*.
NEW CALEDONIA. New Caledonia, 1899. On *Acacia* sp.
NIUE. 1970. On *Albizia falcataria, Eucalyptus grandis, Passiflora edulis, Plumeria rubra, Schefflera* sp.
PAPUA NEW GUINEA. C.P.: 1959. E.H.P.: 1959. E.N.B.P.: 1923. Morobe P.: 1979. On *Casuarina* sp., *Ceiba pentandra, Cocos nucifera, Coffea canephora,* Sapotaceae, *Timonius* sp.
SOLOMON IS. C.I.P.: Florida Is, Tulaghi, 1932. G.P.: Guadalcanal, 1954. W.P.: New Georgia Is, Gizo 1984, Shortland Is, Maleai 1984. On *Aleurites fordii, Cocos nucifera, Prosopis insularum, Psidium guajava*.
TOKELAU. Nukunonu, 1971. On *Cocos nucifera*.
TONGA. Ha'apai Group, Foa, 1977, Lifuka, 1977, Nomuka, 1977; Tongatapu Group, 'Eua, 1977, Tongatapu, 1975; Vava'u Group, Pangaimotu, 1977. On *Aleurites moluccana, Capsicum frutescens, Ceiba pentandra, Chrysanthemum* sp., *Elaeocarpus tonganus, Leucaena glauca, Myristica hypargyraea, P. guajava, Rosa indica*.

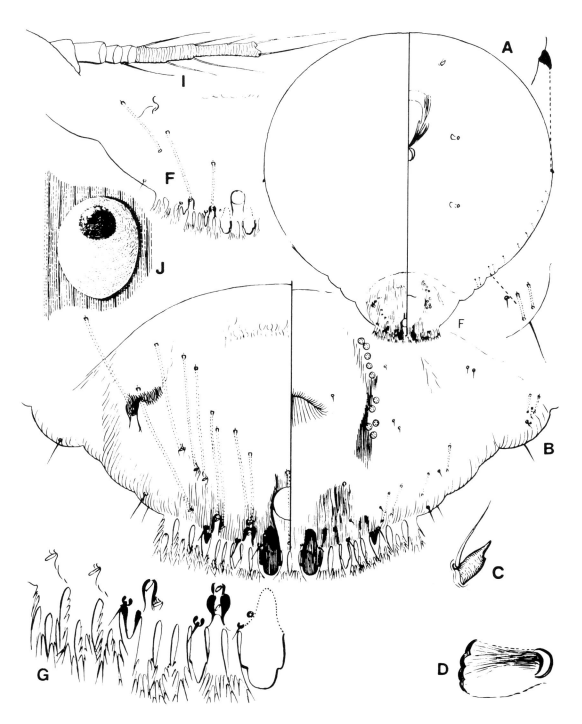

Fig. 65. *Hemiberlesia palmae* (Cockerell).

TUVALU. Funafuti, 1976; Vaitupu, 1976. On *Canavalia* sp., *Chloris* sp., *Cordia* sp., *Cyrtosperma chamissonis*.
VANUATU. Efate, 1983; Espiritu Santo, 1983. On *Citrus aurantifolia*, *Piper methysticum*.
WESTERN SAMOA. Savai'i, 1977; Upolu, 1987. On *Nerium oleander*, *Passiflora edulis*.

Comments
This is probably one of the commonest cosmopolitan species and is found on numerous plants. In addition to the material listed, *H. lataniae* has been recorded from Easter I. on *Melia azederach*, *Persea americana*, *Prunus persica*, and *Phoenix* sp. (Charlin, 1973); from Fiji on *Musa sapientum* by Veitch & Greenwood (1921), and on *Elephantopus scaber*, *Musa* spp., *Pithecellobium saman* and *Veitchia joannis* by Hinckley (1965); from French Polynesia on *Averrhoa carambola*, *Psidium cattleianum*, *P. guajava* (Reboul, 1976), and from Tahiti on *P. guajava* (Doane & Hadden, 1909); and from Irian Jaya on *Artocarpus* sp. by Reyne (1961). Manser (1974) recorded it from Kiribati on *Musa paradisiaca*; it has been recorded from New Caledonia on *Cordyline* sp., *C. neo-caledonica*, *Cycas* sp., *Psidium cattleianum*, *P. guajava*, *Solanum* sp., *Yucca aloifolia* and *Zygogynum* sp. (Cohic, 1956, 1958a; Brun & Chazeau, 1980). Szent-Ivany *et al.* (1956) recorded it from Papua New Guinea damaging *Cycas* sp., while in Solomon Is it has been mentioned occurring on *Dioscorea* spp. (Dumbleton, 1954), citrus (Lever, 1968) and *Ficus glandulifera* (Bigger, 1985). Doane & Ferris (1916) recorded it from Samoa on *Citrus sinensis*, and Reddy (1970) mentioned its presence in Western Samoa on *Balaka* sp. *H. lataniae* has often been intercepted at quarantine inspection of plant material from the South Pacific region.

Hemiberlesia palmae (Cockerell) (Fig. 65)

Aspidiotus rapax var. *palmae* Cockerell, 1892a: 5; Williams, D.J., 1970: 34.
Hemiberlesia palmae (Cockerell), Ferris, 1938: 242; Balachowsky, 1956: 114.
Borchseniaspis palmae (Cockerell), Zahradník, 1959: 65.

Description
Female scale round to oval, white; the exuviae subcentral, dark to almost black. Male scale not known.
A species with 3 pairs of parallel lobes; the median pair well developed, separated by a space equal to the width of one of them, each lobe with a short basal sclerosis. Second and third lobes elongate and much smaller than median lobes. Plates distinctive, just longer than median lobes, wide, present between all the lobes and ending abruptly with 3 lateral to each third lobe, giving the pygidial margin a wide appearance. Other characters are the long but few slender ducts, the 4 elongate groups of perivulvar pores, marginal thoracic spurs, and a conspicuous anus larger than a median lobe.

Material examined
FIJI. Taveuni, 1975; Vanua Levu, 1975; Viti Levu, 1913. On *Alpinia purpurata*, *Annona muricata*, *A. squamosa*, *Artocarpus altilis*, *A. heterophyllus*, *Barringtonia* sp., *B. asiatica*, *B. racemosa*, *Bixa orellana*, *Calophyllum* sp., *C. inophyllum*, *Camellia sinensis*, *Citrus limon*, *Cocos nucifera*, *Curcuma longa*, 'dabi', *Gliricidia sepium*, *Graptophyllum pictum*, *Heliconia* sp., *Hibiscus tiliaceus*, *Kigelia pinnata*, 'lasova', 'Leguminosae', *Macadamia tetraphylla*, *Mangifera indica*, *Miscanthus* sp., *Musa* sp., *M. sapientum*, *Persea americana*, *Pinus caribaea*, *Piper aduncum*, *Pithecellobium saman*, *Plumeria* sp., *P. rubra*, *Psidium cattleianum*, *P. guajava*, *Pyrus communis*, *Roystonea regia*, *Spondias dulcis*, *Theobroma cacao*, *Tripsacum laxum*, 'uiniglu', *Xanthosoma sagittifolium*.
KIRIBATI. Abemama, 1976; Beru, 1976; Tarawa, 1956. On *Artocarpus altilis*, *Guettarda speciosa*, *Musa* sp., *Pandanus odoratissimus*.
PAPUA NEW GUINEA. C.P.: 1959. E.N.B.P.: 1941. E.S.P.: 1984. Madang P.: 1964. Manus P.: Manus, 1959. M.B.P.: Louisiade Archipelago, Misima, 1985; Trobriand Is, Kiriwina, 1985. Morobe P.: Lasanga, 1959. W.H.P.: 1984. W.N.B.P.: 1956. On *Annona* sp., *Bauhinia monandra*,

Calophyllum inophyllum, Canavalia sp., *Casuarina* sp., *Citrus* sp., *Cocos nucifera, Coffea canephora, Cordyline* sp., *Decaspermum* sp., *Elettaria cardamomum, Grammatophyllum papuanum, Hevea brasiliensis,* 'kakamura tree', *Malus* sp., *Musa* sp., *Piper* sp., *Plumeria rubra, Psidium guajava, Schuurmansia henningsii, Theobroma cacao, Vanilla* sp., *V. planifolia.*
SOLOMON IS. G.P.: Guadalcanal, 1954. On *Cocos nucifera, Ficus glandifera, Hibiscus* sp., *Plumeria rubra.*
TONGA. Ha'apai Group, Foa, 1977, Lifuka, 1975, Nomuka, 1975; Tongatapu Group, Tongatapu, 1973; Vava'u Group, Pangaimotu, 1977, Vava'u, 1974. On *Aleurites moluccana, Alpinia purpurata, Annona muricata, A. reticulata, Artocarpus altilis, Bauhinia* sp., *Calophyllum inophyllum, Ceiba pentandra, Chrysophyllum cainito, Citrus limon, Cocos nucifera, Cycas* sp., *C. circinalis, Dioscorea* sp., *D. bulbifera, Eugenia malaccensis, Euphorbia heterophylla, Hibiscus* sp., *Mangifera indica, Morinda citrifolia, Musa* sp., *Persea americana, Plumeria rubra, Psidium guajava.*
WESTERN SAMOA. Savai'i, 1987; Upolu, 1975. On *Acalypha hispida, Anacardium occidentale, Annona muricata, Mangifera indica, Passiflora edulis, Persea americana, Piper* sp.

Comments
Although originally described from Jamaica, this species has now been reported from many tropical countries on numerous host-plants. It is usually common in the South Pacific region, and often occurs in large numbers on the leaves, especially on palms. The white scales with almost black central exuviae make them particularly conspicuous in life.
In addition to the material listed, literature records include mention of *H. palmae* from Fiji on *Colocasia* sp. (Lever, 1945a), *Barringtonia* sp., *Bruguiera gymnorhiza, Epipremnum pinnatum, Heliconia ?bahai* and *Passiflora edulis* (Hinckley, 1965); from Irian Jaya on *Cocos nucifera* (Reyne, 1961); from Kiribati on *Musa paradisiaca* (Manser, 1974); and from the Solomon Is on *Persea americana* (Dumbleton, 1954) and *Ficus glandulifera* (Bigger, 1985). It has been intercepted frequently at quarantine inspection on plant material from the South Pacific area.

Hemiberlesia rapax (Comstock) (Fig. 66)

Aspidiotus rapax Comstock, 1881: 307.
Hemiberlesia rapax (Comstock), Ferris, 1938: 244; Balachowsky, 1956: 116.

Description
Colour of female scale grayish; convex with the exuviae towards the edge. Male scale not known. Occurring on leaves and bark.
Adult female, on slide, similar in appearance to *H. lataniae*, from which it differs in lacking perivulvar pores, and having a larger and more conspicuous anal opening situated nearer the apex of pygidium. In other respects, the lobes, plates and dorsal ducts have a similar arrangement.

Material examined
NEW CALEDONIA. New Caledonia, 1979. On *Drimys pauciflora.*

Comments
Originally described from U.S.A. and commonly known as 'The Greedy Scale', this injurious species is now widespread throughout the tropics and some temperate areas on numerous species of plants. The only material at hand from the South Pacific area is a single specimen, listed above, from the Katickoin region of New Caledonia. It has been listed by Doane & Hadden (1909) from the Society Is, but the record is actually from Flint, part of Kiribati. Beardsley (1966) recorded it from Micronesia, and Borchsenius (1966) mentions its occurrence in Samoa. It is well established in New Zealand.

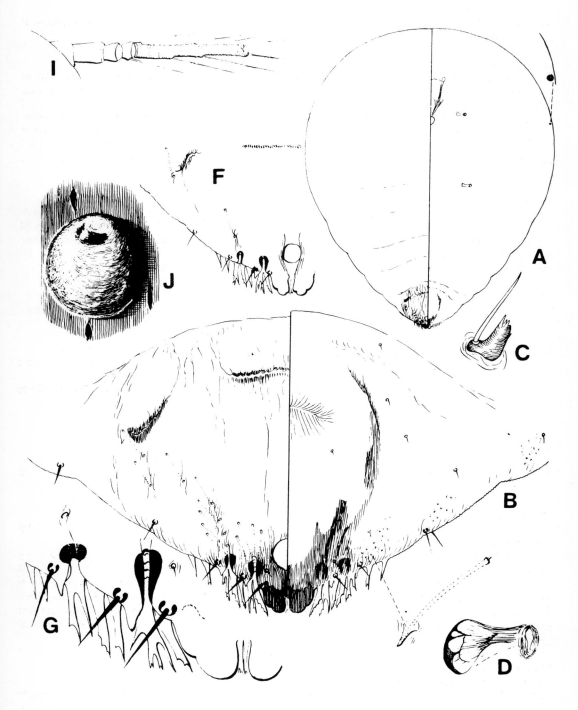

Fig. 66. *Hemiberlesia rapax* (Comstock).

Genus **HOWARDIA** Berlese & Leonardi

Howardia Berlese & Leonardi, 1896: 347. Type-species *Chionaspis biclavis* Comstock, by subsequent designation.

Description
   Body of adult female large, turbinate or fusiform, usually heavily sclerotized; free abdominal segments strongly lobed laterally. Median lobes well developed, set close together; each lobe with short inner margin and longer outer margin, narrow transverse paraphyses present at each basal corner, and a large club-shaped sclerosis arising from inner basal corner. Second lobes reduced or absent. Gland spines represented by a short pair between median lobes; those present forward to segment 2 longer, in marginal groups of 4 or 5. Dorsal ducts 2-barred, all the same size, present on pygidial margins only, or extending also in submarginal rows. Lateral spurs or tubercles present on segments 2 to 4 of abdomen. Perivulvar pores absent.

Comments
   This is a genus that is normally easy to recognize by the large sclerotized body of the adult female, and by the conspicuous clavate scleroses arising from the median lobes. In the formation of the scale the insects usually mine under the epidermal layers of the leaves and stems, and become inconspicuous. The type-species has become well established in the South Pacific area, but *H. stricklandi* Williams, the only other known species, is a new record. The species can be separated as follows:

Key to species of *Howardia*

1	Body turbinate, dorsal ducts present on margins and in submarginal rows on pygidium .... *biclavis* (Comstock)
--	Body fusiform or broadly oval, dorsal ducts confined to margins of pygidium .... *stricklandi* Williams

Howardia biclavis (Comstock) (Fig. 67)

Chionaspis biclavis Comstock, 1883: 98.
Howardia biclavis (Comstock), Berlese & Leonardi, 1896: 348; Ferris, 1937a: 65; Balachowsky, 1954: 252; Takagi, 1970: 32.

Description
   Female scale circular with the exuviae at margin; normally white, but colour obscured by overlying plant material. Apparently parthenogenetic.
   Adult female large, broadly turbinate; head, thorax and anterior abdominal segments heavily sclerotized, pygidium broadly rounded. Important diagnostic characters are the large prominent median lobes set close together, outer margins longer than inner margins, serrate and slanting; each lobe with a prominent clavate sclerosis arising from inner corner. Second lobes each reduced to a small conical process. Gland spines present between median lobes short; long gland spines present in segmental groups of 4 or 5 forward to abdominal segment 5. Dorsal ducts numerous on margins of pygidium and present in submarginal and submedian groups. Perivulvar pores absent.

Material examined
COOK IS. Rarotonga, 1975. On *Gardenia* sp.
FIJI. Viti Levu, 1953. On *Acacia mellifera*, *Camellia sinensis*, *Carica papaya*, *Citharexylum spinosum*, *Citrus sinensis*, *Hibiscus syriacus*, *Ixora coccinea*, *Litchi chinensis*, *Lycopersicon esculentum*, *Macaranga* sp., *Myristica macrantha*, *Parinari laurinum*.

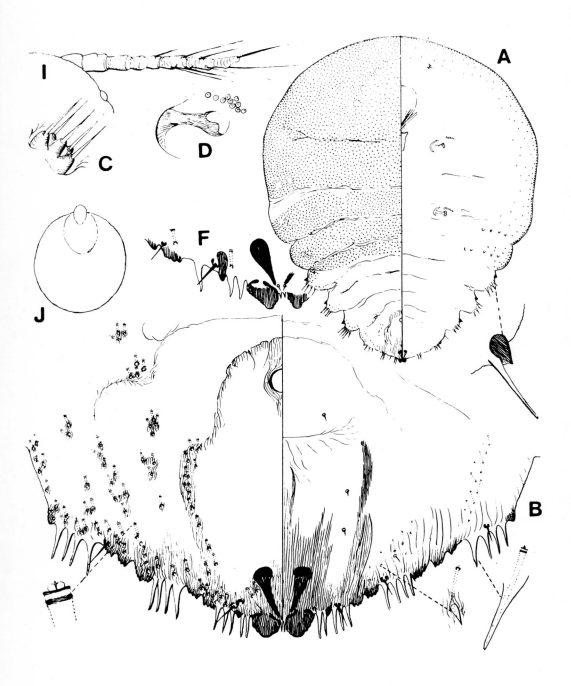

Fig. 67. *Howardia biclavis* (Comstock).

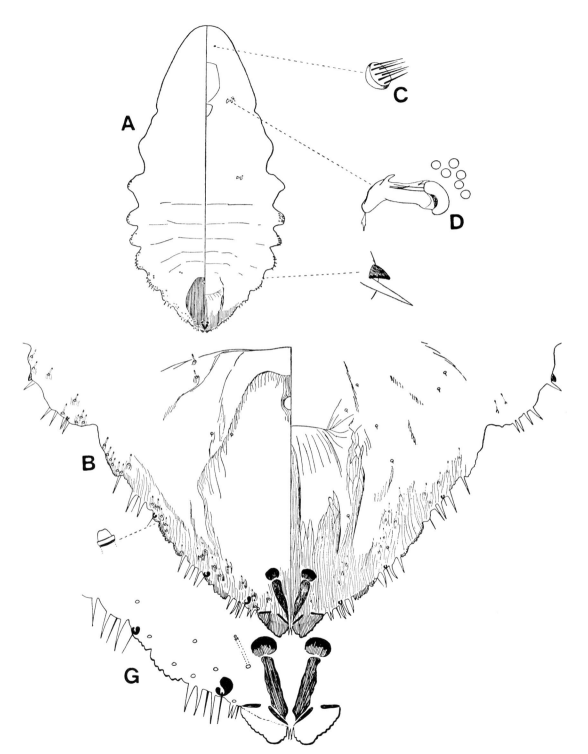

Fig. 68. *Howardia stricklandi* Williams.

NEW CALEDONIA. On ?
NIUE. 1975. On *Acacia spirorbis, Annona muricata, Bixa orellana, Carica papaya, Chrysophyllum cainito, Macadamia tetraphylla, Terminalia catappa.*
PAPUA NEW GUINEA. Morobe P.: 1959. N.P.: 1985. W.H.P.: 1978. On *Albizia stipulata, Camellia sinensis, Citrus* sp., *Neonauclea* sp.
TONGA. Ha'apai Group, Lifuka, 1977, Navea, 1977; Tongatapu Group, 'Eua, 1977, Tongatapu, 1975; Vava'u Group, Pangaimotu, 1977, Vava'u, 1974. On *Allemanda cathartica, Annona squamosa, Carica papaya, Citrus aurantium, C. reticulata, Gardenia* sp., *Grevillea robusta, Indigofera anil, Manilkara zapota, Morus alba, Piper methysticum, Punica granatum, Swietenia macrophylla, Urena lobata.*
VANUATU. Malekula, 1983. On *Piper methysticum.*
WESTERN SAMOA. Savai'i, 1975; Upolu, 1977. On *Carica papaya, Khaya grandifolia,* Malvaceae, *Nephelium lappaceum, Piper methysticum.*

Comments

This is a tropicopolitan species feeding on numerous plant species. It probably has a much wider distribution throughout the South Pacific area. It is fairly easy to recognize by its large size, turbinate shape, and heavily sclerotized cephalothorax. Apart from the shape, it can be distinguished from *H. stricklandi* by the presence of numerous dorsal ducts.

In addition to the material listed above, *H. biclavis* has been recorded from Fiji on *Calophyllum* sp. and *Gardenia* sp. (Hinckley, 1965); from French Polynesia on *Gardenia* 'taitensis' (= *Randia tahitensis?*) (Reboul, 1976); from New Caledonia on *Allemanda schottii, Ervatamia orientalis, Gardenia* sp., *Hibiscus rosa-sinensis, Jasminum sambac, Manilkara zapota, Mundulea suberosa* and *Plumeria acutifolia* (Cohic, 1956, 1958a; Brun & Chazeau, 1980); and from Wallis Is on *Ervatamia orientalis* by Cohic (1959).

Howardia stricklandi Williams (Fig. 68)

Howardia stricklandi Williams, 1960: 391.

Description

Female scale pale brown, fusiform, exuviae terminal. Male scale not known.

Adult female on slide, fusiform, but sometimes broadly oval; up to 3.4 mm long. Important characters are the few short dorsal ducts, confined to the margins and submargins except for one or two near base of pygidium.

Material examined
WESTERN SAMOA. Savai'i, 1975. On *Khaya grandifolia.*

Comments

This species was described originally from Ghana on *Theobroma cacao,* and has not been recorded since. Specimens from Western Samoa were associated with *H. biclavis,* on mahogany, and both species were probably introduced at the same time from West Africa.

Although original material is fusiform, specimens from Western Samoa range in shape from fusiform to broadly oval. One or two specimens in this material also possess minute conical second lobes, but all lack the numerous dorsal ducts of *H. biclavis.*

Genus ISCHNASPIS Douglas

Ischnaspis Douglas, 1887b: 21. Type-species *Ischnaspis filiformis* Douglas = *Mytilaspis longirostris* Signoret, by monotypy.

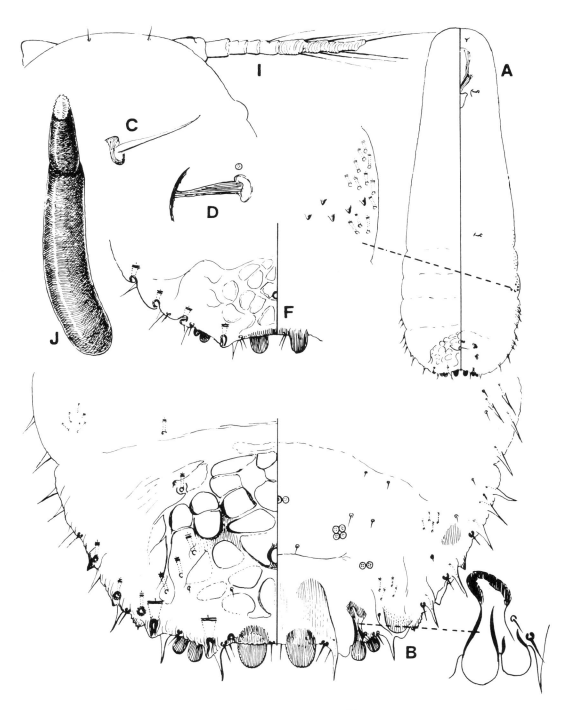

Fig. 69. *Ischnaspis longirostris* (Signoret).

This genus, at present comprising five species, seems to be of African origin. Only the type-species, *I. longirostris*, has been introduced to the Pacific area, where it is widespread. For purposes in the South Pacific region, the characters discussed for the species serve well for the genus. In recent years the genus has been revised, and characters assessed, by Ben-Dov (1974) and Matile-Ferrero (1982).

Ischnaspis longirostris (Signoret) (Fig. 69)

Mytilaspis longirostris Signoret, 1882: xxxv.
Ischnaspis filiformis Douglas, 1887b: 21.
Ischnaspis longirostris (Signoret), Cockerell, 1896: 336; Ferris, 1937a: 67; Balachowsky, 1954: 135; Ben-Dov, 1974: 20.

Description
    Scale of adult female, elongate and narrow, often thread-like, sometimes reaching 3.0 mm long; black, with terminal exuviae yellow. Male scale not known.
    Adult female on slide elongate, widest at anterior abdominal segments; membranous, but dorsum of pygidium somewhat sclerotized in a conspicuous reticulate pattern.
    Pygidium with a pair of prominent, rounded median lobes, well separated. Second lobes rounded, the inner and outer lobules smaller than median lobes. Gland spines absent between median lobes, present singly as far forward as segment 5, and then progressively more numerous to segment 2. Marginal macroducts 2-barred, 1 present between each median and second lobe, and 1 lateral to each second lobe. Other dorsal ducts smaller, few, in submarginal and submedian areas. Anus situated near centre of pygidium. Ventral surface with perivulvar pores present in 5 small groups. Microducts present around submargins as far forward as head. Duct tubercles present on mesothorax, metathorax and abdominal segment 1.

Material examined
COOK IS. Rarotonga, 1937. On *Citrus* sp.
FIJI. Taveuni, 1977; Viti Levu, 1975. On *Coffea* sp., *C. arabica*, *C. canephora*, 'dabi', *Elaeis guineensis*, *Inocarpus fagifer*, *Swietenia macrophylla*.
PAPUA NEW GUINEA. C.P.: 1959. E.N.B.P.: 1980. E.S.P.: 1984. Morobe P.: 1968. N.S.P.: 1965. On *Aloe* sp., *Cocos nucifera*, *Coffea canephora*, *Diospyros kaki*, *E. guineensis*, *Grammatophyllum papuanum*, *Ixora* sp., *Musa* sp., *Strychnos nux-vomica*.
SOLOMON IS. G.P.: Guadalcanal, 1964. On ?
TONGA. Ha'apai Group, Foa, 1977, Nomuka, 1975; Tongatapu Group, Tongatapu, 1979; Vava'u Group, Vava'u, 1977. On *Ficus* sp., *Inocarpus fagifer*, *Justicia* sp., *Mangifera indica*, *Plumeria rubra*.
VANUATU. Efate, 1983. On *Coffea canephora*, *Litchi chinensis*.
WESTERN SAMOA. Upolu, 1977. On *C. canephora*, *C. arabica*.

Comments
    There should be no difficulty in identifying this species from the characters and illustration given. Ben-Dov (1974) has given a detailed description, and has mentioned the presence of dorsal submarginal bosses on segments 1 and 3 of the abdomen.
    The species is often known as the black-thread scale, and despite its small size, it is often noticeable on leaves when in great numbers. It is tropicopolitan, although it is common in greenhouses in temperate areas. The host-plant list is extremely long; in addition to material mentioned above, Hinckley (1965) records it from Fiji on *Inocarpus fagifer* and *Ligustrum japonicum*, and Swaine (1971) mentions it on Arecaceae; from French Polynesia, Reboul (1976) mentions it on *Coffea arabica*, *Jasminum* sp., *J. officinale*, *J. sambac* and *Mucuna bennettii*, while Cohic (1955, 1958a) records it from Tahiti on *Monstera deliciosa* and *Coffea* sp.; it is known in New Caledonia on *M. deliciosa* (Cohic, 1956, 1958a; Brun & Chazeau, 1980); in Papua New Guinea it has been found

on *Ixora* sp. (Anon, 1969); Lever (1968) recorded it from the Solomon Is on *Cocos nucifera*; and in Tonga it has been collected on *Coffea* sp. (O'Connor, 1949) and *Citrus* spp. (Dumbleton, 1954).

Genus **LEPIDOSAPHES** Shimer

Lepidosaphes Shimer, 1868: 372, 373; Ferris, 1937a: 70; Balachowsky, 1954: 28; Takagi 1970: 1.
Type-species *Coccus conchiformis* Gmelin = *C. ulmi* Linnaeus, by monotypy.

Description

Adult female elongate, fusiform, usually membranous except for pygidium, but a few species sclerotized on thorax; lateral lobes of free abdominal segments well developed. Median and second lobes well developed, the second lobes bilobed; third and fourth absent, or represented by serrations. Ventral paraphyses to lobes slender and usually vertical. Gland spines present, always a pair between median lobes; replaced by duct tubercles when present anterior to segment 2. Tubular ducts 2-barred; marginal macroducts usually 6 in number on each side, enlarged, each with orifice surrounded by a thick sclerotized rim, at right angles to pygidial margin. Dorsal ducts smaller than marginal ducts, or represented as microducts. Lateral tubercles or spurs often present between some abdominal segments. Dorsal submarginal bosses present or absent. Anus situated near base of pygidium. Perivulvar pores present in 5 groups.

Comments

Over 150 species have been included in this world-wide genus at one time or another. Borchsenius (1963) recognized eight other genera as being related to *Lepidosaphes*, describing some as new, and resurrecting others that had been regarded as identical. Williams (1971) discussed these genera, based mainly on the presence or absence of certain characters such as bosses, lateral tubercles, duct tubercles on the mesothorax, and ventral paraphyses on the lobes. The genera have not been recognized wholeheartedly by workers on scale insects, mainly because most of the species originally in *Lepidosaphes* have still not been studied critically, and many species assigned to genera described by Borchsenius are world-wide pests of economic crops and are widely recognized under *Lepidosaphes* in the agricultural literature. Takagi (1970) sank most of the names as synonyms of *Lepidosaphes*, and Takagi's action is followed here.

Fourteen species are recognized from the South Pacific area. These include important pests of citrus, cosmopolitan species, and others that may be intercepted at quarantine inspection. The number probably represents a small proportion of those awaiting discovery. Some other species have been studied, but the specimens are too poor to describe or illustrate. *L. laterochitinosa* Green is not represented, but it has been intercepted at quarantine at U.S.A. from the area, on orchids. This is a polyphagous species, known from southern Asia and Micronesia, and is probably widespread in the South Pacific area. Despite the record by Reyne (1961) of *L. pinnaeformis* (Bouché) in New Guinea, no material has been seen.

The name *Lepidosaphes* has been regarded as masculine or feminine by various authors. Although Shimer (1868) defined the name well, he gave no indication of the gender. Kirkaldy (1902) seems to have been the first to use adjectival combinations, and he regarded the name as feminine. This decision (first reviser rule) is followed here.

Key to species of *Lepidosaphes*

1	Bosses present on dorsal submargins of abdomen. Double boss present on dorsal margin of cephalothorax almost opposite first spiracle .... *beckii* (Newman)
--	Bosses absent from dorsum 2

2	With 5 marginal macroducts, same size, present on each side of pygidium. Single marginal macroduct on fifth segment narrower or absent	3
--	With 6 marginal macroducts, same size, present on each side of pygidium	4
3	Duct tubercles present on metathorax, lateral tubercles present on at least 1 abdominal segment. Eye spot modified into a minute spur	*securicula* sp. n.
--	Duct tubercles absent from metathorax, lateral tubercles absent from abdominal margins. Eye spot not modified into a minute spur	*geniostomae* sp. n.
4	Dorsal ducts large to small, much wider than ventral microducts. Duct tubercles present on metathorax	5
--	Dorsal ducts minute, resembling ventral microducts. Duct tubercles absent from metathorax	13
5	Eye spots modified into conspicuous thorn-like spurs	*elmerrilleae* sp. n.
--	Eyes either not discernible or not modified into thorn-like spurs	6
6	With a wide distinct sclerotized band around margins of third and posterior abdominal segments	*marginata* Ferris
--	Without a wide sclerotized band around posterior margins of abdomen, although apex of pygidium often sclerotized to some extent	7
7	Head with large or minute spicules	8
--	Head without large or minute spicules	9
8	Spicules minute, sparse on anterior head margin, large concentrations on lateral angles of head	*rubrovittata* (Cockerell)
--	Spicules large and thorn-like, occupying anterior margin of head	*stepta* sp. n.
9	Thorax and first abdominal segment sclerotized at maturity	*gloverii* (Packard)
--	Thorax and first abdominal segment not sclerotized at maturity	10
10	Head lightly sclerotized, expanded laterally to form lobes or projections	*tokionis* (Kuwana)
--	Head membranous, rounded, not expanded laterally	11
11	Space between median pygidial lobes as wide as or wider than width of a median lobe. Median lobes diverging slightly	*esakii* Takahashi
--	Space between median pygidial lobes narrower than width of a median lobe. Median lobes parallel	12
12	A duct present just anterior to each second lobe, minute, resembling ventral microducts but longer. A similar duct usually present between median lobes. Orifices indistinct	*tapleyi* Williams
--	A duct present just anterior to each second lobe, about half width of a dorsal duct, with conspicuous sclerotized orifice. Duct absent from between median lobes	*eurychlidonis* sp. n.
13	Lateral tubercles between anterior abdominal segments large and conspicuous, 3 present on each side. Space between median lobes almost as wide as width of a median lobe. Microduct present between median lobes	*karkarica* sp. n.
--	Lateral tubercles present only between segments 2 and 3, small, barely perceptible. Space between median lobes about half as wide as a median lobe. Microduct absent from between median lobes	*pometiae* sp. n.

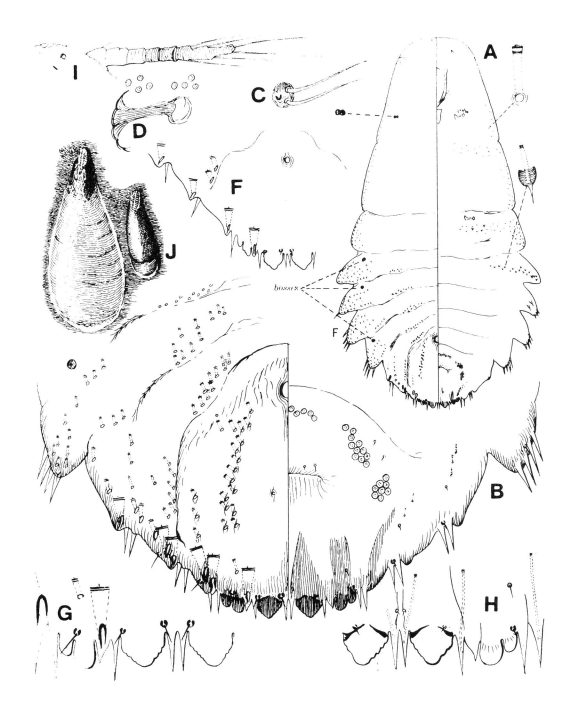

Fig. 70. *Lepidosaphes beckii* (Newman).

Lepidosaphes beckii (Newman) (Fig. 70)

Coccus beckii Newman, 1869: 217.
Aspidiotus citricola Packard, 1869: 527.
Lepidosaphes beckii (Newman), Fernald, 1903: 305; Ferris, 1937a: 71; Balachowsky, 1954: 61.
Cornuaspis beckii (Newman), Borchsenius, 1963: 1168; Borchsenius, 1966: 57.

Description
　　Scale of adult female light to dark purplish brown, fusiform, about 2.5-3.0 mm long. Male scale smaller than female scale, usually about 1.0 mm long.
　　Adult female, when prepared on microscope slide, about 1.25 mm long, fusiform, widest at about first abdominal segment. Lateral lobes of abdominal segments well developed and directed postero-laterally. Head and pygidium rounded. The main distinguishing features of this species are submarginal bosses on abdominal segments 1 to 5, and a double boss on dorsum of prothorax, almost opposite each anterior spiracle, the latter is one of the most distinctive characters of this species, and is present in only a few species of *Lepidosaphes*. Other important characters are the 6 macroducts on each side of pygidium, the numerous dorsal ducts on segment 6 forming an almost continuous line or group and the lack of lateral tubercles or spurs between the abdominal segments.

Material examined
COOK IS. Aitutaki, 1954; Atiu, 1977; Mangaia, 1954; Rarotonga, 1957. On *Citrus* sp., *C. sinensis*.
FIJI. Taveuni, 1977; Viti Levu, 1940. On *Citrus* sp., *C. limon*, *C. paradisi*, *C. sinensis*.
KIRIBATI. Abemama, 1976. On *C. aurantifolia*.
NEW CALEDONIA. Loyalty Is, Ouvea, 1963; New Caledonia, 1942. On *Citrus* sp.
NIUE. 1975. On *C. aurantifolia*, *C. reticulata*.
NORFOLK ISLAND. 1947. On *Citrus* sp., *C. reticulata*.
PAPUA NEW GUINEA. C.P.: 1978. M.B.P.: Trobriand Is, Kiriwina, 1985. On *Citrus* sp., *C. limon*.
SOLOMON ISLANDS. C.I.P.: Bellona, 1974; Florida Is, 1935. On *Citrus* sp., *C. sinensis*.
TONGA. Ha'apai Group, Lifuka, 1977, Nomuka, 1977; Tongatapu Group, 'Eua, 1977, Tongatapu, 1974; Vava'u Group, Pangaimotu, 1977, Vava'u, 1974. On *Citrus* sp., *C. aurantium*, *C. limon*, *C. maxima*, *C. reticulata*, *C. sinensis*.
VANUATU. Efate, 1956; Espiritu Santo, 1945; Malekula, 1978, Tanna, 1978. On *Citrus* sp., *C. aurantifolia*, *C. limon*, *C. maxima*, *Hibiscus* sp.
WESTERN SAMOA. Upolu, 1924. On *C. grandis*, *C. limon*, *C. maxima*, *C. sinensis*.

Comments
　　This species has been reported on numerous host-plants, but it is usually regarded as one of the most important pests of citrus wherever it is grown. It occurs on the fruits, leaves and stems. When collected on citrus it is fairly easy to identify by the characters given.
　　In addition to the material listed above, *L. beckii* has been recorded from the Cook Is on *Citrus aurantifolia*, *C. grandis* and *C. limon* (Maddison, 1976); from Easter I. on *C. limon* and *C. sinensis* (Charlín, 1973); from Fiji, Hinckley (1965) mentions that it sometimes damages citrus significantly, and it has been recorded on *C. aurantifolia* (Lever, 1945b), *C. aurantium* (Greenwood, 1929) and *C. grandis* (Maddison, 1976). Dumbleton (1954) mentions its presence in French Polynesia, and Reboul (1976) calls it the most important citrus pest there; Cohic (1955) describes how the scale prefers hot humid conditions such as occur in Tahiti, and is found in the shady inner parts of the canopy until sheer numbers make it spread outwards on to the leaves and fruit, causing chlorosis of the leaves, defoliation, disfiguration and poor maturation of the fruit, and desiccation and weakening of the branches; dense colonies are formed, particularly on *Murraya exotica*. Borchsenius (1966) mentions its presence in the Marquesas Is and Tuamotu Is. Dumbleton (1954) recorded it from Irian Jaya on citrus, on which it is a major pest according to Maddison (1976); Cohic (1950a, 1956, 1958a) recorded *L. beckii* from New Caledonia as the main pest on *C. aurantium*, *C. grandis*, *C. limon*, *C. paradisi*, *C. reticulata*, *C. sinensis* and *Murraya exotica*, and mentioned its association with the

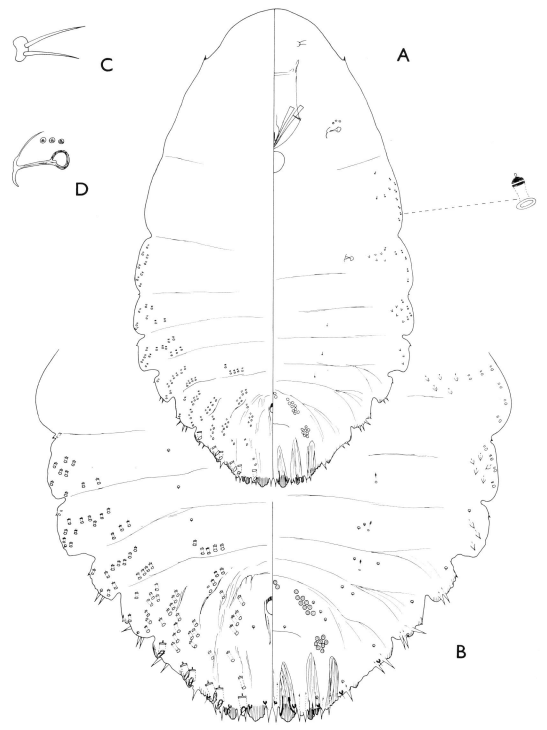

Fig. 71. *Lepidosaphes elmerilleae* sp. n.

symbiotic fungus *Septobasidium bogoriense*. It has been described as a pest of citrus in the Solomon Is (Stapley, 1976), Tonga (Maddison, 1976) and Wallis Is (Cohic, 1959), and Risbec (1937) recorded it from Vanuatu on *C. sinensis*. The record of *L. beckii* from Western Samoa on *Codiaeum variegatum* (Reddy, 1970) probably refers to *L. tokionis* (Kuwana). *L. beckii* has been intercepted at quarantine inspection in New Zealand and Cook Is on citrus fruit from other parts of the region.

Lepidosaphes elmerrilleae sp. n. (Fig. 71)

Description
Adult female, when prepared on microscope slide, oval, almost 1.0 mm long, widest at second abdominal segment; head rounded, apex of pygidium straight; minute sclerotized pointed tubercles present between metathorax and abdominal segment 1 and on anterior edges of lateral lobes of segments 2 to 4, each tubercle with an accompanying duct; conspicuous ocular spurs, directed anteriorly, also present.

Dorsal surface of pygidium with median lobes well developed, parallel, the space between less than width of a single lobe; each lobe with 2 notches on either side. Second lobes with inner lobules narrow, less than half width of median lobes; outer lobules half as wide as inner lobules; ventral paraphyses of lobes well developed. Gland spines about same length as lobes, in pairs between median lobes, between median and second lobes, and lateral to second lobes. A single gland spine present on each side of segment 5, and either 2 or 3 on each side anteriorly to segment 3. Marginal macroducts numbering 6 on each side. Dorsal ducts, mostly about half width and length of marginal macroducts, present as a submarginal widely separated pair and a submedian group of 4 on each side of segment 6; anteriorly, present in marginal groups as far forward as metathorax and in submedian groups to abdominal segment 3. Anus situated near base of pygidium.

Ventral surface with about 40 perivulvar pores in 3 groups. Duct tubercles present on metathorax and segments 1 and 2 of abdomen. Small marginal ducts situated on mesothorax, metathorax and abdominal segment 1. Antennae each with 2 long setae. Anterior spiracles each with 2 or 3 disc pores.

Material examined
Holotype female. **PAPUA NEW GUINEA.** Morobe P., Buso, on *Elmerrillea papuana*, 13.x.1979 (*J.H. Martin*) (BMNH).
Paratype female. **PAPUA NEW GUINEA.** Same data as holotype, 1 (BMNH).

Comments
This species is related to *Eucornuaspis pseudomachili*, which Borchsenius described from Korea on *Magnolia*, but differs in possessing fewer duct tubercles. In the new species these are present on the metathorax and the first and second abdominal segments, but in *E. pseudomachili* they are present also on the mesothorax and third abdominal segment.

Lepidosaphes esakii Takahashi (Fig. 72)

Lepidosaphes esakii Takahashi, 1939: 266; Beardsley, 1966: 539.
Parainsulaspis esakii (Takahashi), Borchsenius, 1966: 60.

Description
Slide-mounted adult female 1.75 mm long, tending to be broadly oval, widest at mesothorax; lateral lobes of abdominal segments only moderately developed; head and pygidium rounded. Other important characters are the rather narrow diverging median lobes, the space between as wide as or wider than a single lobe. Second lobes narrow. Gland spines only slightly longer than lobes, present in pairs on pygidium, but 3 or 4 present on margins of segments 3 and 4. Marginal macroducts 6 on each side. Dorsal ducts small, smaller than half width and length of a marginal

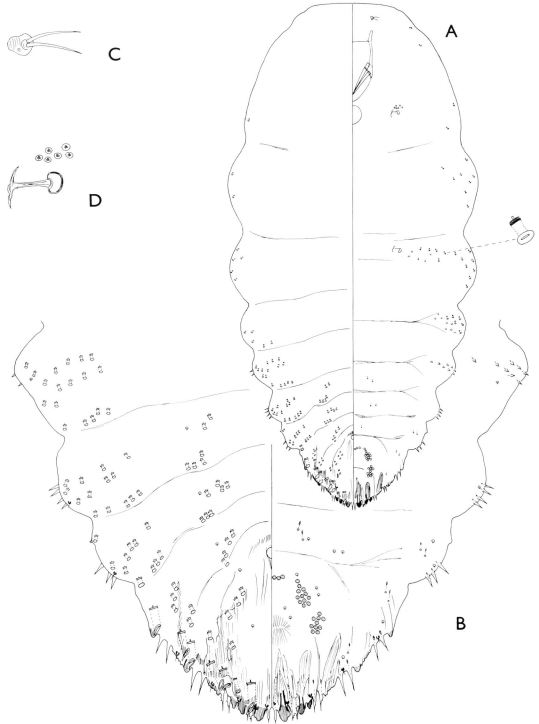

Fig. 72. *Lepidosaphes esakii* Takahashi.

macroduct; segment 6 usually with a single submarginal duct and 2-3 submedian ducts; anteriorly, forming groups on margins of abdomen, and occurring sparsely on margins of thorax; submedian groups present as far forward as segment 3. A single conspicuous duct present just anterior to each second lobe, about half width of a dorsal duct but about twice as long. A single blunt to pointed tubercle present, with accompanying duct, on each margin of segment 4.

Ventral surface with 50-55 perivulvar pores in 5 groups. Microducts sparse, present on pygidial margins and median areas of segments 3 to 5. Duct tubercles present near posterior spiracles and on lateral lobes of segments 1 and 2. Small ducts present on margins of segment 1 and anteriorly to head. Antenna with 2 long setae. Anterior spiracles each with 5-8 disc pores (17 in original description).

Material examined
KIRIBATI. Tarawa, 1976. On *Pandanus odoratissimus*.

Comments
This species was described from the Caroline Is on *Pandanus* sp. Beardsley (1966) recorded it from various localities in Micronesia on *Pandanus* spp. and *Cocos nucifera*, plants to which it may be confined. The illustration is based on specimens collected in the Mariana Is on coconut, kindly made available for this study by J.R. Beardsley. The record from Kiribati is based on a single specimen, but the species is probably common there.

Lepidosaphes eurychlidonis sp. n. (Fig. 73)

Description
Scale of adult female elongate, about 2.5 mm long, dark brown, exuviae pale brown to pale yellow. Male scale about 1.5 mm long, fusiform, pale yellow.

Adult female, on slide, membranous except for pygidium, up to 1.2 mm long, elongate, fusiform, widest at metathorax; head rounded; apex of pygidium straight; lateral lobes of abdominal segments moderately produced. Pygidium with median lobes well developed, parallel, the space between slightly narrower than width of a single lobe; usually with 2 notches on inner edge and 1 notch on outer edge of each. Second lobes each with inner lobule same shape as median lobe but half as wide, outer lobule narrower than inner lobule. Ventral paraphyses of lobes well developed. Gland spines about same length as lobes or slightly longer; 1 pair present between median lobes, and present in pairs on each anterior segment as far forward as segment 3. Marginal macroducts numbering 6 on each side, all the same size. Dorsal macroducts each about half width and length of a marginal macroduct. Segment 6 with 2-4 submedian ducts, forming a group rather than a line when more than 2 present. Anterior ducts present on abdomen in marginal groups as far forward as segment 1, either in definite submedian groups, or forming rows across the segments. A single duct present anterior to each second lobe, this about half as wide as a dorsal duct but longer, with a conspicuous sclerotized orifice.

Lateral sclerotized spurs present between segments 2 and 3, and either blunt or pointed projections present between segments 3 and 4, these spurs or projections with accompanying ducts. Anus situated towards base of pygidium.

Venter of pygidium with 22-36 perivulvar pores in 5 groups. Microducts few, present anterior to vulva, present in median areas of segments 3 to 5, around margins of posterior abdominal segments, near posterior spiracles and on head. Duct tubercles not numerous, situated on metathorax near spiracles, and laterally on segments 1 and 2. Antennae each with 2 long setae. Anterior spiracles each with 1-3 disc pores.

Material examined
Holotype female. **IRIAN JAYA.** Biak, on upper surface of leaf of herbaceous vine, 24.v.1959 (BPBM).

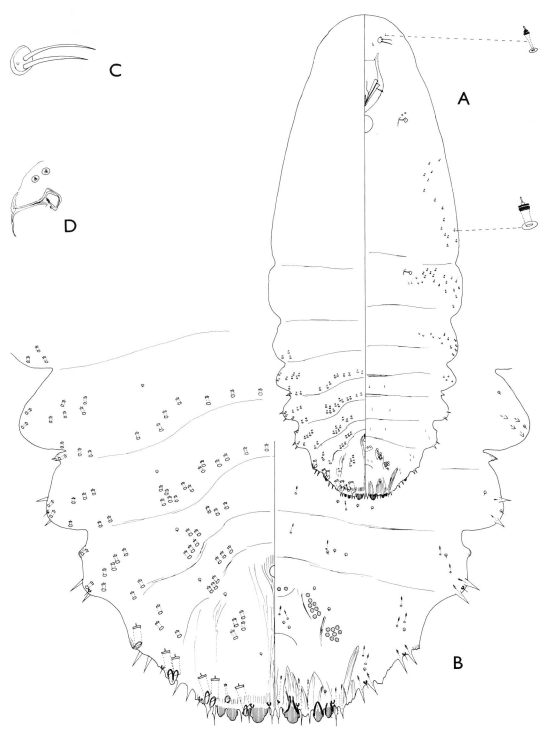

Fig. 73. *Lepidosaphes eurychlidonis* sp.n.

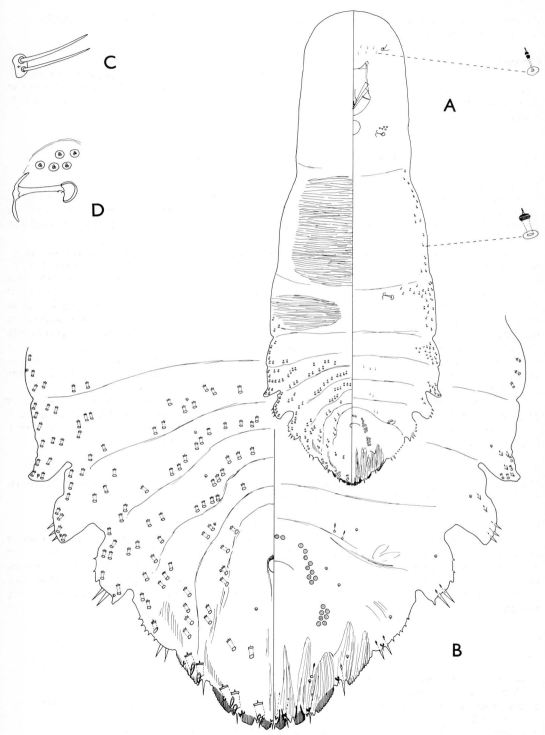

Fig. 74. *Lepidosaphes geniostomae* sp. n.

Paratypes female. **IRIAN JAYA.** Same data as holotype. 7 (BPBM), 4 (BMNH), 2 (DSIR).
Non-type material. **VANUATU.** Efate, 1957, on *Hibiscus* sp. (*J.L. Gressitt*).

Comments
This species resembles *L. tapleyi*, but differs in lacking the lateral spurs between segments 2 and 3; furthermore, the duct immediately above each second lobe is about half the width of a dorsal duct, and has a conspicuous sclerotized orifice. In *L. tapleyi* the duct in this position is filamentous, and the orifice is barely perceptible.

Lepidosaphes geniostomae sp. n. (Fig. 74)

Description
Slide-mounted specimens elongate, about 1.25 mm long; head and thorax narrow, subparallel, expanding to wider anterior abdominal segments; widest at second segment. Head and pygidium rounded. Thorax tending to be sclerotized on dorsum. Lateral lobes of abdomen well produced and directed posteriorly.

Dorsal surface of pygidium with median lobes short, wider than long, inner edges diverging, with single notches on each side. Second lobes each with inner lobule larger than a median lobe, hatchet-shaped, outer lobule reduced to a minute sclerotized point. Position of each third lobe occupied by a long, well-developed sclerotized area, with dentate margin. Gland spines represented by a small pair between median lobes; present singly lateral to each second lobe and lateral to position of third lobes and in pairs on segments 3 and 4. Small but conspicuous points situated on margins of segments 3 and 4. Marginal macroducts numbering 5 on each side. Dorsal ducts mostly about half length and width of marginal macroducts, decreasing in size anteriorly; segment 6 with 2 or 3 present submarginally; other submarginal ducts present in groups anteriorly to metathorax; submedian and median ducts present in more or less single rows as far forward as segment 2. Anus situated at base of pygidium.

Ventral surface with about 35 perivulvar pores in 5 groups. Microducts sparse across median areas of abdominal segments 2 to 4 and between antennae. Duct tubercles present on segments 1 to 3 of abdomen. Small ducts situated on margins of mesothorax, metathorax and abdominal segments 1 and 2. Antennae each with 2 long setae. Anterior spiracles each with 2-5 disc pores.

Material examined
Holotype female. **FIJI.** Viti Levu, Nadarivatu, on *Geniostoma* sp., 1.ix.1955 (*R. Morwood*) (BMNH).
Paratypes female. **FIJI.** Same data as holotype, 2 (BMNH).

Comments
It is difficult to assess this species. It lacks duct tubercles on the metathorax, a character often associated with the presence of dorsal microducts, but the ducts in this species are much larger than microducts. Furthermore there are only 5 marginal macroducts on each side instead of 6, an unusual character but by no means unique. With the elongate sclerotization in the normal positions of the third lobes and the combination of characters given above, this species is easily recognizable.

Lepidosaphes gloverii (Packard) (Fig. 75)

Aspidiotus gloverii Packard, 1869: 527.
Mytilaspis gloverii (Packard), Comstock, 1881: 323.
Lepidosaphes gloverii (Packard), Kirkaldy, 1902: 111; Fernald, 1903: 309; Ferris, 1937a: 74; Balachowsky, 1954: 51.
Insulaspis gloverii (Packard), Borchsenius, 1963: 1172; Borchsenius, 1966: 62.

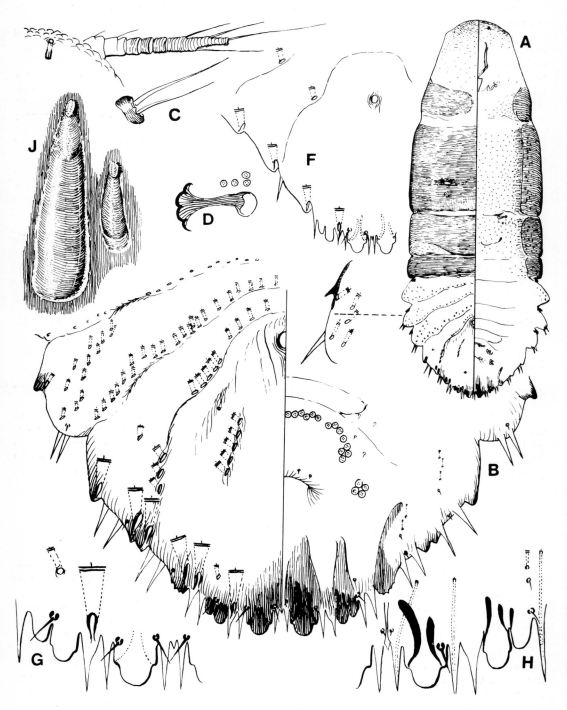

Fig. 75. *Lepidosaphes gloverii* (Packard).

Description
      Scale of adult female elongate, brown, about 2.5-3.0 mm long, the exuviae yellow-brown. Male scale similar to that of female, about 1.5 mm long.
      Adult female, slide-mounted, about 1.35 mm long; widest at about first or second abdominal segments, sides parallel; mature specimens heavily sclerotized on thorax and parts of first abdominal segment; abdomen membranous except for lightly sclerotized pygidium; head and pygidium rounded. Pygidium with median lobes parallel, the space between being less than the width of 1 lobe; second lobes similar shape to median lobes but smaller. Gland spines present in pairs on most abdominal segments. Marginal macroducts numbering 6 on each side. Dorsal ducts small but conspicuous. Segment 6 with a submedian row of 5 or 6 ducts present; anteriorly on abdomen they are numerous across the segments: a small duct also present above each second lobe. Lateral tubercles spur-like, present between first and second segments and between second and third segments; may be blunt on fourth segment.
      Ventral surface with 5 groups of perivulvar pores present. Duct tubercles present on metathorax and first abdominal segment.

Material examined
COOK IS. Rarotonga, 1949. On *Erythrina* sp.
FIJI. Viti Levu, 1944. On *Alocasia macrorhiza, Citrus* sp., *C. aurantium, C. limon, Fortunella japonica.*
FRENCH POLYNESIA. Society Is, Tahiti, 1976. On *C. aurantifolia, C. grandis, C. reticulata.*
NIUE. 1975. On *C. sinensis.*
PAPUA NEW GUINEA. E.S.P.: 1985. M.B.P.: 1985. On *Citrus* sp.
TONGA. Tongatapu Group, Tongatapu, 1974. On *C. limon, C. sinensis.*
WESTERN SAMOA. Upolu, 1975. On *C. aurantifolia, C. limon, C. maxima, C. reticulata, C. sinensis.*

Comments
      This species has been recorded on numerous host-plants, but it seems to favour citrus and is usually found on the fruit, leaves and stems wherever citrus is grown. Slide-mounted specimens are usually easy to identify by the parallel sides and the thick, heavily sclerotized thorax.
      Additional records in the literature mention *L. gloverii* as a major pest of citrus in Fiji (Maddison, 1976); on *Cocos nucifera* in the Society Is (Doane, 1909; Doane & Hadden, 1909; Reboul, 1976); on *Citrus grandis* in Irian Jaya (Reyne, 1961); on *Citrus* sp. in New Caledonia (Brun & Chazeau, 1980); as a minor pest of citrus in the Solomon Is (Stapley, 1976) and as occurring in Samoa on *Codiaeum* sp. The species has been intercepted at quarantine inspection on citrus from the area.

Lepidosaphes karkarica sp. n. (Fig. 76)

Description
      Slide-mounted adult female about 1.15 mm long; elongate; thorax subparallel; abdominal segments with lateral lobes moderately developed; head and pygidium rounded. Median area of head with numerous minute spicules present.
      Pygidium with median lobes wider than long, the space between just narrower than the width of a single lobe, each lobe with a single notch on each side. Second lobes narrower than median lobes, each inner lobule about half as wide as a median lobe, outer lobule slightly narrower than inner lobule; ventral paraphyses of lobes well developed. Gland spines present in pairs on pygidium; segments 2 and 3 each with 3 or 4. Marginal macroducts numbering 6 on each side. Dorsal ducts slender, slightly longer than ventral microducts, tending to be wider anteriorly; present on segment 7 as a small group of 3 or 4, and on segment 6 as an elongate group; submarginal groups present as far forward as segment 1, and submedian groups present as far forward as second abdominal segment. A single microduct situated on midline above gland spines and another just anterior to each second

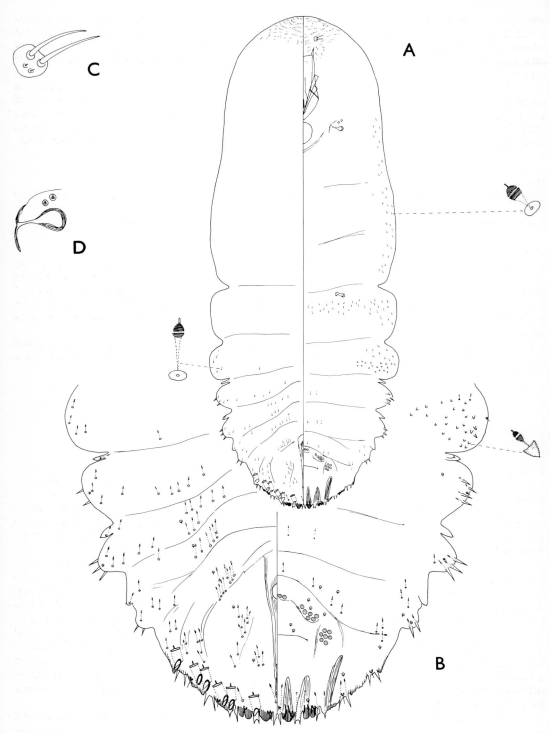

Fig. 76. *Lepidosaphes karkarica* sp. n.

lobe. Anus situated at base of pygidium. Well-developed lateral tubercles present between segments 1 and 2 and between segments 2 and 3, these tubercles elongate and narrower, often with 1 or 2 minute points present at apex or ventrally. Another lateral tubercle present on anterior edge of each lobe of segment 4, wider than anterior tubercles, and with minute point at apex; each tubercle with accompanying duct.

Ventral surface with 24-31 perivulvar pores in 5 groups. Microducts sparse on free abdominal margins and in median areas. Duct tubercles absent from metathorax, present in marginal groups of 18-26 on abdominal segment 1, increasing in size laterally. Minute ducts, slightly wider than abdominal microducts but shorter, present anterior to duct tubercles on segment 1; numerous across metathorax and around margins of metathorax and mesothorax. Antennae each with 2 long setae. Anterior spiracles each with 2-4 disc pores.

Material examined

Holotype female. **PAPUA NEW GUINEA.** Madang P., Karkar Is, on *Cocos nucifera*, 11.x.1979 (*G. Young*) (BMNH).

Paratypes female. **PAPUA NEW GUINEA.** Same data as holotype. 5 (BMNH).

Comments

This species is extremely similar to *L. laterochitinosa* Green, but differs in the form of the lateral tubercles, which are longer and narrower, and in the shorter distance between the median lobes. *L. cocculi* (Green) also has well-developed lateral tubercles, but has a conspicuous group of 6 or 7 gland spines on segment 2, which are fewer in *L. karkarica*; furthermore, there tend to be more gland spines generally in *L. cocculi*, but further research is needed on a greater range of material.

When collected, this species was recorded as severely infesting coconut after volcanic ash had settled, the leaves and nuts being encrusted with the scale. After some months the infested palms started to shed immature nuts and later stopped flowering.

Lepidosaphes marginata Ferris (Fig. 77)

Lepidosaphes marginata Ferris, 1939: 130; Borchsenius, 1966: 50.

Description

Scale of adult female elongate, pale yellow, about 2 mm long. Male scale same colour as female scale, but smaller.

Adult female, when prepared on microscope slide, up to 1.5 mm long; oval, widest at about first abdominal segment; membranous, except for lightly sclerotized pygidium and heavily sclerotized marginal zone on third and posterior segments. Lateral lobes of abdominal segments moderately developed.

Pygidium with median lobes almost parallel, the space between about equal to width of a single lobe; each lobe with single notches on inner and outer margins. Second lobes with inner lobule narrower than a median lobe but similar shape; outer lobule about half as wide as inner lobule. Gland spines longer than lobes; present in pairs between median lobes, between median and second lobes and on margins of segments 5 and 6; and in groups of 3-4 on margins of segments 2 to 4. Marginal pygidial macroducts numbering 6 on each side. Dorsal ducts much smaller than marginal ducts, there being on each side 1 or 2 in submedian position on segment 7: on segment 6, usually a submarginal pair and a submedian group of 4 ducts present, often coalescing. Anteriorly, around submargins, there are groups of ducts as far forward as mesothorax, and submedian groups forward to abdominal segment 2. A single duct present anterior to each second lobe, each duct narrower than a dorsal duct but longer. Anus situated near base of pygidium.

Ventral surface with 43-55 perivulvar pores in 5 groups. Microducts present mainly around pygidial margins and across median areas of abdominal segments. Duct tubercles situated laterally on metathorax and segments 1 and 2 of abdomen. Small tubular ducts present on margins of abdominal

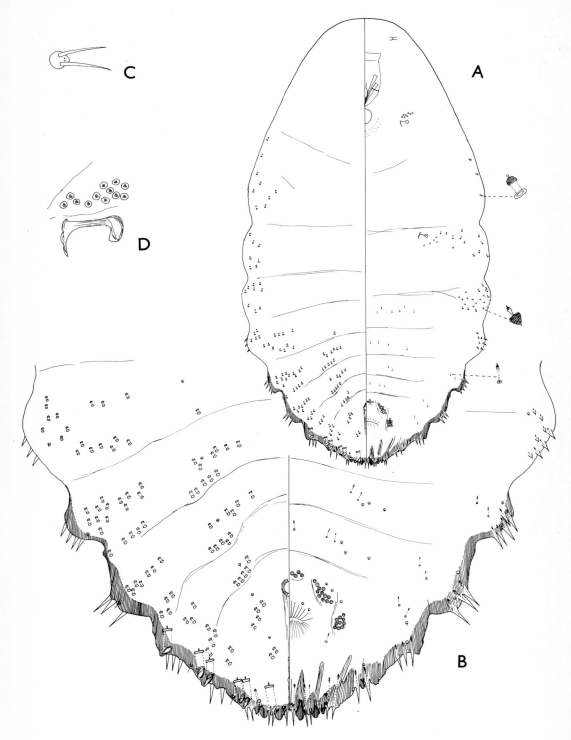

Fig. 77. *Lepidosaphes marginata* Ferris.

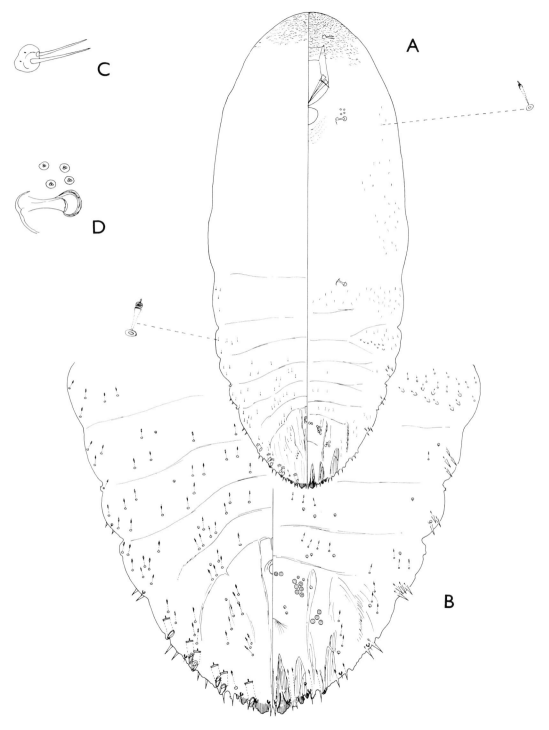

Fig. 78. *Lepidosaphes pometiae* sp. n.

segment 1 and laterally on metathorax and mesothorax. Antennae each with 2 long setae. Anterior spiracles each with 8-15 disc pores.

Material examined
FRENCH POLYNESIA. Marquesas Is, Hivaoa, 1930. On *Cheirodendron marquesense, Reynoldsia marchionensis*.

Comments
   As Ferris mentioned in his original description, this is a typical species of the genus, but differs from all other species in possessing a marginal sclerotized zone around the posterior abdominal segments. It has not been collected outside the Marquesas Is; all records so far are from endemic plants at altitudes above 1000 m.
   There have been available two paratypes and two other original specimens, kindly made available by R.O. Schuster, University of California, Davis. In addition, specimens have been prepared from original dry material held at the Bernice P. Bishop Museum, Honolulu. There is distinct pitting of the leaves where scales in close proximity have fed (Ferris, 1939). One of the original samples is labelled '*Cheirophyllum*' sp. nr *platyphyllum*, while Mumford & Adamson (193?) give the host as *Cheirodendron* sp. nr *platyphyllum*; this plant was subsequently named *Cheirodendron marquesense*. The other original sample was recorded on *Reynoldsia* sp. nr *tahitensis*, which was later named *Reynoldsia marchionensis*.

Lepidosaphes pometiae sp. n. (Fig. 78)

Description
   Slide-mounted specimens about 1.0 mm long, elongate-oval; head and pygidium rounded. Head profusely covered with minute spicules.
   Pygidium with median lobes wider than long, each with a single notch present on inner margin and usually 2 notches on outer margin; the space between less than half width of 1 lobe. Second lobes with inner lobules approximately half width of median lobes, about same shape; outer lobules minute, less than half width of inner lobules. Gland spines present in pairs on margins of pygidium as far forward as segment 5. Margins of segments 2 to 4 each with 3 or 4 gland spines. Marginal macroducts numbering 6 on each side. Dorsal ducts slender, slightly wider and longer than ventral microducts; a group of 3 or 4 present on segment 7; a row or group present on segment 6; submarginal groups present as far forward as metathorax; submedian groups present anteriorly to segment 2, but the groups often merging across the segments to median areas. A single slender duct present just anterior to each second lobe. Anus situated at base of pygidium. Lateral tubercles weak, represented by 1 on each side between segments 2 and 3, each usually tipped with 2 minute points; a swelling also present between segments 3 and 4, the tubercles and swellings each having an accompanying duct.
   Ventral surface with 24-26 perivulvar pores in 5 groups. Microducts present on margins of pygidium and across median areas of free abdominal segments; numerous on margins of mesothorax and metathorax, and across entire metathorax; another group present on abdominal segment 1. Duct tubercles, numbering 7-12, present submarginally on each side of segment 1 posterior to the microducts. Antennae each with a single seta. Anterior spiracles each with 2-4 disc pores.

Material examined
Holotype female. **SOLOMON ISLANDS.** G.P., Guadalcanal, Mt Austen, on bark of *Pometia pinnata*, 11.i.1984 (*M. Bigger*) (BMNH).
Paratypes female. **SOLOMON ISLANDS.** Same data as holotype. 4 (BMNH), 2 (DSIR), 1 (ARSDC).

Comments
   This species shares with *L. karkarica* many of the characters of *L. laterochitinosa*. It differs from both species in having the median lobes separated by a space less than half the width

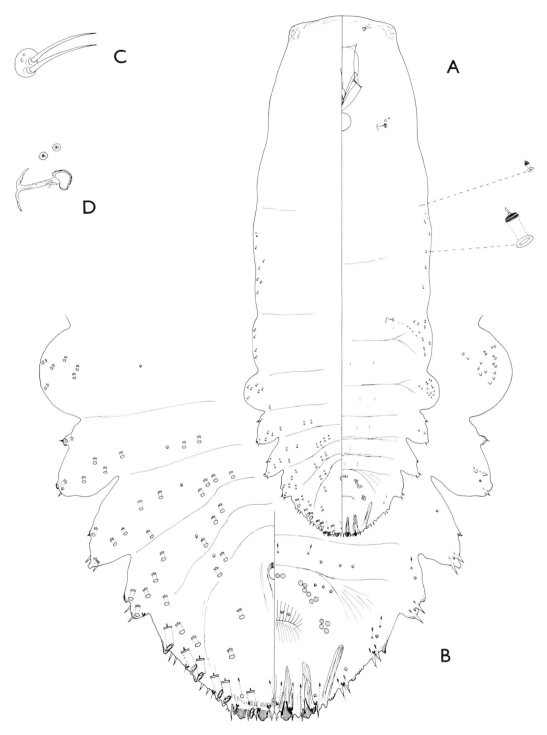

Fig. 79. *Lepidosaphes rubrovittata* Cockerell.

of 1 lobe. Weak lateral tubercles, each usually tipped with 2 minute points, are present only between segments 2 and 3, whereas in the other two species, tubercles are present between three abdominal segments; furthermore, the duct tubercles on segment 1 number only 7-12 on each side, whereas in the other two species they number 17-30. *L. pometiae* also lacks the minute dorsal duct just anterior to the median gland spines. This species was listed as *Lepidosaphes* sp. by Bigger (1985).

Lepidosaphes rubrovittata Cockerell (Fig. 79)

Lepidosaphes rubrovittatus Cockerell, 1905b: 135.
Mytilaspis fasciata Green, 1911: 31. Lectotype female, Sri Lanka (BMNH), here designated [examined]. Syn. n.
Lepidosaphes ulapa Beardsley, 1966: 546. [Synonymized by Beardsley, 1975: 661.]
Insulaspis rubrovittata (Cockerell), Borchsenius, 1966: 66.

Description
    Length of largest slide-mounted specimens about 1.0 mm; sides sub-parallel, widest at about first abdominal segment. Margins of anterior abdominal segments well produced laterally. Head with anterior margin almost straight, and with pronounced lateral angles. Derm of anterior margin with minute spicules, lateral angles with heavier concentrations of spicules so that the angles appear to be sclerotized.
    Pygidium with apex almost straight. Median lobes well developed, the space between about same width as a single lobe; each lobe usually with single notches present on either side. Second lobes each with inner lobule one half to two thirds width of median lobe and of a similar shape; outer lobule narrower than inner lobule. Ventral paraphyses of median and second lobes well developed. Gland spines only slightly longer than lobes, 1 pair present between median lobes, 1 pair between each of second and third lobes, and a single wide gland spine present lateral to each second lobe. Other gland spines present in pairs on each segment, anteriorly to about third segment. Marginal macroducts numbering 6 on each side. A single microduct, about same length as marginal macroduct, present between, and anterior to, median gland spines, and a similar duct present anterior to each second lobe. Dorsal macroducts mostly about half length and width of a marginal macroduct. Segment 6 with a pair of ducts present on each side, usually well separated on submarginal and submedian areas, only rarely closer together; occasionally 3 ducts present on one or more sides. Forward from segment 6, on dorsum, ducts usually form small submarginal and submedian groups to second segment, and marginal ducts are often present as far forward as mesothorax. Distinct sclerotized spurs present singly on anterior edges of lateral lobes of segments 2 and 3; segment 4 with a blunt sclerotized projection; each of these processes with an accompanying duct. Anus situated near base of pygidium.
    Venter of pygidium with 5 groups of perivulvar pores, each group rarely with more than 6 pores, total of 20-27. Microducts few across abdominal segments, and next to posterior spiracles. Duct tubercles present in small numbers laterally on metathorax and on segment 1 of abdomen, accompanied by small marginal ducts. Similar ducts also present on margins of mesothorax. Antennal tubercles each with 2 long setae and 2 spicule-like setae. Anterior spiracles each with 1 or 2 disc pores.

Material examined
FIJI. Beqa, 1949; Viti Levu, 1945. On *Barringtonia* sp., *Ficus* sp., *Inocarpus* sp., *I. fagifer*, Orchidaceae.
IRIAN JAYA. Biak, 1959. On Asteraceae.
PAPUA NEW GUINEA. C.P.: 1959. Morobe P.: 1974. On *Araucaria* sp., *A. hunsteinii*, *Ficus* sp., Sapotaceae.
TONGA. Vava'u Group, Vava'u, 1977. On *Eugenia malaccensis*.

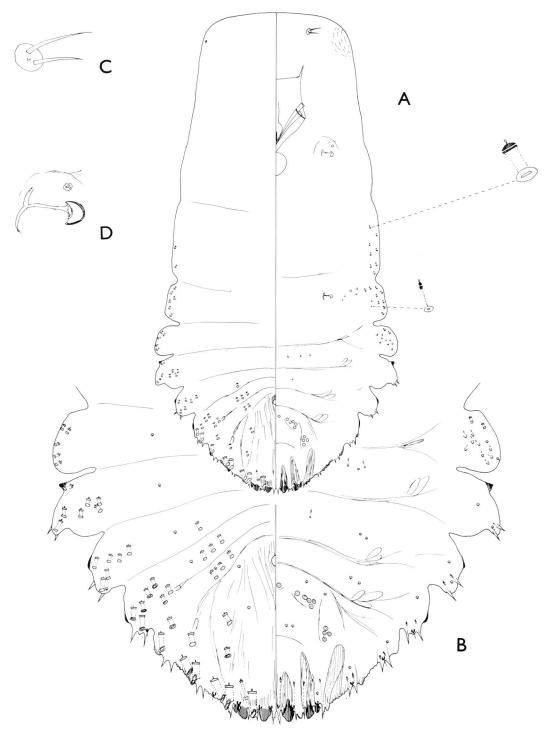

Fig. 80. *Lepidosaphes securicula* sp. n.

Comments

Superficially this species resembles *L. tokionis*; however, in *L. rubrovittata* the lateral angles of the head have concentrations of spicules, whereas in *L. tokionis* the lateral angles of the head are sclerotized and lack spicules.

There have been available a few original specimens of *L. rubrovittata* collected from Manila, Philippine Islands (BMNH), and two paratypes of *L. ulapa*, kindly made available by J.W. Beardsley. *Mytilaspis fasciata* was described from Sri Lanka, Heneratgoda, on *Hevea brasiliensis*. The only original slide available contains seven adult females and one of these has been selected as lectotype and clearly labelled. Shortly after describing this species, Green must have seen the original material of *L. rubrovittata* mentioned above, and altered the name on his single slide. It is one of only a few of Green's type slides not labelled as such. There is apparently no previous formal synonymy.

The distribution of this species has still not been determined, but it is probably widespread throughout southern Asia on a wide range of plants. Beardsley (1975) has recorded it from the Caroline Is and the Mariana Is.

The illustration has been prepared from specimens on *Araucaria hunsteinii* collected at Bulolo, Papua New Guinea.

Lepidosaphes securicula sp. n. (Fig. 80)

Description

Adult female, when mounted on microscope slide, up to 1.0 mm long; elongate-oval, but sides often straight; widest at second abdominal segment; pygidium rounded; head curved to straight; lateral lobes of abdominal segments well developed.

Pygidium with median lobes well developed, each with a notch on either side; the space between about the same as width of one lobe or less. Second lobes each with inner lobules hatchet-shaped, narrower than median lobes; outer lobules rounded, narrower than inner lobules. Ventral paraphyses of all lobes well developed. Gland spines slightly longer than lobes, there being 2 present between median lobes, 2 between median and second lobes, 1 present lateral to each second lobe, and 2 laterally on each segment as far forward as segment 2 of abdomen. Marginal macroducts on each side numbering 5, all the same size; the single duct in normal position of the sixth, on segment 5, about half as wide as the others. Dorsal ducts normally about half length and width of marginal macroducts. Segment 6 usually with 2 widely separated ducts on each side; submarginal groups present as far forward as mesothorax, small submedian groups also present on abdominal segments 3-5. A narrow duct, about half as wide as dorsal ducts, or narrower, but often longer than marginal macroducts, situated above each second lobe. Segment 2 with an antero-lateral tubercle, blunt and heavily sclerotized. Segments 3 and 4 with conspicuous sclerotized thickenings present in same positions, each process with an accompanying duct. Each lateral angle of head with eye spot modified into a posterior projecting spur, occasionally absent or represented by 1 only.

Ventral surface with perivulvar pores numbering 14-20 in 5 groups. Microducts sparse, present in median areas of some abdominal segments, and lateral to posterior spiracles. Duct tubercles present laterally on metathorax and segment 1 of abdomen. Small marginal ducts present on mesothorax, metathorax and first abdominal segment. Antennae each with 2 long setae. Anterior spiracles each with 1 or 2 disc pores.

Material examined

Holotype female. **PAPUA NEW GUINEA.** E.H.P.: Kratke Mts, Kassam, on leaves of Sapotaceae, 15.xi.1959 (*T. Maa*) (BPBM).
Paratypes female. **PAPUA NEW GUINEA.** Same data as holotype, 3 (BPBM), 5 (BMNH).

Comments

This is a distinctive species with 5 macroducts the same size on each side, but the sixth present is narrower. The inner lobule of each second lobe is distinctly hatchet-shaped, and the eye

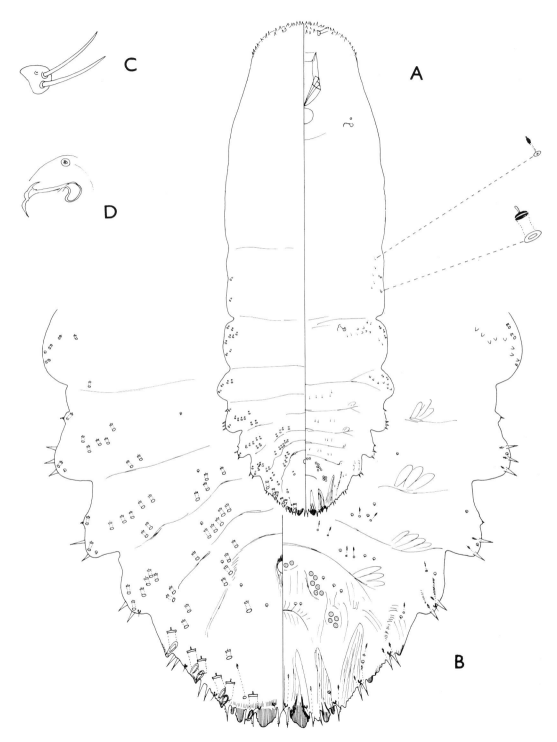

Fig. 81. *Lepidosaphes stepta* sp. n.

spot is modified into a small spur. It differs from *L. pinnaeformis* (= *L. machili* (Maskell)), another species with 5 marginal macroducts and spurs on head, in possessing larger, but much fewer, dorsal ducts.

Lepidosaphes stepta sp. n. (Fig. 81)

Description
   Slide-mounted specimens about 1.25 mm long; narrow, elongate, sides almost parallel, widest at first abdominal segment; lateral lobes of abdominal segments moderately developed; pygidium rounded; head rounded, but jagged because of numerous quite large pointed tubercles or spurs which become sclerotized, occupying anterior edge and area between antennae.
   Pygidium with median lobes well developed, parallel, but inner margins slightly divergent, each with 2 or 3 small notches; outer margins usually with 2 notches; space between lobes narrower than width of single lobe. Second lobes each with inner lobule less than half width of median lobes; outer lobules small and narrow. Ventral paraphyses of lobes well developed. Gland spines about same length as lobes, 1 pair present between median lobes and 1 pair between each median and second lobe; present singly lateral to second lobes, and in pairs forward to third abdominal segment. Segment 2 usually with 3 gland spines present. Gland spines with jagged edges, associated with the lobes. Macroducts numbering 6 on each side. Dorsal ducts much smaller, those on segments 5 and 6 about half length and width of a marginal macroduct, but ducts further forward decreasing in size. Segment 6 usually with single submarginal and submedian ducts, but sometimes pairs present; rarely 3 in total. Submarginal ducts present in groups as far forward as mesothorax; submedian groups present up to segment 3. A single microduct, longer than a marginal macroduct, situated just anterior to each second lobe. Spur-like lateral tubercles present on anterior edges of lateral lobes of segments 2 and 3.
   Ventral surface with 23-26 perivulvar pores in 5 groups. Microducts present on pygidial margins near posterior spiracles, and in single median rows on abdominal segments. Duct tubercles present lateral to posterior spiracles, and on lateral lobes of abdominal segment 1. Small ducts few, present on margins of mesothorax, metathorax and segment 1 of abdomen. Antennae each with 2 long setae. Anterior spiracles each with either 1 or 2 disc pores.

Material examined
Holotype female. **FIJI.** Viti Levu, Naduruloulou, on *Stachytarpheta* sp., x.1949 (*B.A. O'Connor*) (BMNH).
Paratypes female. **FIJI.** Viti Levu, Naduruloulou, on *Stachytarpheta* sp., 15.i. 1950 (*T.S. Pillai*), 3 (BMNH). **NIUE.** Vaipapahi Farm, on *Aleurites moluccana*, 31.i.1977, (*P.A. Maddison*), 3 (DSIR), 1 (BMNH).
Non-type material. **FIJI.** Colo-i-Suva, on unidentified plant, 28.ii.1941 (*R.A. Lever*). Intercepted in New Zealand, on *Croton* sp., 18.viii.1975 (*R.G. Holland*).

Comments
   This distinctive species is easily separated from other Pacific species in possessing numerous large tubercles or spurs on the head margin. It is related to *L. okitsuensis* Kuwana, described from Japan, but this species has more numerous ducts posterior to segment 5 on the pygidium. *L. stepta* is very close to the species currently accepted as *Scobinaspis serrifrons* (Leonardi), redescribed as *Velataspis serrifrons* by Balachowsky (1954). It was described originally from specimens collected in a botanical garden in Italy on *Croton* sp. Specimens have not been available for this work, but according to Balachowsky's description, it differs from *L. stepta* in lacking the sclerotized tubercles on segments 2 and 3 of the abdomen, and the long microduct above each second lobe.

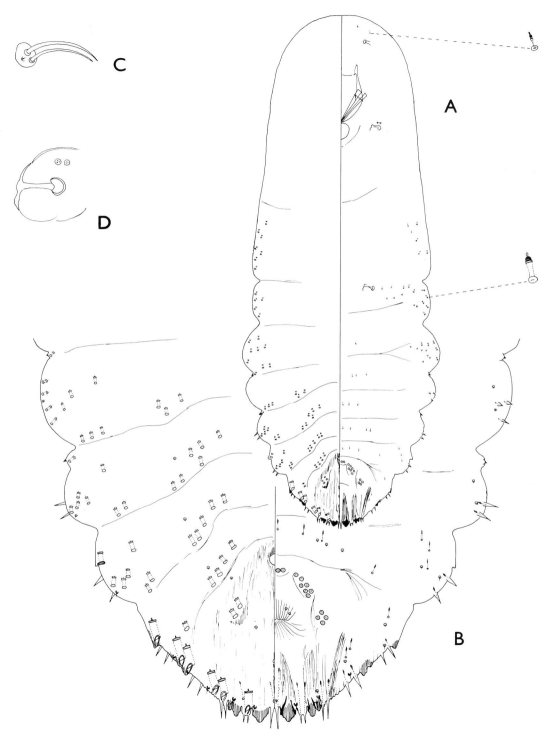

Fig. 82. *Lepidosaphes tapleyi* Williams.

Lepidosaphes tapleyi Williams (Fig. 82)

Lepidosaphes tapleyi Williams, 1960: 393.
Insulaspis tapleyi (Williams), Borchsenius, 1963: 1173.

Description
   Scale elongate, pale to reddish brown, about 2.5 mm long. Slide-mounted specimens about 1.0 mm long; elongate-oval, widest at first or second abdominal segment; membranous except for pygidium; lateral lobes of abdominal segments moderately developed. Head straight to rounded.
   Pygidium with apex straight to slightly rounded. Median lobes well developed, parallel, the space between a little narrower than width of 1 lobe; each lobe with 2 notches present on inner edge and with 1 or 2 notches on outer edge. Second lobes each with inner lobule about half width of a median lobe, outer lobule about half width of inner lobule. Ventral paraphyses of lobes well developed. Gland spines usually surpassing length of lobes, there being 2 between median lobes, 2 between each median and second lobe; pairs of gland spines present on margins of segments forward to third abdominal segment; rarely 3 on segment 3. Marginal macroducts numbering 6 on each side. A single microduct, about same length as a marginal macroduct, usually situated on midline just anterior to median gland spines, and a similar duct present just lateral to each first marginal macroduct; orifices of the microducts sometimes difficult to detect. Dorsal macroducts about half width and length of marginal macroducts. Segment 6 usually with 2 submedian ducts set close together present on each side, occasionally with 3 and rarely with 4. Anterior ducts arranged in small submarginal and submedian groups as far forward as segment 2. Other ducts present marginally to mesothorax. Lateral tubercles small, sclerotized and pointed, present between first and second and second and third segments; tubercles on fourth segment blunt and not sclerotized; each tubercle with an accompanying duct. Anus situated at base of pygidium.
   Venter of pygidium with 5 groups of perivulvar pores, totalling 25-30. Microducts present in more or less single rows across abdominal segments; a few also present near posterior spiracles, near margins of mesothorax and on head. Duct tubercles few, present in groups on metathorax and first abdominal segments, there being also 2 or 3 on margins of second abdominal segment. Small ducts present on margins of mesothorax, metathorax and first abdominal segment. Antennae each with 2 long setae. Anterior spiracles each with 2 or 3 disc pores.

Material examined
IRIAN JAYA. Biak, 1959. On Asteraceae.
KIRIBATI. Butaritari, 1976; Tarawa, 1972; Phoenix Is, Enderbury, 1964. On *Capsicum frutescens*, *Cocos nucifera*, *Hibiscus* sp., *Lycopersicon esculentum*, *Plumeria rubra*.
TUVALU. Vaitupu, 1976. On *Hibiscus* sp.

Comments
   This species was described from Sudan and Tanzania, and has since been collected throughout East and West Africa. It is now known from southern Asia, and may be widespread on many plant species. In Africa and Asia it seems to be particularly common on mango. Specimens collected on tomato at Tarawa, Kiribati, were reported as causing damage. Most of the specimens examined from Enderbury in the Phoenix group have dorsal ducts on the sixth segment, numbering 4 or sometimes 5 on each side, but the specimens are here regarded as *L. tapleyi*.

Lepidosaphes tokionis (Kuwana) (Fig. 83)

Mytilaspis newsteadi var. *tokionis* Kuwana, 1902: 81.
Mytilaspis auriculata Green, 1907: 204. [Synonymized by Ferris, 1942: 398].
Lepidosaphes gloverii (Packard), Laing, 1927: 42 (Misidentification).
Lepidosaphes tokionis (Kuwana), Takahashi, 1935: 21; Ferris 1942: 398; Zimmerman, 1948: 426.
Lepidosaphes lasianthi (Green), Ferris, 1938: 145 (Misidentification).

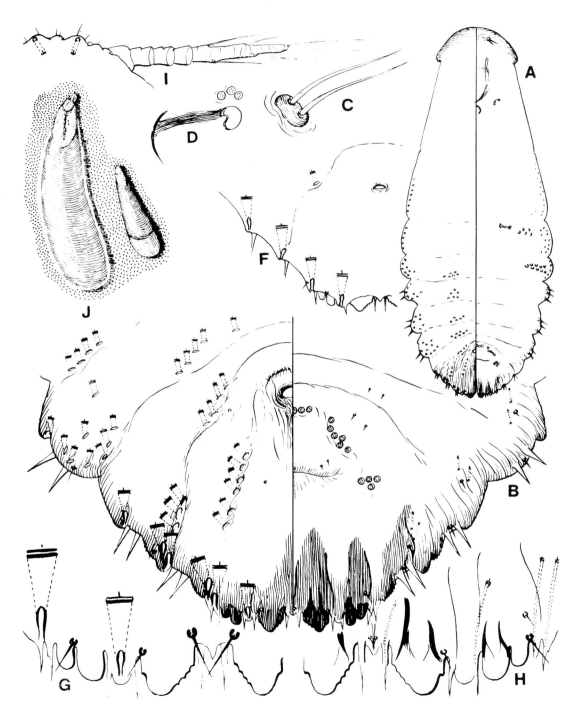

Fig. 83. *Lepidosaphes tokionis* (Kuwana).

Description
    Scale of adult female pale brown, elongate. Male scale about same colour as female scale, smaller.
    Adult female, when mounted on slide, elongate; head of mature specimen sclerotized and expanded laterally to form sclerotized lobes. Other important characters are the median lobes with space between either same width as a single lobe, or narrower; each lobe with a few notches present on inner edge; the second lobes narrower than median lobes. Gland spines present in pairs. Marginal macroducts represented by 6 on each side; dorsal ducts each about half width and length of a marginal duct, present in a submedian line of 6 or more on segment 6, in marginal groups as far forward as mesothorax, and in submedian groups forward to segment 1 of abdomen.
    Ventral surface with perivulvar pores in 5 groups. Duct tubercles present on metathorax and abdominal segment 1. Antennae each with 2 long setae. Anterior spiracles each with 2 or 3 disc pores.

Material examined
FIJI. 1911; Viti Levu, 1942. On *Citrus maxima, Codiaeum* sp., *C. variegatum, Gossypium* sp.
PAPUA NEW GUINEA. C.P.: 1958. On *C. variegatum.*
TONGA. Tongatapu Group, Tongatapu, 1975. On *C. variegatum.*
WESTERN SAMOA. Upolu, 1925. On *Codiaeum* sp., *C. variegatum.*

Comments
    This species is usually easy to identify by the lateral sclerotized lobes on the head. It has been recorded in various parts of the world on numerous species of plants, but these records need to be verified. The main host-plant is *Codiaeum*, on which it is often found throughout the tropics and some temperate areas.
    In Fiji, Swaine (1971) states that occasional heavy infestations may damage *Codiaeum* by drying the tissues. This species has been intercepted from the area at quarantine inspection in New Zealand.

Genus **LINDINGASPIS** MacGillivray

Lindingaspis MacGillivray, 1921: 388; McKenzie, 1950: 98; Williams, 1963: 3. Type-species *Melanaspis samoana* Lindinger, by original designation.

Description
    The scale of the adult female is usually deeply coloured, often brown or blackish, circular and flat, with the exuviae subcentral. Male scale similar to female scale, but smaller and elongate.
    Important characters of the adult female are 1-barred ducts; fringed plates; 3 pairs of pygidial lobes, not bilobed, and a series of tooth-like projections anteriorly to fourth abdominal segment. Slender paraphyses are numerous; those arising from the angles of the lobes usually the longest; others, that become shorter, extend forward to fourth abdominal segment. The dorsal ducts are long and are of two types; the larger type occurring in the interlobular spaces, and exceedingly slender ducts occurring further forward.

Comments
    The genus is closely related to *Chrysomphalus*, but differs in possessing the long marginal series of paraphyses. It differs from *Melanaspis* Cockerell, in which the type-species was originally described, in having 2 sizes of dorsal ducts, the larger type being absent in *Melanaspis*. About 25 species are recognized, from southern Asia, Australia and Africa; but three species appear to be endemic to Samoa, and further species may occur there. *L. rossi* (Maskell) has become widely distributed in the tropics.

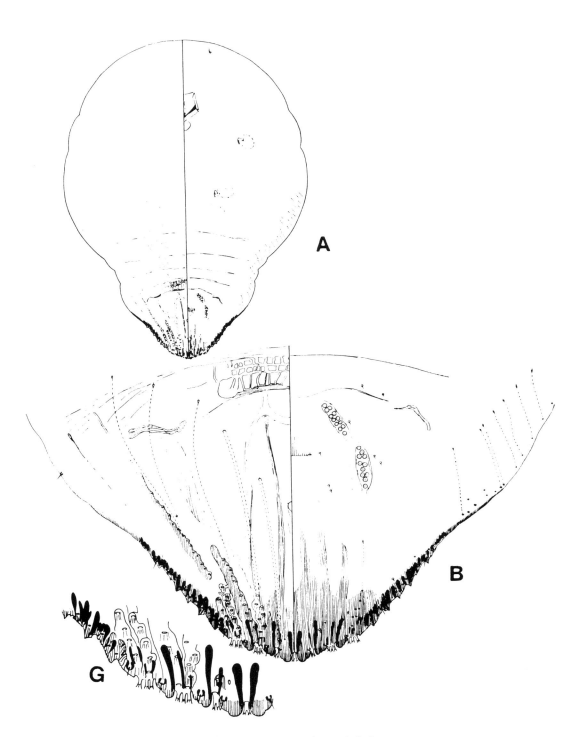

Fig. 84. *Lindingaspis buxtoni* (Laing).

Key to species of *Lindingaspis*

1	Perivulvar pores present in 2 groups	*samoana* (Lindinger)
--	Perivulvar pores present in 4 groups	2
2	With a series of medium-sized dorsal pygidial ducts extending forward from near seta of abdominal segment 4	*similis* McKenzie
--	Without this series of dorsal ducts	3
3	Paraphysis arising from lateral angle of each second lobe minute, shorter than neighbouring paraphyses	*buxtoni* (Laing)
--	Paraphysis arising from lateral angle of each second lobe long and slender, about same size as neighbouring paraphyses	*rossi* (Maskell)

Lindingaspis buxtoni (Laing) (Fig. 84)

Chrysomphalus buxtoni Laing, 1927: 40.
Lindingaspis buxtoni (Laing), McKenzie, 1939: 53; Williams, 1963: 4.

Williams (1963) redescribed and illustrated this species. Although the paraphyses are large and tend to be thick, the species is easily distinguishable from others in the Pacific area by the minute paraphysis arising from the lateral angle of each second lobe, as shown in the accompanying diagram.

Material examined
WESTERN SAMOA. Upolu, 1924. On ?

Comments
Since the original record, this species has apparently not been collected again, although it may be common in Samoa.

Lindingaspis rossi (Maskell) (Fig. 85)

Aspidiotus rossi Maskell, 1891: 3.
Lindingaspis rossi (Maskell), Ferris, 1938: 246; McKenzie, 1950: 57.

This species was described originally from Australia, and has since been reported from many tropical countries and temperate areas on a wide variety of plants. It should be easily identifiable from the key, but other important characters are the presence of 3 plates between the second and third lobes; a row of about 12 dorsal ducts arising from between the second and third lobes, increasing in width with distance from the margin. In other species from the Pacific area these ducts remain the same size or decrease in width. The specimens from New Caledonia on *Araucaria* have a shorter row in this position, containing only about 8 ducts; this is assumed to be a host effect.

Material examined
NEW CALEDONIA. New Caledonia, 1970. On *Araucaria brasiliana*.
NORFOLK I. 1984. On *Elaeodendron curtipendulum*, 'a Monocot', *Olea apetala, O. verrucosa, Schinus terebinthifolius*.

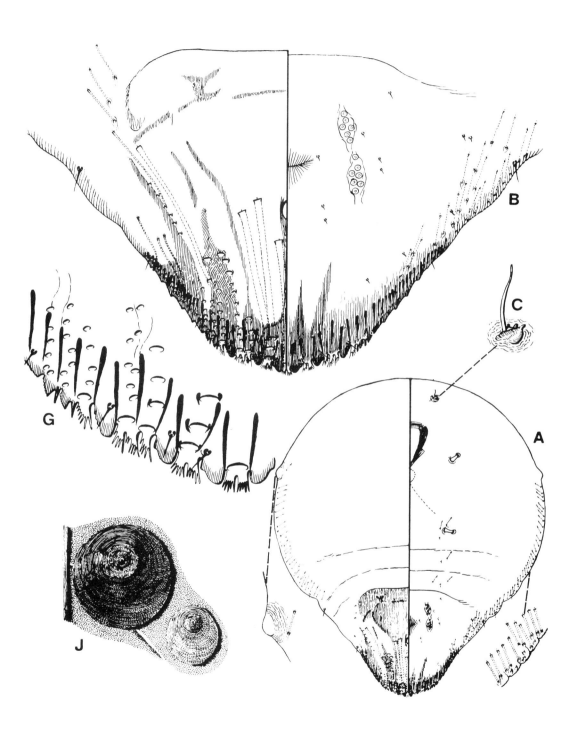

Fig. 85. *Lindingaspis rossi* (Maskell).

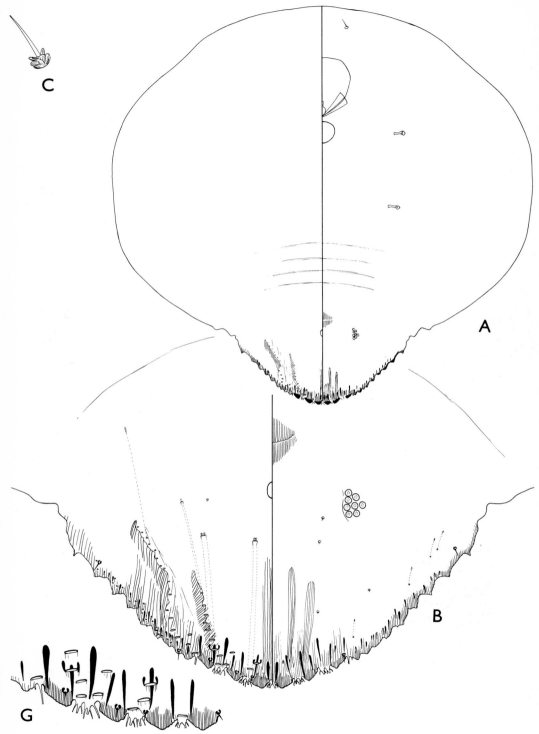

Fig. 86. *Lindingaspis samoana* (Lindinger).

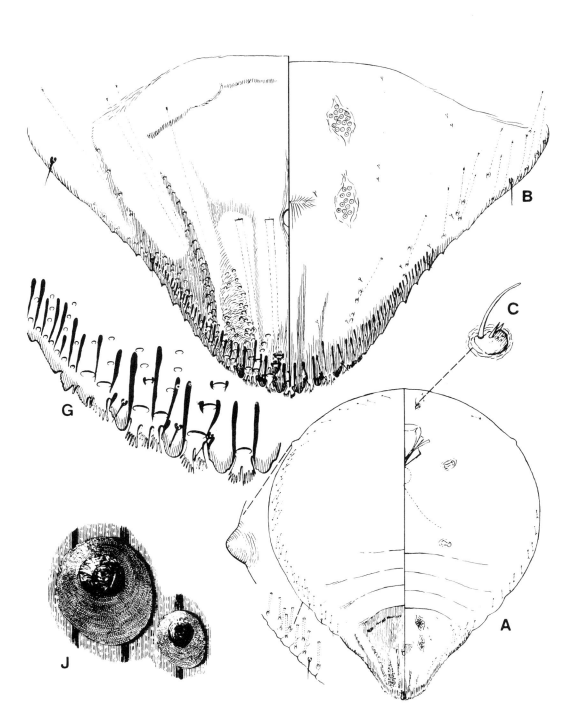

Fig. 87. *Lindingaspis similis* McKenzie.

Comments
    L. rossi was recorded from one locality in New Caledonia, on *Avicennia officinalis*, by Cohic (1958a), who mentions it had been intercepted at quarantine on *Macadamia ternifolia* several times. Borchsenius (1966) mentions its occurrence in Samoa, but no material has been seen from there in this study.

Lindingaspis samoana (Lindinger) (Fig. 86)

Melanaspis samoana Lindinger, 1911: 177.
Chrysomphalus samoana (Lindinger), Sasscer, 1912: 93.
Lindingaspis samoana (Lindinger), MacGillivray, 1921: 422; Williams, 1974: 217.

   After examining original material, Williams (1974) redescribed and illustrated this species, and the illustration is reproduced here. The species differs from all known species of *Lindingaspis* in possessing only 2 groups of perivulvar pores. In other species they may be absent, or are represented by 4 or 5 groups. The species is endemic to Samoa, but has not been collected since the original record.

Material examined
WESTERN SAMOA. Savai'i. On *Myristica hypargyraea*.

Lindingaspis similis McKenzie (Fig. 87)

Chrysomphalus rossi (Maskell), Doane & Ferris, 1916: 401. (Misidentification)
Lindingaspis similis McKenzie, 1950: 105. Lectotype female, Western Samoa (USNM), here designated [examined].

   The type material of this species was described from Upolu, Western Samoa, on an unidentified host and from specimens collected later from the same locality on *Pandanus* sp. Material examined by McKenzie (1950) from American Samoa, recorded by Doane & Ferris (1916) on *Cocos nucifera* as *C. rossi*, also proved to be this species. The species has not been recorded from elsewhere. Among the South Pacific species, it can be distinguished in possessing a series of dorsal pygidial ducts extending forward from the margins of abdominal segment 4.

Material examined
SAMOA. 1838-42. On *Hernandia sonora*.
AMERICAN SAMOA. Tutuila, 1913. On *Cocos nucifera*.
WESTERN SAMOA. Upolu, 1940. On *C. nucifera*.

Comments
    The lectotype selected is one of 2 specimens on a slide labelled 'type', and is clearly marked. The other specimen, and 2 others on another slide labelled 'type' (USNM), are here designated paralectotypes. Doane & Ferris (1916) misidentified *L. similis* from American Samoa on coconut as *Chrysomphalus rossi*; Swezey (1924) and Dale (1959) refer to this misidentification.

Genus LOPHOLEUCASPIS Balachowsky

Lopholeucaspis Balachowsky, 1953: 153; Balachowsky, 1958: 335; Takagi, 1969: 30. Type-species *Leucaspis japonica* Cockerell.

Description
This is a genus containing pupillarial species, the adult female remaining within the second instar exuviae after moulting. Body elongate with almost parallel sides. Pygidial lobes projecting, elongate to conical; present as 2 pairs, each lobe well separated. Plates elongate, fimbriate apically, a pair present between median lobes and between each median and second lobe; replaced anteriorly by broader plates, and then by a row of duct tubercles that curves to ventral surface beyond anterior spiracles. Perivulvar pores present, usually in 5 groups; supplementary submarginal groups present on 2 preceding segments.

Comments
About ten species are now recognized in the genus. Balachowsky erected the genus because it differed from *Leucaspis* Targioni in the adult female possessing a continuous row of duct tubercles from the abdomen forward to thorax, and in lacking the group of duct tubercles lateral to the anterior spiracles. This definition holds for all described species, but the new species described herein has an interrupted row of duct tubercles, which are lacking between the prothorax and third abdominal segment; nevertheless, the species lacks the group lateral to the anterior spiracles, and is here referred to *Lopholeucaspis*.

Key to species of *Lopholeucaspis*

1 Lateral row of duct tubercles interupted, absent from between prothorax and abdominal segment three .... *baluanensis* sp. n.
-- Lateral row of duct tubercles complete, not absent from between prothorax and abdominal segment three .... *cockerelli* (Grandpré & Charmoy)

Lopholeucaspis baluanensis sp. n. (Fig. 88)

Description
Female pupillarial; scale elongate, with longitudinal ridge on mid-line, dark brown, but covered in white wax. Male scale similar to female scale but much smaller.

Adult female, on slide, about 1.55 mm long; elongate-oval, widest at thorax; membranous, except for sclerotized patches on pygidium; head and pygidium rounded.

Pygidium with 2 pairs of lobes present. Median lobes well separated, elongate, projecting, bluntly pointed at apices. Second lobes similar in shape to median lobes, but often with a notch present on inner margins. Plates elongate, about same length as lobes; 2 present between median lobes, slender and fimbriate at apices; 2 present between median and second, and 3 lateral to second lobes. Anteriorly to segment 5 they increase in width and are notched on each side, and forward to segment 3 on margins of venter they are replaced by prominent truncate duct tubercles. Dorsal ducts slender; 1 marginal duct present between median plates, more present forward to segment 3 of abdomen. Other dorsal ducts arranged irregularly on pygidium, associated with sclerotized patches on the derm; some ducts with long slender filament at inner end, longer than the duct. Anus situated at base of pygidium.

Ventral surface with perivulvar pores in 5 groups; supplementary lateral groups also present on the 2 preceding segments. Microducts sparse on abdomen, but a few present between spiracles. Duct tubercles present in a single row anterior and posterior to each anterior spiracle. Antennae each with 2-3 setae of different lengths. Anterior spiracles curving postero-laterally, each with a group of disc pores. An oval, lightly sclerotized patch present lateral to each antenna.

Material examined
Holotype female. **PAPUA NEW GUINEA.** Manus P., Baluan I., on *Citrus sp.*, 11.iv.1985 (*T. Mala*) (BMNH).
Paratypes female. **PAPUA NEW GUINEA.** Same data as holotype. 6 (BMNH), 1 (DSIR), 1 (USNM).

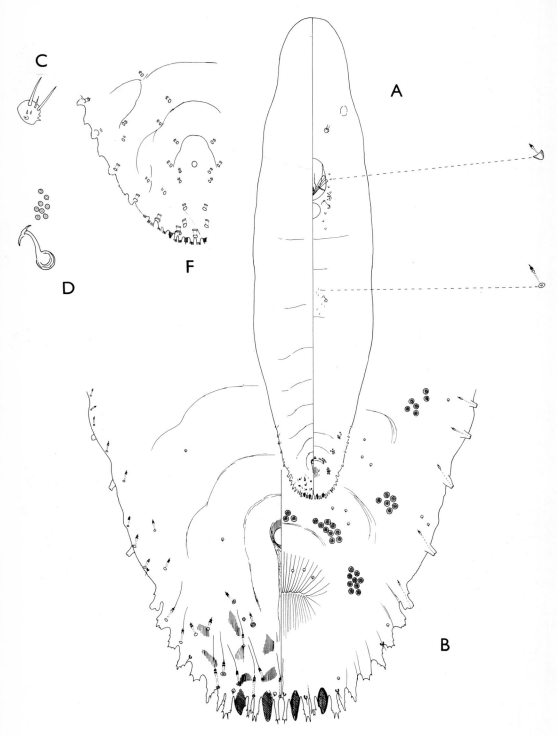

Fig. 88. *Lopholeucaspis baluanensis* sp. n.

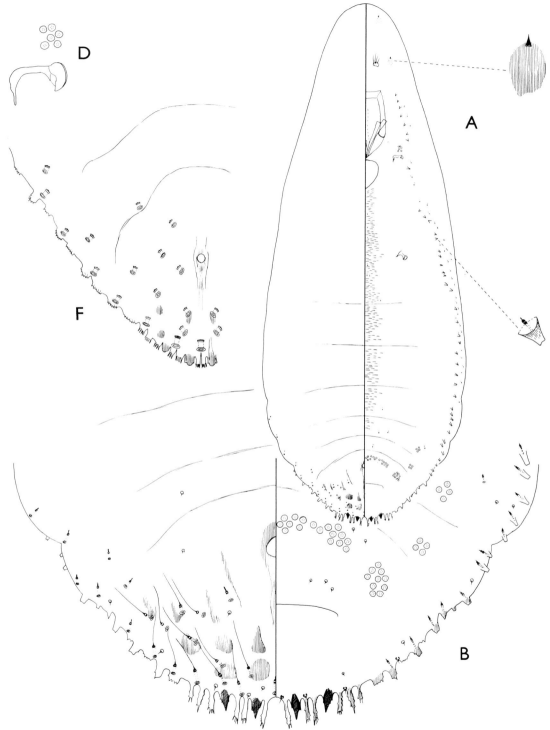

Fig. 89. *Lopholeucaspis cockerelli* (Grandpré & Charmoy).

Comments

The complete lack of duct tubercles between the prothorax and third abdominal segment, forming an interrupted row on each side, is a character that separates this species from any other in the genus.

Lopholeucaspis cockerelli (Grandpré & Charmoy) (Fig. 89)

Fiorinia cockerelli Grandpré & Charmoy, 1899: 35.
Leucaspis cockerelli (Grandpré & Charmoy), Green, 1905: 344; Ferris, 1941a: 289.
Lopholeucaspis cockerelli (Grandpré & Charmoy), Balachowsky, 1953: 155; Balachowsky, 1958: 336.

Description

Female pupillarial; scale elongate, ridged and triangular in cross-section, brown, but covered in a thin secretion of white wax. Male scale similar to female scale but narrower and much smaller.

Adult female elongate-oval, widest at about third abdominal segment; membranous except for parts of pygidium. Head and pygidium rounded.

Pygidium with 2 pairs of lobes present. Median lobes wide apart, each elongate and pointed, often with single notches present on each side. Second lobes similar to median lobes in shape. Plates either same length as lobes, or longer, dentate or notched on sides, and more deeply fringed at apices; 2 present between median lobes, 2 between median and second lobes, and 3 present lateral to second lobes. Anteriorly, plates sclerotized and blunt, then replaced by duct tubercles, at first projecting from the margins; but on the free abdominal segments they continue in a ventral submarginal row, curving to anterior spiracle, and terminating towards anterior margin of clypeolabral shield. Dorsal ducts slender, there being 1 present between median plates, and more present marginally forward to third abdominal segment; others present irregularly on submargins of pygidium, often associated with sclerotized patches; many ducts with a long filament at inner end, this much longer than duct. Perivulvar pores in 5 groups, but median and antero-lateral groups often merging; supplementary groups also present on submargins of 2 preceding segments. Microducts present in median areas of abdomen, and lateral to the duct tubercles. Antennae each with 4 setae. Anterior spiracles each with a group of disc pores. Small spurs present on head directed anteriorly, 1 lateral to each antenna, and 1 on either side of head margin, all at anterior ends of oval sclerotized areas.

Second instar with 2 pairs of lobes and fringed plates. Marginal ducts with heavily sclerotized rims to orifices; dorsal ducts present irregularly as far forward as segment 5.

Material examined

COOK IS. Aitutaki, 1977; Mangaia, 1977. On *Citrus maxima*, *Heliconia* sp., *Inocarpus fagifer*, *Persea americana*.
FIJI. Taveuni, 1977; Vanua Levu, 1974; Viti Levu, 1956. On *Barringtonia* sp., *B. racemosa*, *Calophyllum inophyllum*, *Pinus caribaea*, *Piper aduncum*, *Schefflera* sp., *Theobroma cacao*.
KIRIBATI. Butaritari, 1976. On *Citrus aurantifolia*.
NIUE. 1975. On *C. aurantifolia*, *Pinus caribaea*, *Terminalia catappa*.
TONGA. Ha'apai Group, Nomuka, 1977; Tongatapu Group, 'Eua, 1977; Vava'u Group, Vava'u, 1974. On *Aleurites moluccana*, *C. aurantium*, *C. limon*, *C. maxima*.
WESTERN SAMOA. Upolu, 1975. On *C. aurantifolia*, *Passiflora edulis*.
VANUATU. Efate, 1983; Espiritu Santo, 1983. On *Theobroma cacao*.

Comments

This species is now known to be tropicopolitan and probably has a wide distribution, although it has still not been reported from some tropical countries. It is polyphagous and is often found in large numbers on the leaves. Outbreaks have occurred on *Pinus caribaea* in Fiji giving rise to concern. In addition to the material listed above, Maddison (1976) recorded *L. cockerelli* from Fiji

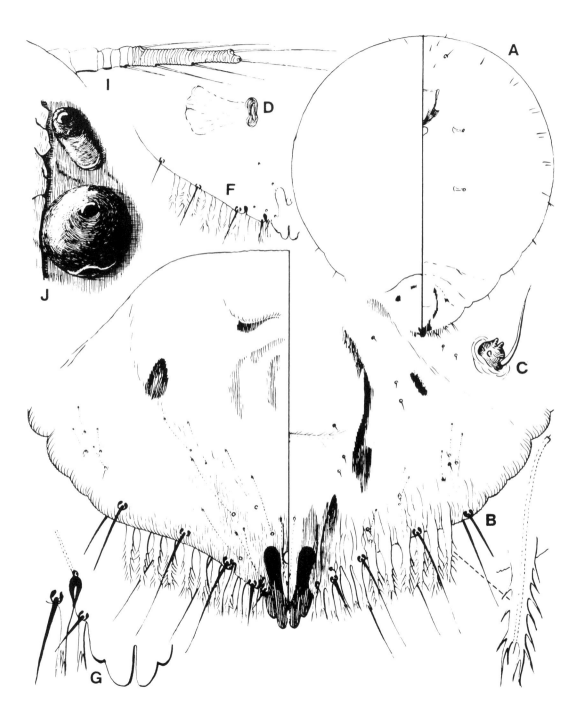

Fig. 90. *Morganella longispina* (Morgan).

and Niue as a pest of *Citrus aurantifolia* and *C. aurantium*, and on other hosts; Brun & Chazeau (1980) recorded it from New Caledonia on *Citrus* sp.; and Maddison (1976) recorded it from Tonga as a pest of *C. aurantium*. Specimens have been intercepted at quarantine inspection in New Zealand on produce from the Pacific area.

L. cockerelli was described originally from Mauritius on citrus, and there have been available, for comparison, a few specimens of the original material sent by de Charmoy to Green.

Genus **MORGANELLA** Cockerell

Morganella Cockerell, 1897a: 22; Ferris, 1938: 247; Balachowsky, 1956: 120. Type-species *Aspidiotus (Morganella) maskelli* Cockerell = *Aspidiotus longispinus* Morgan, by monotypy.

Few characters separate this genus from *Oceanaspidiotus*. *Morganella* contains 6 species; although most are African, the type-species and one other seem to be New World in origin. Main characters are the presence of a single pair of well-developed median lobes, the space between very narrow, each lobe with a well-defined basal sclerosis. The second and third lobes are lacking entirely. In the type-species the plates are elongate and fringed, but in other species they may be simple, pointed or rounded. In all but one African species there are no plates between the median lobes. Small paraphyses are present on margins of segments 6 and 7. The dorsal ducts are slender and 1-barred. The anus is situated near the apex of the pygidium.

In *Oceanaspidiotus* paraphyses are absent; otherwise the genera are closely related.

Morganella longispina (Morgan) (Fig. 90)

Aspidiotus longispina Morgan, 1889: 352.
Aspidiotus (Morganella) maskelli Cockerell, 1897a: 24. Lectotype female, Hawaii (USNM), here designated [examined].
Morganella longispina (Morgan), Fernald, 1903: 282; Ferris, 1938: 249; Balachowsky, 1956: 126.

Material examined
COOK IS. Mangaia, 1970; Rarotonga, 1969. On *Citrus* sp., *C. limon*.
FIJI. Viti Levu, 1977. On *Carica papaya*.
FRENCH POLYNESIA. Society Is, Tahiti. On 'kukui tree'.
PAPUA NEW GUINEA. E.H.P.: 1959. Morobe P.: 1959. N.P.: 1985. On *Citrus* sp., *Eugenia* sp.
TONGA. Tongatapu Group, Tongatapu, 1980. On *Citrus maxima*.
WESTERN SAMOA. Savai'i, 1975. On *Persea americana*.

Comments
This species is normally tropicopolitan, but it does extend to some temperate areas, and has been recorded on numerous plant species. In the South Pacific area, it should be recognizable fom the generic description and key. Although it comes close to some species of *Oceanaspidiotus*, it always lacks completely any second and third lobes. These lobes are always present in *Oceanaspidiotus*, even though they are sometimes reduced to points. *M. longispina* can also be distinguished in lacking plates between the median lobes.

The lectotype selected above is one of 3 specimens, clearly marked, labelled *Aspidiotus (Morganella) maskelli* Ckll. Ohia tree, Kailua, N. Kona, Hawaii, Dec. 22 1893, from H.G. Wait. The other 2 specimens are here designated paralectotypes.

In addition to the material listed, *M. longispina* has been recorded from French Polynesia, Society Is and Tuamotu Archipelago, on *Psidium guajava* (Doane & Hadden, 1909), *Averrhoa carambola*, *Citrus* sp., *Ficus carica* and *Tecoma stans* (Cohic, 1955). The latter mentions damage to grapefruit, lemon and fig; on fig, feeding punctures cause tumerous cankers on twigs, and branches become desiccated and tend to break off. Reboul (1976) recorded damage to the above hosts and *Carica papaya, Lagerstroemia flos-reginae, Mangifera indica* and *P. cattleianum*. In New Caledonia, Cohic (1956, 1958a) mentions *M. longispina* on *Bauhinia variegata, C. papaya, Citrus aurantium, C. grandis, C. limon, C. paradisi, C. reticulata, Ficus* sp., *F. carica, Hibiscus rosa-sinensis, Jasminum* sp., *J. sambac,* and *Tecoma stans*; he describes it as particularly damaging to *Bauhinia* and *Jasminum*, which it may kill, and mentions it is often associated with the fungus *Fusarium juruanum*.

Genus **OCEANASPIDIOTUS** Takagi

Oceanaspidiotus Takagi, 1984: 16. Type-species *Octaspidiotus araucariae* Adachi & Fullaway, by original designation.

Description
  Prepygidium expanded, ovate or round, remaining membranous in type species; pygidium broad. Ducts 1-barred. Lobes present as 3 or 4 pairs, well developed, never bilobed. Median lobes more or less elongate, well sclerotized, subapically notched on outer or both edges; apical margins slightly slanting outwards, sometimes serrate; median lobes set quite close together, with or without basal scleroses. Second and third lobes varying in size and shape; fourth lobes pointed if present. Plates well developed, occurring as far forward as the fifth segment; 2 present between median lobes, 2 between median and second lobes, 3 between second and third lobes (and third and fourth lobes when the latter are present); number present lateral to outermost lobe variable. Median pair of plates narrow, often simple; other plates more or less deeply and finely fringed; plates present lateral to outermost lobe simpler, often forming slender processes. Dorsal marginal seta present at outer basal corner of median lobe sometimes elongate; those present on second and third lobes more or less thickened, widest just before or at base, tapering gradually to apex. Anus usually smaller than median lobe, located subapically, or at posterior third to two fifths of pygidium. Dorsal macroducts slender (not filiform), with sclerotized rims on orifices; marginal ducts arranged as 1 present between median lobes (absent in type species), 1 between median and second lobes, 1-2 between second and third lobes, and up to 5 present lateral to third lobe on each side. Dorsally, 3 pairs of clear furrows present on each side towards apex of pygidium, containing most of the submarginal macroducts; innermost furrow usually stopping before anterior edge of anus. Marginal macroducts present on segments 2 and 3 of each side. Dorsal submarginal bosses present on either side of segments 1 and 3, sometimes quite prominent and slightly sclerotized. Perivulvar pores present or absent. An area of dermal granulation, a smooth tubercle sometimes slightly sclerotized, and 1-2 microducts associated with each anterior spiracle. Antennae each bearing 1 robust seta and more or less sclerotized spurs. Often several long setae present on anterior margin near midline.

Comments
  Takagi (1984) placed *O. araucariae, O. caledonicus* and *O. spinosus* in *Oceanaspidiotus*. To these are here added *Aspidiotus pangoensis* Doane & Ferris, and *Oceanaspidiotus nendeanus* sp. n.; both fit the generic description well, except for the gradual sclerotization of the expanded prosoma in mature females. This sclerotization also occurs in paratype material of *O. caledonicus*.
  This genus is remarkable for its morphological variation. Takagi (1984) commented on the characteristic variation in size and shape of the lateral lobes within species, and in the number of lobes, the presence or absence of perivulvar pores, and position of the anus between the species. Generally, specimens with reduced lateral lobes tend to have rather long marginal segmental setae present on the pygidium.
  Material is at hand with data on the labels corresponding with those of *Aspidiotus suvaensis* Green *nomen nudum*, mentioned by Lever (1945a, 1947), this material appears to belong to the genus *Oceanaspidiotus*. Further study is necessary to decide whether it consists of one or two new species.

Key to species of *Oceanaspidiotus*

1 Anus situated near posterior quarter of pygidium, or subapical; submarginal (other than marginal) prepygidial macroducts absent; prosoma remaining membranous at maturity 2
-- Anus always situated anterior to posterior quarter of pygidium; submarginal and marginal prepygidial macroducts present; prosoma becoming more or less sclerotized at maturity 3

2 4 pairs of lobes present; median marginal macroduct and perivulvar pores absent
 .... *araucariae* (Adachi & Fullaway)
-- 3 pairs of lobes present; median marginal macroduct and perivulvar pores present
 .... *spinosus* (Comstock)

3 With more than 18 perivulvar pores and over 120 pygidial macroducts present
 .... *nendeanus* sp. n.
-- With 0-12 perivulvar pores and fewer than 95 pygidial macroducts present 4

4 4 pairs of lobes present; 44-68 pygidial macroducts present; vulva situated near centre of pygidium; 1 microduct present near anterior spiracle
 .... *caledonicus* (Matile-Ferrero & Balachowsky)
-- 3 pairs of lobes present; 70-92 pygidial macroducts present; vulva situated near anterior third of pygidium. With 2-4 microducts present near anterior spiracle
 .... *pangoensis* (Doane & Ferris)

Oceanaspidiotus araucariae (Adachi & Fullaway) (Fig. 91)

Octaspidiotus araucariae Adachi & Fullaway, 1953: 89.
Oceanaspidiotus araucariae (Adachi & Fullaway), Takagi, 1984: 17.

Description
 Scale of adult female subcircular, flat, whitish to pale fawn with darker particles included; pale yellow exuviae subcentral. Male scale similar to female scale, but elongate-oval.
 Adult female, slide-mounted, 0.5-0.85 mm long; prosoma becoming swollen, almost oval at maturity; narrowing at caudal end; membranous except for some parts of pygidium.
 Pygidium broad, with dorsal sclerotized patches present at bases of second, third and fourth lobes. Median lobes well sclerotized, longer than wide, each notched subapically on either side; lobes separated by less than one third lobe width; basal scleroses absent, or very small and present on inner basal angle of each median lobe. Second lobes of variable size, rounded to almost pointed; third lobe with subapical notch present on outer margin; third and fourth lobes well sclerotized, pointed. Plates extending well beyond lobes, deeply and finely fringed; 4-5 plates present lateral to fourth lobe less fringed, outermost plates often forming long slender processes. Dorsal marginal seta present at outer basal corner of each median lobe minute. Anus oval, slightly smaller than median lobe, situated at or behind posterior quarter of pygidium. Dorsal pygidial macroducts 13-20 times as long as wide, 11-22 present. Median marginal macroduct absent; submarginal macroducts present in 3 distinct membranous furrows on each side. Short marginal macroducts numbering 6-7 present on each of abdominal segments 2 and 3 on each side; occasionally 1 present on segment 1. Submarginal prepygidial macroducts absent. A few rather long microducts present on prosoma.
 Venter of pygidium lacking perivulvar pores. Microducts few, rather long, mostly present submarginally. Single microduct associated with each anterior spiracle. Antennal bases each with 3 short sclerotized spurs and 1 robust seta.

Material examined
NEW CALEDONIA. New Caledonia, 1978. On *Araucaria* sp., *A. cookii*.

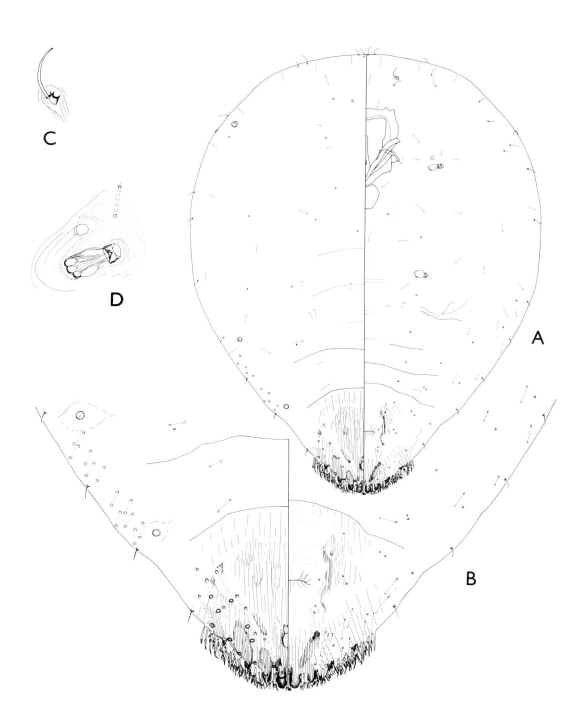

Fig. 91. *Oceanaspidiotus araucariae* (Adachi & Fullaway).

Comments

O. araucariae was described from Hawaii, and has also been recorded from Florida, Puerto Rico and the Caroline Is. It is specific to *Araucaria*, particularly *A. heterophylla*, native to New Caledonia and Norfolk Island; Takagi (1984) considers it likely to have been introduced to elsewhere from these islands with the host. In addition to the material examined, it has been recorded from French Polynesia on *A. cookii* (Reboul, 1976), and has been described as a serious pest on the same host from Wallis Is, causing desiccation of the extremities of the branches (Cohic, 1959). Similar damage to *A. heterophylla* has been recorded from Florida by Hamon (1985). In countries where *Araucaria* is grown for timber, this species could be a potential pest.

Oceanaspidiotus caledonicus (Matile-Ferrero & Balachowsky) (Fig. 92)

Octaspidiotus caledonicus Matile-Ferrero & Balachowsky, 1973: 239.
Oceanaspidiotus caledonicus (Matile-Ferrero & Balachowsky), Takagi 1984: 18.

Description

Scale of adult female circular, flat, whitish with pale brown subcentral exuviae. Male scale similar to female scale but more elongate.

Adult female, slide-mounted, 0.6-1.1mm long; prosoma swollen, becoming almost oval, with some sclerotization of margins at maturity; narrowing at caudal end.

Pygidium broad, particularly in mature specimens, with dorsal sclerotized patches present at bases of second, third and fourth lobes. Median lobes well developed and sclerotized, longer than wide, subapically notched on lateral edges; apices slightly slanting outwards, sometimes finely serrate; separated by less than one third lobe width, each lobe with a distinct sclerosis present at base. Second lobes of variable size and shape; third lobes may have notched lateral margins; third and fourth lobes well sclerotized, pointed. Plates extending well beyond lobes; 2 present between median lobes simple; those present between other lobes deeply fringed; 0-4 simpler plates present beyond fourth lobe, often forming slender processes. Dorsal marginal seta present at outer basal corner of each median lobe, long, reaching beyond lobes and plates. Anus ovoid, usually smaller than median lobe, situated at posterior third to two fifths of pygidium. Dorsal pygidial macroducts 8-25 times as long as wide, 44-70 present. Median marginal macroduct present; submarginal macroducts rather scattered, but approximately arranged in 3 membranous furrows on each side; present as far forward as anterior edge of pygidium, continuing as 2-8 submarginal macroducts on segment 3, 0-4 on segment 2, and 0-3 on segment 1. Marginal macroducts present, 4-7 on segment 3 on each side, 3-7 on segment 2, and occasionally 1 present on segment 1. Dorsal microducts few, quite long, mostly submarginal, present on prosoma. Dorsal submarginal bosses situated on segments 1 and 3 quite prominent, sometimes lightly sclerotized.

Venter of pygidium with 0-7 perivulvar pores in up to 4 groups; 0-2 in each antero-lateral group and 0-1 in each postero-lateral group. Few microducts present, quite long, mostly submarginal, most numerous on pygidium and abdomen. Anterior spiracles quite lightly sclerotized, each with a single associated microduct. Antennae each with 2 prominent sclerotized spurs and 1 robust seta.

Material examined
NEW CALEDONIA. New Caledonia, 1967. On *Erythrina fastigiata, Pittosporum* sp.

Comments

Eleven paratype specimens from the Muséum National d'Histoire Naturelle, Paris, were kindly made available by D. Matile-Ferrero; the original description gives the host as *Erythrina fusca fastigiata*, here regarded as *E. fastigiata*. These differ from the specimens at hand, which have perivulvar pores present, and dorsal macroducts rather longer and more numerous than the paratype material. D. Matile-Ferrero reports that a further 20 paratype specimens examined also all lack perivulvar pores (personal communication). These differences seem likely to represent intraspecific

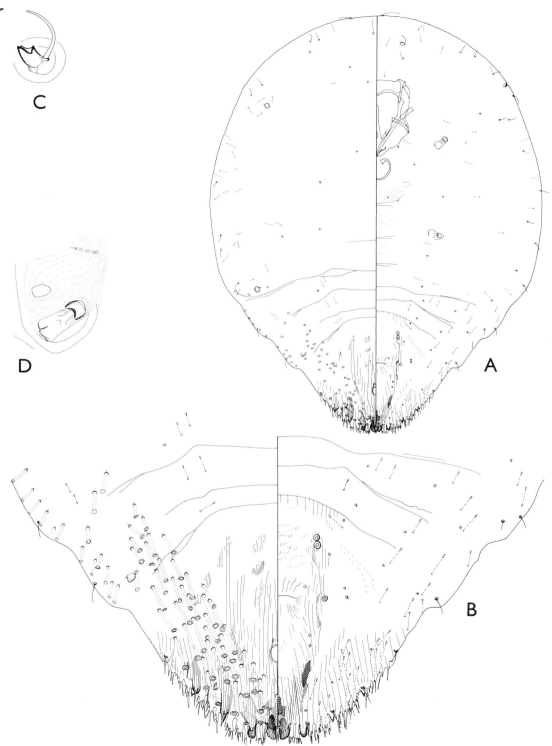

Fig. 92. *Oceanaspidiotus caledonicus* (Matile-Ferrero & Balachowsky).

variation similar to that found in *O. pangoensis* and *O. spinosus*, and may be due to a host effect. The species is apparently native to New Caledonia, and has not been recorded from elsewhere.

Oceanaspidiotus nendeanus sp. n. (Fig. 93)

Description
    Scale of adult female subcircular, convex, light to mid-brown, with paler secretion covering yellow-brown subcentral exuviae. Male scale similar to female scale, but smaller and elongate-oval.
    Adult female, slide-mounted, 0.8-1.7 mm long; prosoma swollen, mature specimens becoming almost circular, with narrower caudal end. Dorsum and ventral edges of prosoma becoming moderately sclerotized at maturity, except for 4 small dorsal fenestrations sometimes visible on posterior margin of mesothorax.
    Pygidium broad, particularly in mature specimens; with dorsal sclerotized patches present at bases of second and third lobes, and 4 dorsal sclerotized submedian scars on anterior third of pygidium. Median lobes well developed and sclerotized, about twice as long as wide, with subapical notches present on either side; apices slightly slanting outwards, finely serrate; separated by less than half median lobe width, each with a distinct sclerosis present at base. Second lobes slightly smaller than median lobes, notched subapically on either side; third lobes smaller again, each with prominent subapical notch present on lateral margin. Fourth lobes absent. Plates extending well beyond tips of lobes; 2 present between median lobes; the plate lateral to each second lobe simple or only slightly fringed; other plates present between lobes fringed; 9-10 plates present beyond third lobe, the outer ones simpler, often forming slender processes. Dorsal marginal seta at outer basal corner of each median lobe long, reaching to tips of plates. Anus ovoid, smaller than median lobe; situated anterior to, or on, posterior third of pygidium. Dorsal pygidial macroducts 10-28 times as long as wide, with lightly sclerotized rims: 120-138 present. Single median marginal macroduct present, 0-1 present between median and second lobes on each side, 2 between second and third lobes, 2 between third lobe and dorsal marginal seta of fifth segment, and 1-2 present on fourth segment. Submarginal macroducts present in 3 membranous furrows on each side; more scattered towards anterior edge of pygidium, continuing as 9-16 submarginal macroducts on segment 3, and 0-2 submarginal macroducts present on segment 2. Marginal prepygidial macroducts present, 0-4 on segment 4 on each side, extending in a row forward to posterior edge of segment 1; total of 20-26. A few quite long dorsal microducts present scattered on prosoma; short submarginal microducts present, some forming loose groups. Dorsal submarginal bosses situated on either side of segments 1 and 3 may be lightly sclerotized.
    Venter of pygidium with 20-44 perivulvar pores in 4 (occasionally 5) groups, 0-2 in median group, 9-11 in each antero-lateral group and 5-10 in each postero-lateral group. Vulva situated between centre of pygidium and anterior third. Microducts few, quite long, mostly present submarginally, most numerous on pygidium and abdomen. Spiracle orifices in slight pits; a single microduct associated with each anterior spiracle, and 4-5 microducts associated with each posterior spiracle. Antennal bases each with 3 sclerotized spurs and 1 robust seta.

Material examined
Holotype female. **SOLOMON IS.** T.P.: Santa Cruz Is, on sweet potato vine (*Ipomoea batatas*), iii.1983 (BMNH).
Paratypes female. **SOLOMON IS.** Same data as holotype, 8 (BMNH), 2 (USNM), 2 (DSIR); T.P.: Santa Cruz Is, Nende, on sweet potato, iii.1984, 22 (BMNH); Nende, on sweet potato, bred on pumpkin, 18.ix.1984 (*R. Macfarlane*), 7 (BMNH); Tenotu, on sweet potato vine, iii.1983, 20 (BMNH), 2 (ARSDC).

Comments
    O. nendeanus resembles *O. pangoensis* in having numerous dorsal macroducts, 3 pairs of lobes, and a tendency to develop sclerotization of the prosoma at maturity; it is the only species of *Oceanaspidiotus* which consistently has numerous perivulvar pores, however. It is probably an

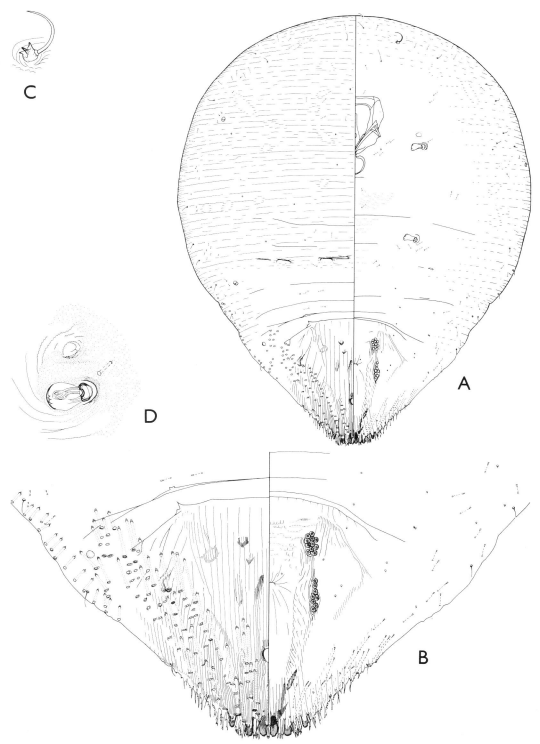

Fig. 93. *Oceanaspidiotus nendeanus* sp. n.

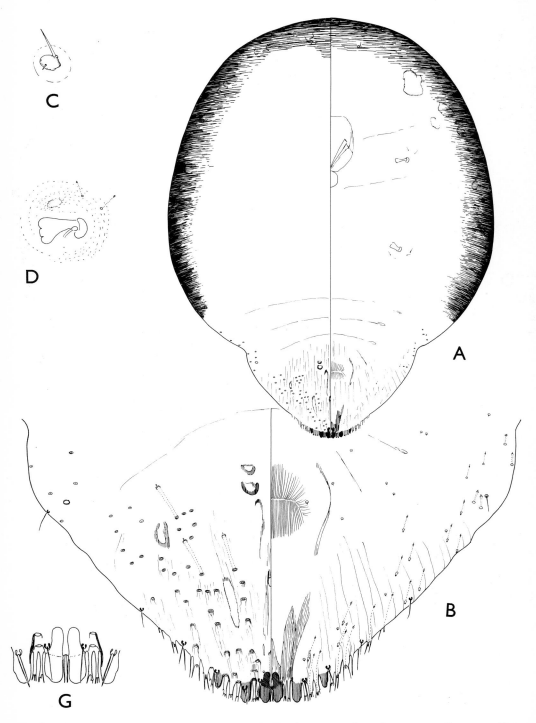

Fig. 94. *Oceanaspidiotus pangoensis* (Doane & Ferris). American Samoa, on *Cocos nucifera*.

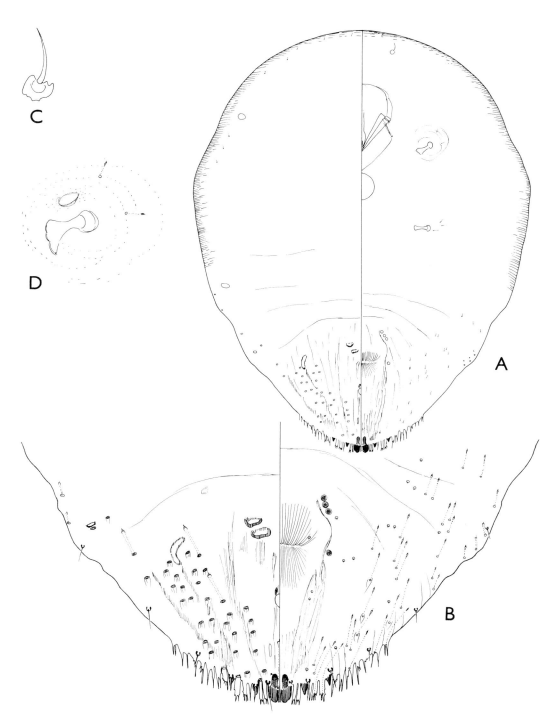

Fig. 95. *Oceanaspidiotus pangoensis* (Doane & Ferris). Fiji, taken from *Hernandia peltata*.

endemic species feeding on non-cultivated host plants: quite possibly it only attacks *Ipomoea* in certain conditions, and may only appear sporadically, but it could be a potential pest.

Oceanaspidiotus pangoensis (Doane & Ferris) comb. n. (Figs 94, 95)

Aspidiotus pangoensis Doane & Ferris, 1916: 400; Green & Laing, 1923: 125.

Description
    Scale of adult female approximately circular, flat, slightly translucent, light to mid-brown; exuviae central, slightly yellower. Male scale similar to female scale but more elongate.
    Adult female, slide-mounted, 0.6-1.3 mm long; prosoma swollen, becoming almost oval and moderately sclerotized at edges at maturity; sometimes constriction present at posterior edge of segment 2.
    Pygidium broad and rounded, particularly in mature specimens; with narrow sclerotized patches present at bases of second and third lobes and dorsal marginal seta of fifth segment, and 4 heavily sclerotized submedian scars on anterior third of pygidium. Median lobes well developed and sclerotized, each about twice as long as wide, with subapical notch present on lateral margin; apex may be finely serrate; both lobes separated by less than one third lobe width, each with a distinct sclerosis at base. Second and third lobes variable in size and shape (extremes illustrated). Plates extending to tips of lobes or beyond; 2 present between median lobes, and the plate lateral to each second lobe simple or slightly fringed; other plates between lobes sparsely fringed; 5-7 simpler plates present beyond each third lobe, those beyond dorsal marginal seta of each fifth segment usually simple and deeply cleft. Dorsal marginal seta present at outer basal corner of each median lobe slender, length variable from slightly shorter than, to longer than, median lobe. Anus oval, smaller than a median lobe, situated near posterior two fifths of pygidium. Dorsal pygidial macroducts 8-15 times as long as wide, with thick rims to their small orifices; 68-92 present. Marginal median macroduct present; submarginal macroducts often scattered, sometimes approximately forming 3 membranous furrows on each side; reaching nearly to anterior edge of pygidium. Prepygidial macroducts present, segment 3 bearing 1-4 submarginal and 3-9 marginal ducts on each side, and segment 2 with 0-1 submarginal and 3-8 marginal ducts. A few dorsal submarginal microducts present on prosoma. Dorsal submarginal bosses situated submarginally on segments 1 and 3 becoming slightly sclerotised at maturity.
    Venter of pygidium with 0-8 perivulvar pores in up to 4 (rarely 5) groups; occasionally 1 antero-median pore present, and 0-3 in each of the lateral groups. Microducts moderately long, submarginal, most frequent on pygidium and abdominal segments; those opening on the pygidium and plates sometimes with the orifice on a conical projection (Fig. 94). A relatively large area of dermal granulation and 2-4 microducts associated with each spiracle. Antennae each with 3 short sclerotized spurs and 1 robust seta.

Material examined
AMERICAN SAMOA. Tutuila, 1913. On *Cocos nucifera*.
FIJI. Kadavu, 1949; Vanua Levu, 1921; Viti Levu, 1899; Yasawa Group, Yasawa. On *Bruguiera gymnorhiza, Cocos nucifera, Hernandia peltata, Smilax* sp.
NIUE. 1977. On *Cocos nucifera*.
TONGA. Ha'apai Group, Foa, 1977, Lifuka, 1977, Nomuka, 1977; Tongatapu Group, 'Eua, 1977, Tongatapu, 1974; Vava'u Group, Vava'u, 1977. On *Broussonetia papyrifera, C. nucifera, Cordia subcordata*.

Comments
    Takagi (1984) described *Oceanaspidiotus* as having a membranous prosoma, and assigned *O. araucariae, O. caledonicus* and *O. spinosus* to the genus. The mature paratype specimens of *O. caledonicus* examined are obviously sclerotized around the margins of the prosoma. *Aspidiotus pangoensis* develops a similar pattern of sclerotization at maturity, and agrees well with the

generic definition, with expansion of the prosoma, sclerotization of macroduct orifices and variably sized lateral lobes: it is therefore placed in *Oceanaspidiotus*. The cotype specimen was available through the courtesy of R.O. Schuster.

Aspidiotus simmondsi Green & Laing is a *nomen nudum* and was listed by Hinckley (1965) as being *A. pangoensis*. This species was originally described from American Samoa on coconut husks (Doane & Ferris, 1916), and Reddy (1970) records the species from Western Samoa on *Cocos nucifera*.

Oceanaspidiotus spinosus (Comstock) (Fig. 96)

Aspidiotus spinosus Comstock, 1883: 70; Ferris, 1941b: 58; Balachowsky, 1956: 78.
Oceanaspidiotus spinosus (Comstock), Takagi, 1984: 18.

Description
    Scale of adult female subcircular, flat, white to light brown, sometimes including fragments of bark; exuviae subcentral, golden brown. Male scale similar to female scale but more elongate.
    Adult female, slide-mounted, 0.75-0.9 mm long, broadly pyriform, becoming broader with maturity but remaining membranous except for parts of pygidium.
    Pygidium broad, with dorsal sclerotized patches present at bases of second and third lobes and the dorsal marginal seta of the fifth segment, and submarginally on a level with the vulva; occasionally apex slightly recessed between the second lobes. Median lobes quite large and well sclerotized, longer than wide, notched subapically on either side; apices may be serrate; separated by less than one third lobe width, each with a distinct sclerosis at the base. Second lobes variable; at one extreme, each with a well-developed rounded tip and a notch on each margin; at the other extreme, each reduced to a sclerotized rounded tubercle. Third lobes shorter and narrower than second lobes, especially when second lobes are reduced; lateral margin of each sometimes notched; tip may be pointed. Plates at least as long as median lobes; 2 present between median lobes usually simple; those between other lobes deeply fringed; 5-6 plates present lateral to each third lobe increasingly spiniform with distance from the lobe. Dorsal marginal setae present on segments 5 to 7 and ventral seta on segment 7 more or less thickened, often exceeding plates in length; seta at outer basal corner of each median lobe more slender, about as long as lobe. Anus oval, about half as long as a median lobe, situated within posterior fifth of pygidium. Dorsal pygidial macroducts 14-24 times as long as wide, 16-45 present. Median marginal macroduct present. Submarginal macroducts present in 3 membranous furrows on each side, usually posterior to level of vulva. Prepygidial marginal macroducts present, 3-7 on segment 3 and 4-7 on segment 2 on each side; submarginal prepygidial macroducts absent. Many short dorsal microducts present just submarginally on prosoma. Dorsal submarginal bosses present on either side of segment 1 appreciably larger than the bosses on segment 3.
    Venter of pygidium with 4 groups of perivulvar pores, 2-8 in each group, total of 10-27. Microducts moderately long, submarginal and submedian, more common on abdominal segments and pygidium. Quite a large area of dermal granulation and 1-2 microducts associated with each spiracle. Antennae each with 2 short sclerotized spurs and 1 robust seta.

Material examined
COOK IS. Rarotonga, 1977. On *Persea americana*, *Vitis vinifera*.

Comments
    O. spinosus was described from camellias in a glasshouse in the USA. It is now known to occur in North, Central and South America, Canary Is, Turkey, Nepal and Japan (Takagi, 1984), Algeria, and in glasshouses in England. There are no records of it causing serious damage. Specimens have been examined from Cook Is on Arecaceae and from Norfolk I. on *Diospyros* sp., intercepted at quarantine inspection in New Zealand.

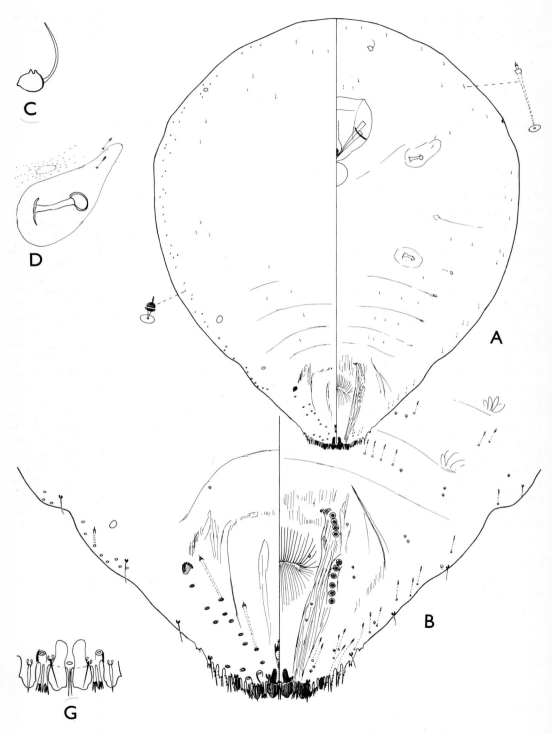

Fig. 96. *Oceanaspidiotus spinosus* (Comstock).

Takagi (1984) comments on the variability of this species, particularly the length of the ventral marginal setae, the occasional presence of a pygidial recess, and the size and shape of the lateral lobes.

Genus **OCTASPIDIOTUS** MacGillivray

Octaspidiotus MacGillivray, 1921: 387; Borchsenius, 1966: 272; Takagi, 1984: 3.   Type-species *Aspidiotus subrubescens* Maskell, by original designation.

Description
Prepygidium oval to circular, membranous or sclerotized; pygidium more or less produced. Ducts 1-barred. Lobes well developed, 3 or 4 pairs present, never bilobed. Median lobes well developed, each lobe usually notched on either side; narrow membranous zone present at base of each lobe, giving a hinged or jointed appearance; median lobes separated by at least one third lobe width; with ventral sclerotized area extending into pygidium from base of each lobe. Second lobes smaller than median lobes, notched subapically on lateral or both margins; third lobes similar, each usually notched only on lateral margins. Fourth lobes, if present, smaller, pointed, notched on lateral margin. Plates as long as or exceeding adjacent lobes, 2 present between median lobes, 2 between median and second lobes on each side, 3 between second and third lobes (and third and fourth lobes on each side if the latter are present); plates present lateral to outermost lobes reaching forward to dorsal marginal setae of fourth segment. Plates between lobes fringed; those present lateral to outermost lobe broad, with sclerotized dentate margins and 1 or a few long fleshy processes, sometimes furcate, present at each mesal angle. Dorsal marginal seta located at outer basal corner of each median lobe slender, usually less than half as long as lobe; setae present on second and third lobes lanceolate (broadened and flattened); that present on fourth lobe, or if this is absent, on fifth segment, also more or less lanceolate. Anus elongate-oval, situated at posterior third of pygidium. Dorsal macroducts elongate, not very slender, with elliptical orifices; marginal ducts arranged as 1 present between median lobes, 1 between median and second lobes on each side, 2 between second and third lobes and between third and fourth lobes (or dorsal marginal seta of fifth segment), and 1-2 present beyond this. Dorsally, 3 pairs of poriferous furrows more or less distinct on each side towards apex of pygidium; innermost furrows usually reaching to anus or slightly beyond. Marginal macroducts present on segments 1 to 3, and sometimes on segment 4 and postero-lateral corners of metathorax on each side. Dorsal submarginal bosses present on segments 1 and 3 may become sclerotized at maturity. Perivulvar pores present in 4 (occasionally 5) groups. An area of dermal granulation, a smooth tubercle and 1 microduct associated with each anterior spiracle. Antennae each with sclerotized spurs and bearing 1 robust seta.

Comments
Takagi (1984) redefined this genus to include species with 3 pairs of lobes, as well as those with 4: of the 9 species he placed in *Octaspidiotus*, only one is known from the South Pacific area.

Octaspidiotus australiensis (Kuwana) (Fig. 97)

Aspidiotus australiensis Kuwana *in* Kuwana & Muramatsu, 1931: 652.
Octaspidiotus australiensis (Kuwana). Borchsenius, 1966: 272; Takagi, 1984: 12.

Description
Scale of adult female broadly oval to subcircular, flat, light brown with slightly darker central exuviae. Male scale similar to that of female, but smaller and more elongate.

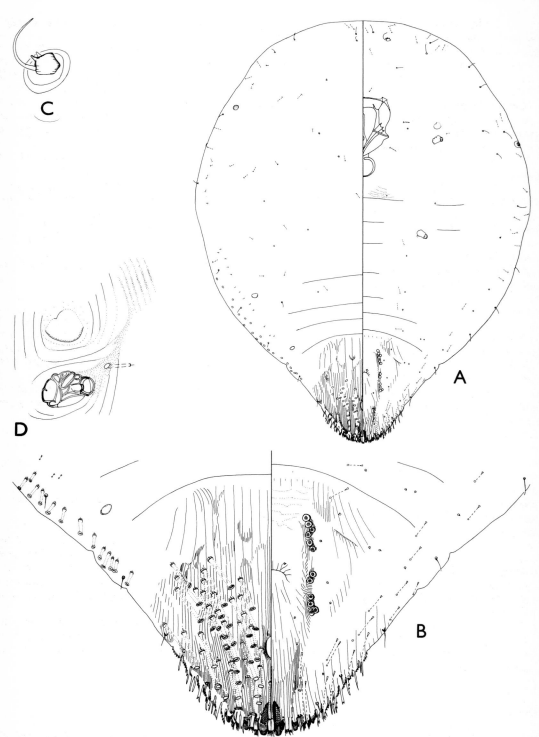

Fig. 97. *Octaspidiotus australiensis* (Kuwana).

Slide-mounted adult female 0.5-1.0 mm long, broadly pyriform; prepygidium becoming enlarged and ovoid at maturity, very occasionally slightly sclerotized at margins.

Pygidium well produced, with 3 pairs of lobes. Median lobes well sclerotized, each longer than wide; with notches usually present on both margins, the notch on the inner margin of each lobe obviously nearer apex than that on the lateral margin; both lobes separated by one third to half lobe width. Second and third lobes of about equal size, half as large as median lobes. Plates between lobes fringed except for single fleshy plate just lateral to second lobe on each side. Anus longer and slightly narrower than a median lobe. Dorsal pygidial macroducts 9-20 times as long as wide, with oval orifices; submedian poriferous furrow on each side extending to anus or beyond. A row of short dorsal submarginal microducts present, running from segment 2 forward to metathorax; a few present further anteriorly. Some longer marginal and submarginal macroducts present mainly on prosoma. Dorsal submarginal boss present on segment 1 larger than that on segment 3.

Venter of pygidium with 4 groups of perivulvar pores, 5-12 in each antero-lateral group, 4-9 in each postero-lateral group, occasionally 1 or 2 submedian anterior pores present; total of 21-40. Microducts quite long, mostly submarginal or marginal, most numerous on pygidium. Spiracles moderately sclerotized. Antennae quite well sclerotized, each with 2 spurs and 1 robust seta.

Material examined
PAPUA NEW GUINEA. C.P.: 1959. E.H.P.: 1959. Manus P.: Manus, 1959. Morobe P.: 1979. On *Araucaria hunsteinii*, *Ficus* sp., Orchidaceae, Sapotaceae.

Comments
This species was described from Japan on an orchid imported from Thursday I., Australia, and Takagi (1984) records it on various host-plants from Nepal, India, and quarantine material from Australia, Philippine Is, New Guinea and Solomon Is. It appears to attack a range of hosts including fruit trees and *Araucaria*, but has not been recorded causing damage. In the dry material at hand it is found on the upper leaf surfaces.

O. australiensis has been intercepted at United States quarantine inspection on Orchidaceae from the Pacific islands, so its distribution in the area is probably more widespread than given above.

Genus **ODONASPIS** Leonardi

Odonaspis Leonardi, 1897: 284; Ferris, 1938: 161; Balachowsky, 1958: 299. Type-species *Aspidiotus secretus* Cockerell, by monotypy.

This genus belongs to the tribe Odonaspidini and, as discussed under *Froggattiella*, all the species apparently feed on grasses or closely related plant groups. The genus differs from *Froggattiella* in possessing perivulvar pores and a single median lobe. Descriptions are here omitted pending the publication of the revision of the genus by Ben-Dov (?1987), who has examined the Pacific material herein listed. It should be possible, however, to identify the species with the aid of the key and illustrations.

As in *Froggattiella*, the adult females have a ventral as well as a dorsal scale in life, giving a clue to the identity of the genus. All the South Pacific species are known also from either Micronesia or Hawaii, but careful consideration should be given to the notes on *O. secreta*. The following key is adapted from Beardsley (1966).

Key to species of *Odonaspis*

1	Posterior spiracles without disc pores	*secreta* (Cockerell)
--	Posterior spiracles each with a group of disc pores	2

2   With lateral groups of perivulvar pores only, anterior median group absent
.... *greenii* Cockerell
--   With an anterior and lateral groups of perivulvar pores present 3

3   With a conspicuous clavate sclerotized process arising from each lateral angle of median lobe
.... *morrisoni* Beardsley
--   A clavate scerotized process absent from each lateral angle of median lobe. If there are
     marginal clavate processes present, they are well lateral to the median lobe 4

4   Ventral duct tubercles present on thorax. Perivulvar pores totalling 200 or more, lateral
     and anterior groups confluent .... *saccharicaulis* (Zehntner)
--   Ventral duct tubercles absent from thorax. Perivulvar pores totalling about 100, present in
     3 groups .... *ruthae* Kotinsky

Odonaspis greenii Cockerell

Odonaspis secretus var. *greenii* Cockerell, 1902b: 25.
Odonaspis secreta Cockerell, Zimmerman, 1948: 428 (Misidentification).

Materal examined
FIJI. Viti Levu, 1955. On bamboo.
WESTERN SAMOA. Upolu, 1977. On *Bambusa vulgaris*.

Comments
    This species has been recorded only from bamboos in China, Sri Lanka and Hawaii.

Odonaspis morrisoni Beardsley

Odonaspis morrisoni Beardsley, 1966: 525.

Material examined
FIJI. Viti Levu, 1899. On Poaceae, [?]*Citrus limon*.

Comments
    Beardsley (1966) described this species from the Caroline Islands and the Philippines, on grasses other than bamboo. It probably has a wider distribution in southern Asia. The specimens listed above from Fiji were labelled *Citrus limon*, but this seems unlikely. The correct host-plant may have been lemon grass.

Odonaspis ruthae Kotinsky (Fig. 98)

Odonaspis ruthae Kotinsky, 1915: 102; Ferris, 1938: 165; Zimmerman, 1948: 428; Balachowsky, 1958: 300.

Material examined
COOK IS. Penrhyn, 1977. On *Euphorbia heterophylla*, Poaceae.
FRENCH POLYNESIA. Society Is, Tahiti, 1934. on *Hibiscus* sp.
KIRIBATI. Gilbert Is, Butaritari, 1976; Line Is, Kiritimati, 1924, Tabueran, 1924. On 'honey grass', *Lepturus repens*, *Panicum* sp.
PAPUA NEW GUINEA. W.N.B.P. On *Nypa fruticans*.

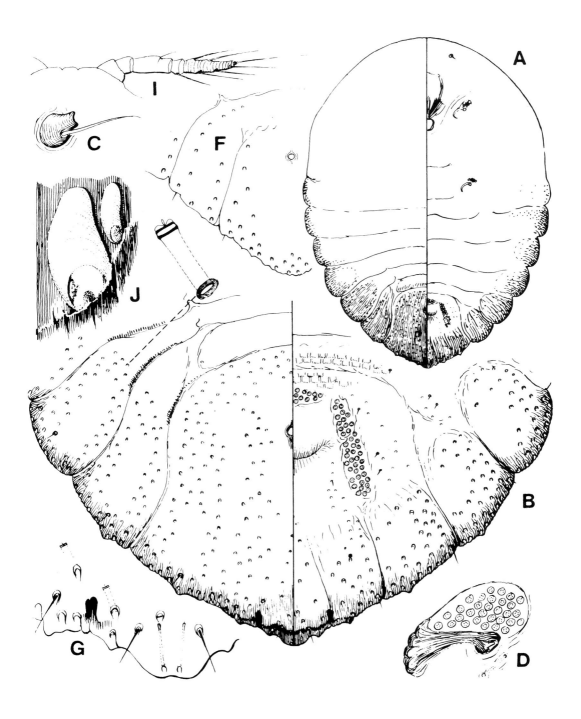

Fig. 98. *Odonaspis ruthae* Kotinsky.

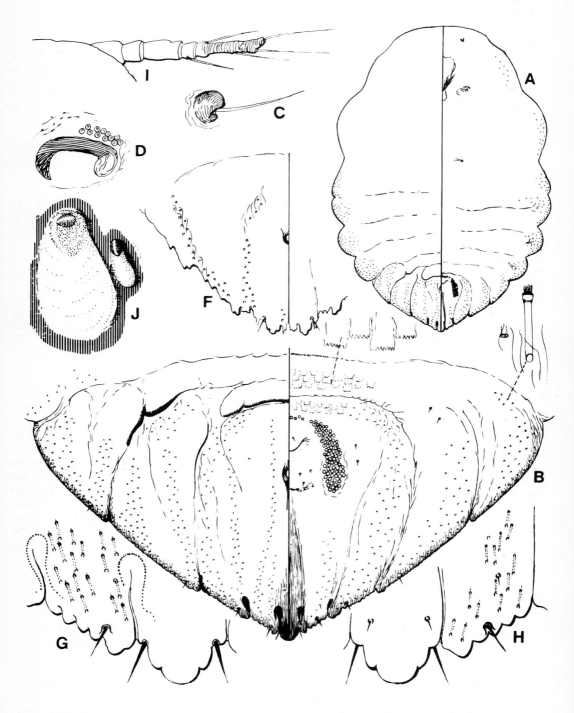

Fig. 99. *Odonaspis secreta* (Cockerell).

TUVALU. Funafuti, 1976. On *Chloris* sp. stem-root junction.
VANUATU. Efate, 1983. On *Chloris* sp.

Comments
   Since this species was described from Hawaii, it has been recorded from many tropical and temperate areas. Borchsenius (1966) listed it on Gramineae [Poaceae] (excluding bamboos), Juncaceae and Crassulaceae. The single specimen from Papua New Guinea is incomplete, so its identity is uncertain. Charlín (1973) recorded *O. ruthae* from Easter I. on *Cymbopogon citratus*, and Cohic (1956, 1958a) and Brun & Chazeau (1980) mention its presence in New Caledonia on *Cynodon dactylon*.

Odonaspis saccharicaulis (Zehntner)

Aspidiotus sacchari-caulis Zehntner, 1897: 560.
Odonaspis saccharicaulis (Zehntner), Green & Laing, 1923: 129; Balachowsky, 1958: 302; Beardsley, 1966: 528.

Material examined
FIJI. Viti Levu, 1977. On ?
PAPUA NEW GUINEA. Madang P.: 1937. On *Saccharum* sp.

Comments
   This species was described from Java, and has been recorded from many tropical and temperate areas since. Although its host-plants are various grasses, it is widely known for infesting sugarcane. Beardsley (1966) recorded it from Palau in Micronesia.

Odonaspis secreta (Cockerell) (Fig. 99)

Aspidiotus secretus Cockerell, 1896: 20.
Odonaspis secreta (Cockerell), Fernald, 1903: 300; Ferris, 1938: 166.

Comments
   This species was described from Japan, and has since been recorded from tropical and temperate areas of the world. It seems to be confined to bamboos. No material has been examined, but it has been recorded from Samoa (Doane & Ferris, 1916) on bamboo; Dumbleton (1954) recorded it from Western Samoa on *Bambusa* sp., while Dale (1959) described it as a pest, and Reddy (1970) also recorded it on bamboo there. Hinckley (1965) recorded it from Fiji on *Schizostachyum glaucifolium*. These records may refer to *O. greenii*, but in case the records of *O. secreta* may prove to be correct, an illustration is provided.

Genus **PARLATORIA** Targioni

Parlatoria Targioni, 1868: 42; Ferris, 1937a: 84; McKenzie, 1945: 50; Balachowsky, 1958: 320; Takagi, 1969: 32. Type-species *Aspidiotus proteus* Curtis, by subsequent designation.

Description
   Female scale of various colours, oval to elongate, with exuviae terminal, often occupying greater part of scale. Male scale elongate and smaller than female scale.
   Body of adult female circular to oval; free abdominal segments not strongly produced; only pygidium becoming sclerotized with maturity. Dorsal ducts 2-barred, the orifice of each marginal

duct, especially, surrounded by a sclerotized lunate rim. Three pairs of prominent lobes present, never bilobed; each usually with a single notch present on each side, but sometimes with additional notches. Fourth and fifth pairs of lobes often present, smaller and sclerotized, or membranous and resembling plates. Plates fimbriate distally, narrow between the 3 pairs of prominent lobes; those plates further forward wider and sometimes rounded or triangular, each with a microduct. Perivulvar pores present, often in 4 groups. Ventral gland spines often present on thorax and abdominal segment 1. Eye spot sometimes modified into sclerotized spur. Derm pocket present between each posterior spiracle and margin in some species. Anus situated near centre of pygidium. Anterior spiracles each with adjacent disc pores, each pore with 5 loculi. Posterior spiracles without disc pores.

Comments

Over 50 species are now recognised: the genus is tropicopolitan, with extensions into the temperate areas. Many species have been introduced throughout the world, but there are so many local species in south eastern Asia that the genus probably originated there.

In the tropical South Pacific area six species are here recorded. These include some important species on citrus. Dumbleton (1954) listed *P. ziziphi* (Lucas) from Irian Jaya on citrus, referring to a record by H.W. Moll in 1953; but *P. ziziphi* is not mentioned at all by Moll, and the record is here regarded as erroneous. Reyne (1961) has, however, recorded it from Merauke in Irian Jaya on citrus, and the species is included in the key.

Key to species of *Parlatoria*

1	With a conspicuous earlike process on each side of prosoma opposite anterior spiracles	*ziziphi* (Lucas)
--	Without such earlike processes opposite anterior spiracles	2
2	With at least 1 or 2 dorsal ducts on pygidium within the frame formed by the perivulvar pores	3
--	Without dorsal ducts within the frame formed by the perivulvar pores	4
3	Second and third lobes each with 2 or 3 lateral notches. Plates tapering between third and fourth lobes	*cinerea* Hadden
--	Second and third lobes each with only single lateral notches. Plates fringed apically between third and fourth lobes	*citri* McKenzie
4	Eyespot not modified to form a stout spur. Minute derm pocket absent from between posterior spiracles and body margin	*pergandii* Comstock
--	Eyespot modified to form stout spur. Minute derm pocket present between each posterior spiracle and body margin	5
5	Fourth pygidial lobe sclerotized and spur-like	*crotonis* Douglas
--	Fourth pygidial lobe replaced by a membranous plate-like structure, resembling adjacent plates but smaller	*proteus* (Curtis)

Parlatoria cinerea Hadden (Fig. 100)

Parlatoria cinerea Hadden, *in* Doane & Hadden, 1909: 299; Ferris, 1937a: 86; McKenzie, 1945: 59; Takagi, 1969: 38.

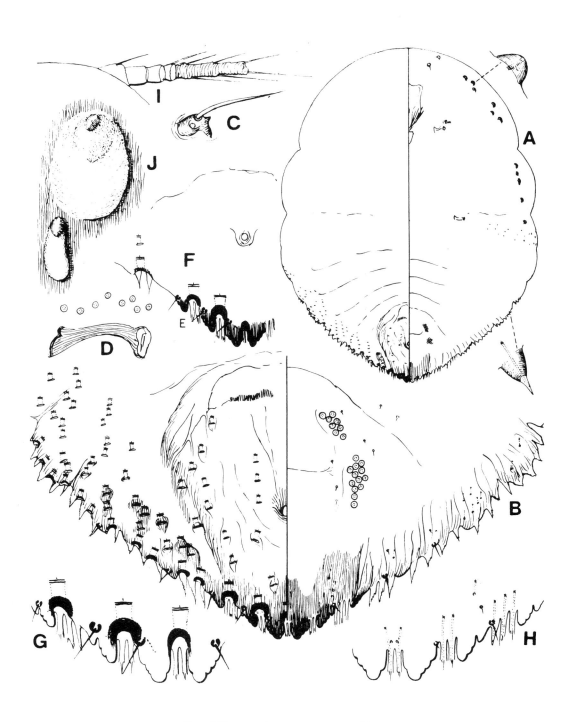

Fig. 100. *Parlatoria cinerea* Hadden.

Description
Female scale whitish, elongate-oval; exuviae terminal, yellow-brown. Male scale similar colour to female scale, but smaller and more elongate.
Adult female, slide-mounted, broadly oval, but pygidium pointed; head rounded; free abdominal segments with lateral lobes poorly developed.
Pygidium with median lobes prominent, appearing to converge apically, separated by a narrow space; each lobe with 3 notches present on the outer margin. Second and third lobes similar to median lobes, but progressively smaller. Plates elongate, almost parallel; those between third lobes and apex of pygidium each pointed at apex; plates anterior to third lobes progressively more tapering and pointed further forward. Marginal macroducts conspicuous by sclerotized lunate orifices, decreasing in size forward to about abdominal segment 2; but species easily identifiable by the vertical row of 4 ducts on either side of anus, and another row lateral to these. Perivulvar pores present in 4 elongate groups. Duct tubercles present around head and thorax.

Material examined
COOK IS. Aitutaki, 1975; Atiu, 1977; Mangaia, 1975; Rarotonga, 1937. On *Citrus* sp., *C. limon*, *C. paradisi*, *C. sinensis*, *Malus sylvestris*, *Vitis vinifera*.
FRENCH POLYNESIA. Society Is, Tahiti, before 1909; Marquesas Is, Hivaoa, 1923. On *Citrus* sp., *C. aurantifolia*, *C. grandis*, *C. reticulata*, *C. sinensis*.
NEW CALEDONIA. New Caledonia. On *C. grandis*.
NIUE. 1978. On *C. aurantifolia*.
PITCAIRN. 1951. On *C. sinensis*.
VANUATU. Efate, 1983; Malekula, 1983. On *C. aurantifolia*, *C. maxima*.
WESTERN SAMOA. Savai'i, 1975; Upolu, 1975. On *Annona muricata*, *C. aurantifolia*, *C. limon*, *C. reticulata*, *C. sinensis*.

Comments
This is a tropicopolitan species found on numerous host-plants. It is usually associated with citrus, and is probably one of the commonest scale insects on citrus in the South Pacific area. Although it was described from Tahiti, it may have been introduced there; its country of origin is still not certain. A few specimens from the original material, collected by R.W. Doane in Tahiti, have been available for study.
P. cinerea usually occurs with *P. pergandii* Comstock on citrus; known as chaff scales, they cover the stems and branches as well as leaves and fruit. Walker & Deitz (1979) found *P. cinerea* forming 94% of the scale cover on bark and 60% on stems of *C. sinensis* in Cook Is, while *P. pergandii* accounted for most of the remaining infestation. In these islands chaff scales are often associated with gumming, flaking and splitting of the bark, causing dieback of whole branches and sometimes killing the tree. Additional information in the literature includes a record from Cook Is and Western Samoa of its pest status on *C. sinensis* (Maddison, 1976); from the Society Is on *C. sinensis* and *Vitis vinifera*, and from French Polynesia on citrus (Dumbleton, 1954); from New Caledonia on *C. aurantium*, *C. limon*, *C. paradisi*, and *C. reticulata* (Cohic, 1956), where it is common but difficult to find underneath colonies of *Lepidosaphes beckii* and *Unaspis citri* (Cohic, 1958a); and from Wallis Is on *Citrus* sp. (Cohic, 1959; Gutierrez, 1981).

Parlatoria citri McKenzie (Fig. 101)

Parlatoria citri McKenzie, 1943: 155; McKenzie, 1945: 61.

Description
Scale of adult female described as pale brown, with exuviae terminal, slightly darker. Male scale smaller and narrower than female scale but similar in colour.
Adult female, when slide-mounted, similar to *P. cinerea*, but it differs in possessing only 1 or 2 dorsal ducts on either side of anus instead of 3 or 4. It differs from *P. crypta* McKenzie,

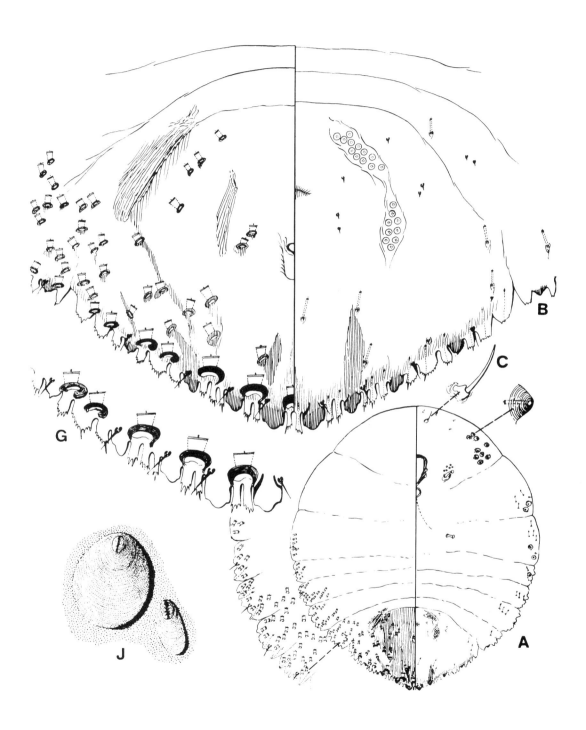

Fig. 101. *Parlatoria citri* McKenzie.

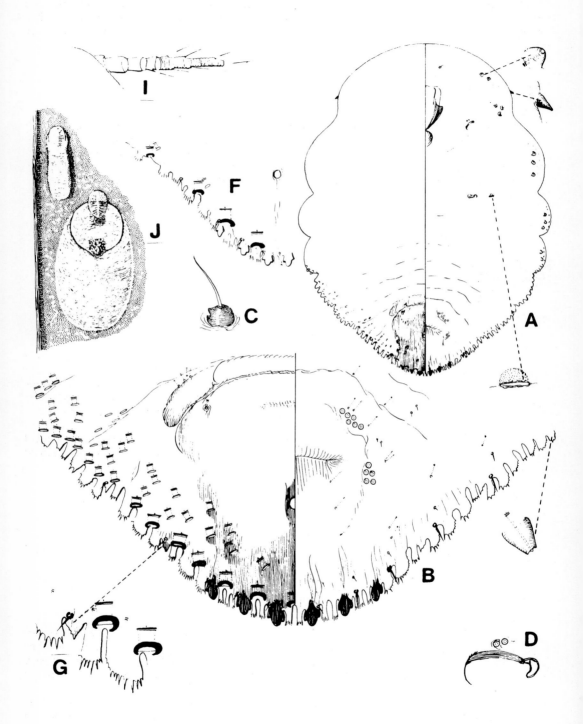

Fig. 102. *Parlatoria crotonis* Douglas.

found throughout southern Asia, in lacking dorsal intermediate ducts on segment 4. Whereas the lobes of *P. cinerea* tend to be pointed medially at apex, because of the long outer margins with a few notches, the lobes of *P. citri* are almost parallel, and have but a single notch on the outer margins.

Material examined
COOK IS. Rarotonga, 1975. On *Citrus sinensis*.

Comments
This species was described from material intercepted from Java at quarantine inspection in California, on leaves of *Citrus* sp. It does not appear to have been recorded from elsewhere, but specimens are at hand from W. Malaysia, collected on *Citrus* spp. at various times, the earliest in 1928. The record above from Cook Is is based on a single specimen. It seems to agree in every respect with the description given by McKenzie. Reyne (1961) refers to the presence of *P. citri* in New Guinea, but no material from there has been seen.

Parlatoria crotonis Douglas (Fig. 102)

Parlatoria proteus var. *crotonis* Douglas, 1887a: 242.
Parlatoria crotonis Douglas, Cockerell, 1902a: 59; Ferris, 1942: 401; McKenzie, 1945: 61; Balachowsky, 1958: 326.

Description
Female scale elongate-oval, yellowish; exuviae similar colour, terminal. Male scale yellowish and elongate, smaller than female scale.
Adult female, slide-mounted, recognizable by fairly distinctive characters. Median, second and third pairs of lobes prominent, successively decreasing only slightly in size, each with a single notch on either side. Fourth lobes much smaller than third lobes, pointed, always sclerotized. Plates fimbriate distally, narrow and parallel between lobes, wider beyond third lobes. Dorsal ducts not numerous, present on margins and submargins forward to abdominal segment 1. Anus situated near centre of pygidium.
Ventral surface with 4 groups of perivulvar pores present. A minute derm pocket located between each posterior spiracle and margin. Duct tubercles present on thorax and abdominal segment 1. Eye spot modified into a stout spur.

Material examined
COOK IS. Mangaia, 1954; Mauke, 1954. On *Codiaeum* sp., *C. variegatum*.
FIJI. Viti Levu, 1949. On *Codiaeum* sp.
IRIAN JAYA. Biak, 1959; Jayapura, 1959. On Arecaceae, Pandanaceae.
KIRIBATI. Butaritari, 1976; Tarawa, 1976. On *C. variegatum*, *Pandanus odoratissimus*.
NIUE. 1977. On *C. variegatum*.
PAPUA NEW GUINEA. E.H.P.: 1959. On Sapotaceae.
SOLOMON IS. C.I.P.: Florida Is, 1939. On *Cocos* sp.

Comments
The modification of eye spots into stout spurs provides a small but useful character for distinguishing this species in the Pacific area. It could be confused with *P. proteus*, another species with small spurs in the same position; but the fourth lobes are membranous in *P. proteus*, resembling the neighbouring plates, whereas in *P. crotonis* the fourth lobes are always sclerotized and pointed.
This is a tropicopolitan and polyphagous species. It is found on leaves and, although it is a general feeder in the Pacific area, it is often found on *Codiaeum* spp. In addition to the material listed above, Dumbleton (1954) recorded it from Fiji on *Inocarpus fagifer*; from French

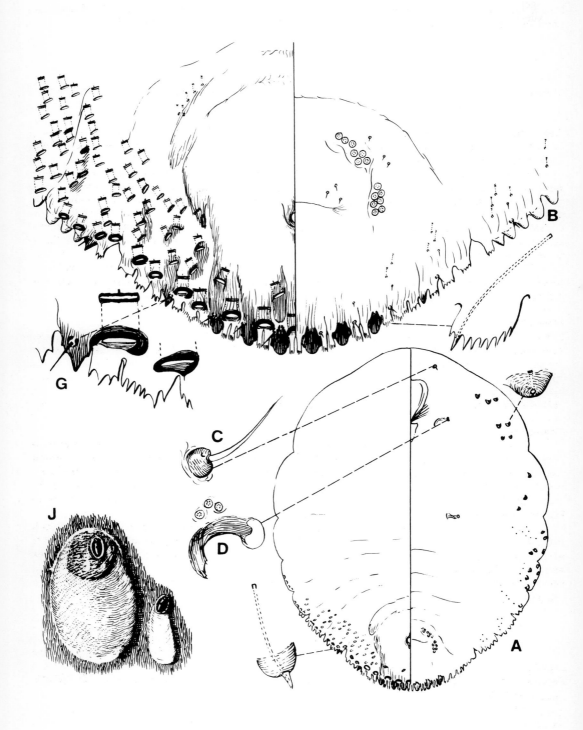

Fig. 103. *Parlatoria pergandii* Comstock.

Polynesia, Reboul (1976) recorded *P. crotonis* on *C. variegatum*, and it has been recorded from New Caledonia on *C. variegatum* and *Croton* (Cohic, 1956, 1958a; Brun & Chazeau, 1980).

Parlatoria pergandii Comstock (Fig. 103)

Parlatoria pergandii Comstock, 1881: 321; Ferris, 1937a: 88; McKenzie, 1945: 70; Balachowsky, 1958: 330.

Description
Scale of adult female oval, usually whitish; exuviae darker, terminal. Male scale smaller and narrower, same colour as female scale.
Adult female, slide-mounted, with 3 pairs of prominent lobes, each lobe notched once on each side; and a fourth pair of lobes, sclerotized, usually about half size of third lobes, but dentate on outer margins, projecting apically, each as prominent tooth. Fifth lobe present but smaller than other lobes. Plates fimbriate apically, narrow between lobes, wider and rounder further forward. Dorsal marginal and submarginal ducts numerous forward to about abdominal segment 1; median ducts absent, but a few small submedian ducts present just above level of perivulvar pores, and sometimes on segment above this.
Ventral surface with perivulvar pores in 4 groups, the anterior groups each usually with 5-8 pores. Duct tubercles present around thorax and abdominal segment 1. Anterior spiracles each with 3-5 pores.

Material examined
COOK IS. 1949; Aitutaki, 1954; Atiu, 1975; Mangaia, 1954; Penrhyn, 1977; Rarotonga, 1954. On *Citrus* sp., *C. aurantifolia*, *C. limon*, *C. maxima*, *C. paradisi*, *C. reticulata*, *C. sinensis*, *Inocarpus fagifer*, *Tournefortia argentea*.
NIUE. 1975. On *C. aurantifolia*, *C. paradisi*.
NORFOLK I. 1947. On *Malus pumila*.
WESTERN SAMOA. Savai'i, 1975. On *C. aurantifolia*, *C. sinensis*.

Comments
This species is commonly known as the chaff scale. In the Pacific area it should be easy to identify from the key and the characters given; elsewhere, it could be confused with related species. It has been recorded on a wide range of plants, but is widely known as a pest of citrus, occurring on the bark, leaves and fruit, and is present in most countries where citrus is grown. Although this species is so far not widely distributed in the South Pacific region, it seems to become common wherever it is introduced.
Walker & Deitz (1979) report *P. pergandii* as the second most numerous scale species from Cook Is on *C. sinensis* stems and bark after *P. cinerea* Hadden. These species frequently occur together, and are often associated with gumming, flaking and cracking of the bark in Cook Is, killing branches and sometimes whole trees. Maddison (1976) records it from Niue as a pest on *C. sinensis*, in addition to the material listed above. According to Bodenheimer (1951), trees over ten years old are particularly prone to attack because the insect has a decided shade preference.

Parlatoria proteus (Curtis) (Fig. 104)

Aspidiotus proteus Curtis, 1843: 676.
Parlatoria proteus (Curtis), Signoret, 1869b: 450; Ferris, 1937a: 89; McKenzie, 1945: 72; Takagi, 1969: 34.

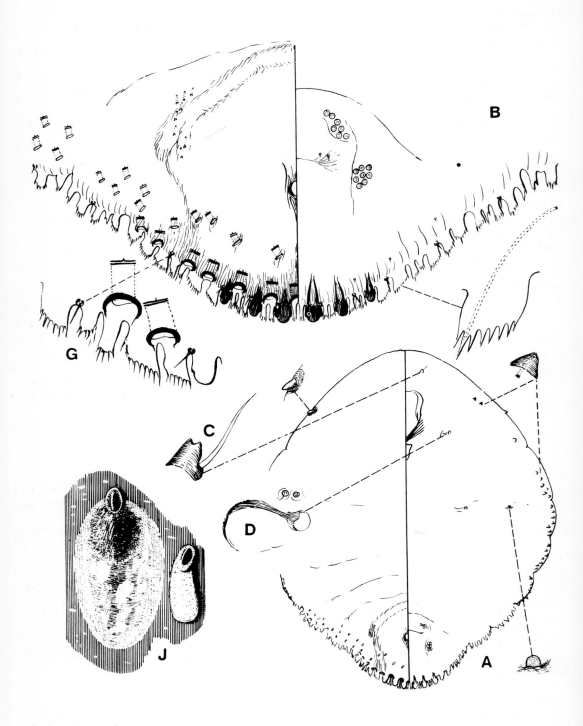

Fig. 104. *Parlatoria proteus* (Curtis).

Description
Female scale almost transparent; general colour yellowish brown, exuviae terminal. Male scale elongate, white, exuviae yellowish bown.

Adult female, on slide, almost circular; membranous, pygidium lightly sclerotized. Pygidium with 3 pairs of well-developed lobes present, successively decreasing only slightly in size, each lobe with a single notch on either side. Fourth and fifth pairs of lobes present, but smaller than other lobes and resembling plates. Plates distally fimbriate, slender between third lobes; wider plates present forward from third lobes to about abdominal segment 2. Dorsal ducts, not numerous, present around margins and submargins as far forward as segment 2. A group of small submedian ducts present at base of pygidium. Eye spot modified into pointed tubercle.

Ventral surface with perivulvar pores in 5 groups. Duct tubercles present on margins of thorax and abdominal segment 1. A small derm pocket present between each posterior spiracle and the margin.

Material examined
FIJI. Viti Levu, 1955. On *Bixa orellana, Derris elliptica, Epipremnum pinnatum, Gardenia* sp., *Macadamia tetraphylla, Theobroma cacao.*
PAPUA NEW GUINEA. C.P.: 1957. E.N.B.P.: 1959. Manus P.: 1959. On *Citrus* sp., *Dendrobium* sp., *Gardenia scandens.*
SOLOMON IS. W.P.: Shortland Is, Maleai, 1984. On *Citrus* sp.
WESTERN SAMOA. Upolu, 1975. On Arecaceae, *Coffea canephora.*

Comments
This is a tropicopolitan and polyphagous species that probably has a wider distribution in the South Pacific area than indicated by the above list. Although it is found on citrus, it is not regarded as one of the normal citrus pests. Despite its considerable host-plant range, it apparently shows a preference for palms and orchids.

In addition to the above material, Hinckley (1965) records *P. proteus* from Fiji on *Citrus* spp., *Freycinetia* sp. and *Eugenia malaccensis*, and it has been intercepted at quarantine inspection in New Zealand on *Ervatamia coronaria* from the area.

Genus **PINNASPIS** Cockerell

Pinnaspis Cockerell, 1892b: 136; Ferris, 1937a: 96; Ferris & Rao, 1947: 25; Balachowsky, 1954: 275. Type-species *Mytilaspis pandani* Comstock = *Aspidiotus buxi* Bouché by subsequent designation.

Hemichionaspis Cockerell, 1897b: 592. Type-species *Chionaspis aspidistrae* Signoret, by original designation.

Description
Species in this genus have 2-barred ducts and gland spines. The most distinctive feature is the pair of median lobes, which are united at the base by a more or less elongate zygosis; the inner edges of the lobes are parallel and very close together to the apices, appearing to be fused. Second lobes, when present, small, often indicated by the ventral paraphyses. Normally the marginal ducts are larger than the dorsal ducts, the latter usually few and represented by submarginal rows, although sometimes there are submedian series. The body is elongate, with the free abdominal segments strongly lobed.

Comments
At present, three cosmopolitan species, *P. aspidistrae* (Signoret), *P. buxi* (Bouche) and *P. strachani* (Cooley), are represented in the South Pacific area, and these may be separated by the key that follows. Some identification difficulties may arise that are still not resolved, and much

research is needed to define the limits of *P. aspidistrae* and *P. strachani*. Other species may be involved. As will be seen in the key, the species are separated mainly by the presence or absence of dorsal preanal scleroses, which are shown in Figs 107 and 108, representing *P. strachani*. It has always been argued that *P. aspidistrae*, a species without these scleroses, has a brown scale and *P. strachani*, with the scleroses, has a white scale. It is the experience here that frequently specimens have been received for identification possessing a brown scale, but with distinct preanal scleroses; and conversely, species with a white scale may show no signs of these scleroses. In the Pacific area the situation is still more complicated because specimens on coconut and other monocotyledons have a definite silvery white scale, but the preanal scleroses are slightly V-shaped and are often indistinct. There has been no possible way in this work to solve these problems. Normally the pygidium of *P. strachani* tends to be pointed, with small second lobes, whereas in *P. aspidistrae* the second lobes are more developed, and the median lobes tend to be smaller; but the characters are not consistent. Similar difficulties in identification were encountered by Beardsley (1966) when examining slide-mounted specimens from Micronesia. The records herein are based on identifications following present concepts, pending much needed research on live material. Normally most specimens can be identified with certainty, and reference should be made to the accompanying figures.

Some records from the South Pacific area refer to *Chionaspis minor* Maskell, to *Hemichionaspis minor*, or even to *Pinnaspis minor*. This species does not belong to *Pinnaspis*, but sometimes *P. aspidistrae* and *P. strachani* have been identified under these names. Specimens examined with the epithet '*minor*' all appear to be *P. strachani*, but not all specimens upon which the records of '*minor*' were based have been examined, and some may refer to *P. aspidistrae*.

Records of *P. strachani* are here presented in a simplified form, listing countries and hostplants separately.

Key to species of *Pinnaspis*

1 Abdominal segment 5 without submarginal macroducts. Second pygidial lobes with the inner lobule large, the apices expanded and truncated. Inner margins of median lobes usually showing a small gap between them .... *buxi* (Bouché)
-- Abdominal segment 5 with 1-3 submarginal macroducts present on each side. Second pygidial lobes smaller, but if the inner lobules rather large, then apices not strongly expanded. Median lobes contiguous for their entire length, the inner margins appearing to be fused 2

2 Dorsal preanal scleroses present. Second lobes usually small. Scale of female white .... *strachani* (Cooley)
-- Dorsal preanal scleroses absent. Second lobes usually as long as median lobes. Scale of female brown .... *aspidistrae* (Signoret)

Pinnaspis aspidistrae (Signoret) (Fig. 105)

Chionaspis aspidistrae Signoret, 1869b: 443.
Hemichionaspis aspidistrae (Signoret), Cockerell, 1897b: 592.
Pinnaspis aspidistrae (Signoret), Lindinger, 1912: 79; Ferris, 1937a: 97; Ferris & Rao 1947: 30; Balachowsky, 1954: 281.

Material examined
COOK IS. Aitutaki, 1975; Rarotonga, 1975. On *Cocos nucifera, Elaeis guineensis, Guettarda speciosa, Inocarpus fagifer, Musa paradisiaca, Pandanus odoratissimus*, 'tuitui pakarangi'.
KIRIBATI. Abemama, 1976; Tarawa, 1975. On *Cocos nucifera, Rhaphidophora acuminata*.

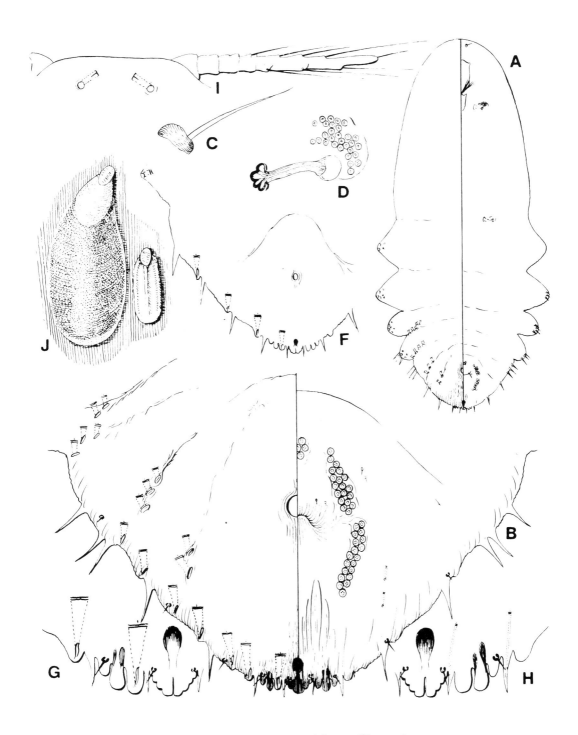

Fig. 105. *Pinnaspis aspidistrae* (Signoret).

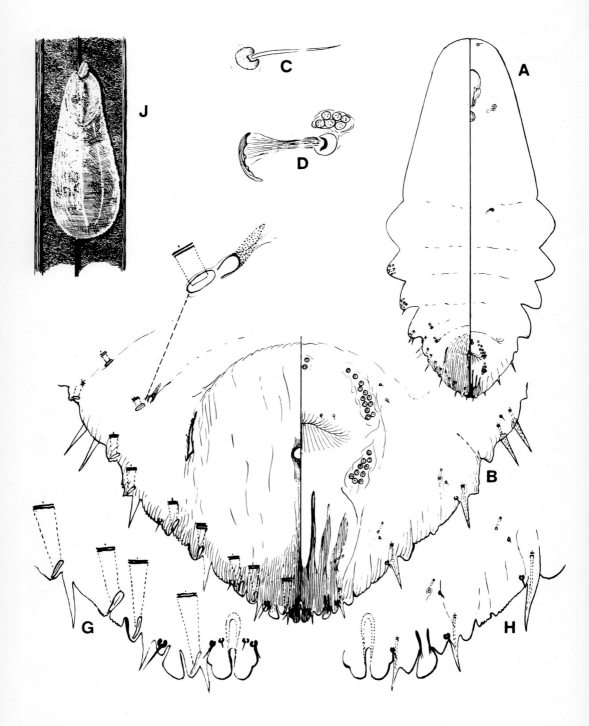

Fig. 106. *Pinnaspis buxi* (Bouché).

NIUE. 1933. On *Cocos nucifera*.
PAPUA NEW GUINEA. C.P.: 1980. Morobe P.: 1979. On *Acalypha godseffiana*, unidentified vine.

Comments

In addition to the above material, *P. aspidistrae* has been recorded from Fiji on *Citrus* spp., *Cocos nucifera, Musa* spp. and *M. sapientum* (Veitch & Greenwood, 1921; Lever, 1940; and Hinckley, 1965); in Society Is, Doane (1909) described it as the most serious pest of coconut after *Aspidiotus destructor;* Cohic (1955) mentioned it drying the tissues of *Cocos*, and damaging Solanaceae including *Lycopersicon esculentum* in French Oceania, where Dumbleton (1954) mentions it on *Citrus* spp., *Cocos nucifera* and *Musa* sp.; Reboul (1976) lists additional hosts as *Cordyline terminalis, Erythrina indica, Plumeria acutifolia, Portulaca lutea, Solanum melongena* and *Terminalia catappa*. It has been recorded from New Caledonia on *Citrus* spp., *Cocos nucifera, Cordyline* sp., *C. neo-caledonica, Crinum* sp., *C. asiaticum, C. pedunculatum, Heliconia brasiliensis, Hibiscus rosa-sinensis, Hippeastrum* sp. and *H. equestre* (Cohic, 1956, 1958a; Brun & Chazeau, 1980). *P. aspidistrae* was recorded from New Guinea [PNG] on *Adiantum* sp. by Froggatt (1940), who also mentioned it as a pest of citrus in Western Samoa; additional hosts there are Arecaceae, *Cocos nucifera, Musa* sp. and *M. sapientum* (Doane & Ferris, 1916; Dumbleton, 1954; and Reddy, 1970).

Pinnaspis buxi (Bouché) (Fig. 106).

Aspidiotus buxi Bouché, 1851: 111.
Mytilaspis pandani Comstock, 1881: 324.
Pinnaspis buxi (Bouché), Newstead, 1901: 207; Ferris, 1937a: 98; Ferris & Rao, 1947: 32; Balachowsky, 1954: 277.
Hemichionaspis pseudaspidistrae Green, 1916: 58. Lectotype female, Australia (BMNH), here designated [examined]. [Synonymised by Ferris & Rao, 1947: 32].

Material examined
COOK IS. Manihiki, 1977; Rarotonga, 1975. On *Coprosma laevigata, Cordyline terminalis, Lycopersicon esculentum, Monstera deliciosa*.
FIJI. Viti Levu, 1920. On *Areca catechu, Cocos nucifera, Colocasia esculenta, Monstera deliciosa*.
FRENCH POLYNESIA. Society Is, Huatine, 1979; Tahiti, 1976. On *Cordyline terminalis, Heliconia* sp., *Maranta* sp., *Momordica charantia, Monstera deliciosa*.
IRIAN JAYA. Biak, 1959. On ?
PAPUA NEW GUINEA. E.H.P.: 1959. E.N.B.P.: 1959. Manus P.: Manus, 1959. M.B.P.: Louisiade Archipelago, Misima, 1979. Morobe P.: 1959. N.I.P.: Lavongai I. (New Hanover), 1934. N.S.P.: Buka, 1959. W.S.P.: 1985. On *Adiantum aethiopicum*, Arecaceae, *Artocarpus heterophyllus, Citrus* sp., *Helicia* sp., *Hevea brasiliensis, Momordica* sp., *Musa* sp., *Schuurmansia* sp.
SOLOMON IS. G.P.: Guadalcanal, 1954. T.P.: Reef Is, 1985. W.P.: New Georgia Is, Kolombangara, 1981; Shortland Is, Maleai, 1985. On *Areca* sp., *Barringtonia edulis, Canarium commune, Morinda citrifolia, Piper* sp., unidentified 'liane'.
TONGA. Ha'apai Group, Nomuka, 1975; Vava'u Group, Vava'u, 1977. On *Inocarpus fagifer, Phaseolus vulgaris*.
WESTERN SAMOA. Savai'i, 1977; Upolu, 1975. On *Cordyline terminalis, Pandanus* sp., *P. upoluensis*.

Comments

Froggatt (1936) listed *Chionaspis simulatrix* Green from New Guinea [PNG] on *Citrus* sp. The name is a *nomen nudum*, but specimens are at hand clearly labelled with the name and information, and they are undoubtedly *P. buxi*, listed above.

The lectotype of *Hemichionaspis pseudaspidistrae* has been selected from 1 of 10 specimens on a slide labelled 'type', Australia, N.T., Koopiniyah, on *Pandanus* (J.F. Hill), and this is clearly marked. The other 9 specimens, and 17 others on 3 slides labelled 'co-type', are here designated

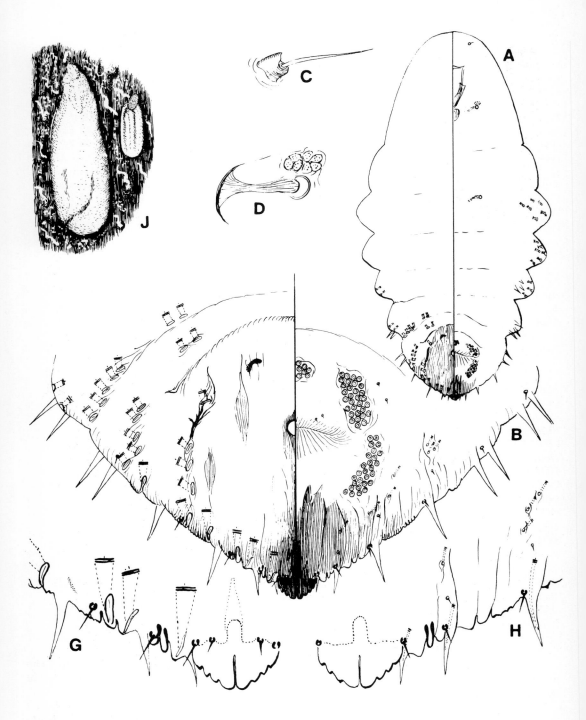

Fig. 107. *Pinnaspis strachani* (Cooley). (From original material).

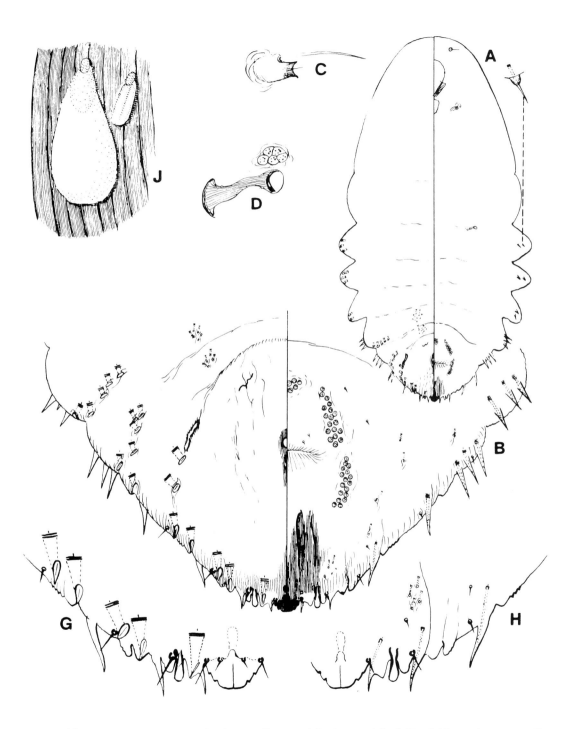

Fig. 108. *Pinnaspis strachani* (Cooley). (From original material of *Hemichionaspis townsendi*).

paralectotypes. In addition to the material listed above, *P. buxi* has been recorded from Fiji on ?*Veitchia joannis* by Hinckley (1965); and from New Guinea [PNG], as *P. pseudoaspidistrae*, on *Cocos nucifera* (Froggatt, 1940).

Pinnaspis strachani (Cooley) (Figs 107, 108)

Hemichionaspis minor var. *strachani* Cooley, 1899: 54.
Hemichionaspis townsendi Cockerell, 1905b: 135.
Pinnaspis temporaria Ferris, 1942: 407.
Pinnaspis strachani (Cooley), Ferris & Rao, 1947: 39; Balachowsky, 1954: 284.

Material examined
COOK IS. Aitutaki, 1975; Atiu, 1970; Mangaia, 1977; Manihiki, 1977; Mauke, 1954; Penrhyn, 1977; Pukapuka, 1969; Rakahanga, 1977; Rarotonga, 1948.
FIJI. Beqa, 1949; Kadavu, 1943; Lau Is, Cicia, 1945, Lakeba, 1977, Mago, 1976, Vanua Balavu, 1945; Taveuni, 1959; Vanua Levu, 1945; Viti Levu, 1914.
FRENCH POLYNESIA. Society Is, Tahiti, 1978.
KIRIBATI. Abemama, 1976; Butaritari, 1976; Marakei, 1976; Tarawa, 1956; Line Is, Kiritimati, 1948.
NEW CALEDONIA. New Caledonia, 1928.
NIUE. 1937.
PAPUA NEW GUINEA. C.P.: 1958. Morobe P.: 1979. N.I.P.: 1962. W.N.B.P.:1923.
SOLOMON IS. C.I.P.: Russell Is, Mbanika, 1962. G.P.: Guadalcanal, 1954. M.P.: Malaita, 1945. W.P.: New Georgia Is, Gizo, 1984, Kolombangara, 1980.
TOKELAU. Fakaofo, 1924.
TONGA. Ha'apai Group, Foa, 1977, Lifuka, 1977, Nomuka, 1977; Tongatapu Group, 'Eua, 1975, Tongatapu, 1974; Baba'u Group, Pangaimotu, 1977, Vava'u, 1977.
TUVALU. Funafuti, 1956; Nanumanga, 1924; Nanumea, 1924; Nukufetau, 1976; Nukulaelae, 1924; Vaitupu, 1976.
VANUATU. Efate, 1983; Espiritu Santo, 1983; Malekula, 1983; Tanna, 1983.
WESTERN SAMOA. Manono, 1977; Savai'i, 1975; Upolu, 1924.
On *Abutilon* sp., *A. hybridum*, *Acacia* sp., *Agave americana*, *Aleurites moluccana*, *Alpinia purpurata*, *Anacardium occidentale*, *Annona muricata*, *A. reticulata*, Arecaceae, *Artocarpus altilis*, *A. heterophyllus*, *Asplenium nidus*, *Barringtonia* sp., *B. asiatica*, *B. butonica*, *Bauhinia* sp., *Cananga odorata*, *Canna indica*, *Capsicum annuum*, *C. frutescens*, *Cassia occidentalis*, *Cassytha filiformis*, *Citrus* sp., *C. aurantifolia*, *C. limon*, *C. maxima*, *C. sinensis*, *Coccoloba* sp., *Cocos nucifera*, *Colocasia esculenta*, *Cordyline* sp., *Cordyline terminalis*, *Crinum* sp., *C. asiaticum*, *Crotalaria* sp., *C. usaramoensis*, *Cucurbita maxima*, *Datura metel*, *Dioscorea* sp., *D. alata*, *D. bulbifera*, *Diospyros kaki*, *Dodonaea viscosa*, *Elaeis guineensis*, *Erythrina indica*, *E. lithosperma*, *E. subumbrans*, *Eucalyptus grandis*, *Euphorbia heterophylla*, *Excoecaria agallocha*, *Fitchia* sp., *Gmelina arborea*, *Gossypium* sp., *Heliconia* sp., *Hernandia ovigera*, *H. peltata*, *Hibiscus* sp., *H. manihot*, *H.* ?*syriacus*, *H. tiliaceus*, *Inocarpus fagifer*, *Justicia* sp., *Laportea* sp., *Leucaena leucocephala*, *Lycopersicon esculentum*, *Malvaviscus arboreus*, mangrove, *Manihot esculenta*, *Maranta* sp., *Micromelum minutum*, *Mimosa pudica*, *Morinda citrifolia*, *Musa* sp., *M. sapientum*, *Ocimum gratissimum*, Orchidaceae, *Pandanus* sp., *P. odoratissimus*, *Pedilanthus* sp., *Persea americana*, *Phaseolus vulgaris*, *Plumeria* sp., *P. rubra*, *Rhaphidophora* sp., *Rhizophora mangle*, *Ricinus communis*, ?*Schefflera* sp., *Solanum melongena*, *S. torvum*, *Sophora tomentosa*, *Stachytarpheta* sp., *Strelitzia* sp., *Terminalia* sp., *T. calamansanay*, *T. catappa*, *T. complanata*, *Thespesia propulnea*, *Tournefortia* sp., *T. argentea*, 'uiniglu', unidentified creeper, unidentified fern, *Urena lobata*, *Vitis vinifera*, *Zingiber officinale*.

Comments
Maddison (1976) recorded *P. strachani* from Cook Is, Fiji, Kiribati, Niue and Tonga as a pest of *Citrus aurantifolia*, and remarked that it is also a pest on the stems of *Dioscorea* spp. and

Persea americana in the area. In addition to the material listed above, it has been recorded from Fiji on *Asparagus plumosus, Bruguierra gymnorhyza, Caesalpinia pulcherrima, Eucharis* sp., *Sophora tomentosa* and *Vietchia joannis*; from Irian Jaya on *Portulaca* sp. by Reyne (1961); from New Caledonia on *Aloe* sp., *Cassia alata, Litchi chinensis, Saintpaulia* sp. (Cohic, 1958a) and *Orchis* sp. (Brun & Chazeau, 1980); and from Wallis Is on many of the hosts listed above, and on *Albizia lebbek, Calophyllum inophyllum, Cycas* sp., *Dioscorea alata, D. esculenta, Erythrina fusca, Hernandia peltata, Hibiscus rosa-sinensis, Musa fehi, M. nana, M. sapientum, Plumeria acutifolia* and *Triumfetta rhomboidea* (Cohic, 1959). Reboul (1976) described it as a serious pest on *Hibiscus rosa-sinensis* in French Polynesia, also occurring there on *Cucumis sativus*, Cucurbitaceae and *Erythrina indica*; Walker & Deitz (1979) recorded it from Cook Is on *Erythrina subumbrans*.

Genus **PSEUDAONIDIA** Cockerell

Pseudaonidia Cockerell, 1897a: 14; Ferris, 1938: 252; Balachowsky, 1958: 268. Type-species *Aspidiotus duplex* Cockerell, by original designation.

Pseudaonidia trilobitiformis (Green)

Aspidiotus trilobitiformis Green, 1896b: 4.
Pseudaonidia trilobitiformis (Green), Cockerell, 1899b: 396; Balachowsky, 1958: 272.

In the South Pacific area this species could only be confused with *Duplaspidiotus claviger*, but as with all species in *Pseudaonidia*, it lacks the well-developed clavate paraphyses of *Duplaspidiotus*. Recognition characters are 1-barred ducts, fringed plates, and 3 pairs of well-developed lobes each about twice as long as wide. It resembles *D. claviger* in having numerous areolations on the dorsum of the pygidium, giving it a mosaic appearance. The female scale is recognizable by its large size, approximately 3.0 mm in diameter, pale or yellow-brown.

Material examined
IRIAN JAYA. Biak, 1959. On *Ficus* sp.
NEW CALEDONIA. Loyalty Is, Ouvea, 1963; New Caledonia, 1897. On *Citrus* sp., *Codiaeum* sp., *Ficus* sp., *Mangifera indica, Nerium oleander, Schinus terebinthifolius*.
VANUATU. Efate, 1983; Espiritu Santo, 1983; Malekula, 1983. On *Anacardium occidentale, Artocarpus* sp., *Citrus maxima, Persea americana, Plumeria rubra, Premna* sp., *Theobroma cacao*.

Comments
 The origin of this species is probably southern Asia, where it is common, but it has spread throughout Africa, the Malagasian area and tropical America on numerous host-plants. About 20 species are known in the genus, but this is the only one that has gained a foothold in the South Pacific area. Although it is often found in large numbers on individual plants, it does not seem to cause much damage. There are no records yet from Micronesia, but Beardsley (1966) recorded *P. manilensis* Robinson from Palau.
 In addition to the material listed above, *P. trilobitiformis* has been recorded from New Caledonia on *Acacia simplicifolia, A. spirorbis, Aleurites moluccana, A. montana, Annona* sp., *A. squamosa, Artocarpus altilis, A. incisa, A. heterophyllus, Barringtonia asiatica, Bauhinia variegata, Calophyllum inophyllum, Capsicum* sp., *C. annuum, C. frutescens, Carica papaya, Catharanthus roseus, Cerbera oppositifolia, Citrus aurantium, C. grandis, C. limon, C. paradisi, C. reticulata, Clitoria terneata, Cordyline* sp., *C. neo-caledonica, Crescentia cujete, Dodonaea viscosa, Eriobotrya japonica, Ficus pumila, Laurus nobilis, Maranta* sp., *Mucuna bennettii, Murraya exotica, Passiflora* sp., *P. edulis, P. laurifolia, P. quadrangularis, Persea* sp., *P. americana,*

Plumeria acutifolia, Pothos aureus, Psidium sp., *P. cattleianum, P. guajava, Pyrostegia venusta* and *Santalum austro-caledonicum* by Cohic (1956, 1958a) and Brun & Chazeau (1980).

Genus **PSEUDAULACASPIS** MacGillivray

Pseudaulacaspis MacGillivray, 1921: 305. Type-species *Diaspis pentagona* Targioni, by original designation.

Description

Adult female turbinate or elongate and fusiform, the free abdominal segments with lateral lobes usually well developed. Median lobes well developed, zygotic at base, often divergent, but always with a pair of basal setae present between lobes. Second lobes present, smaller than median lobes; third lobes present or absent. Lobes varying considerably in shape, depending on whether individuals of any species are bark or leaf feeding. Dorsal ducts about as large as marginal macroducts, present in submarginal and submedian rows. Perivulvar pores usually present in 5 groups. Anterior spiracles each with disc pores. Posterior spiracles each with or without disc pores.

Comments

In the South Pacific area this genus probably contains numerous species, all with white scales, either elongate or circular, the exuviae at one end or at edge, never within the borders of the scale. Nine species in the genus, following the concepts of Takagi & Kawai (1967) and Takagi (1970, 1985), have been studied from the South Pacific region and are recognized here. The genus is closely related to *Chionaspis* and *Aulacaspis*, already discussed, and the presence of a pair of setae between the median lobes clearly separates it from these genera; but concepts may change when further species are described. Species from the South Pacific area may be separated by the following key:

Key to species of *Pseudaulacaspis*

1	Body turbinate, less than twice as long as wide. Gland spines on pygidium each with 2-3 microducts present	2
--	Body elongate, usually fusiform, more than twice as long as wide. Gland spines on pygidium each with 1 microduct present	3
2	Gland spines on pygidium branched or bifid distally. Dorsal ducts absent from median areas of thorax	*pentagona* (Targioni)
--	Gland spines on pygidium thick and blunt. Dorsal ducts present in median areas of thorax	*major* (Cockerell)
3	Perivulvar pores absent	*ponticula* sp. n.
--	Perivulvar pores present	4
4	Dorsal ducts numerous in median areas of metathorax and abdominal segment 1	5
--	Dorsal ducts absent from these areas, although they are present posteriorly	6
5	Dorsal ducts present in median areas of prothorax and mesothorax. Spicules numerous on ventral surface of thorax	*papulosa* sp. n.
--	Dorsal ducts absent from median areas of prothorax and mesothorax. Spicules absent from ventral surface of thorax	*multiducta* sp. n.

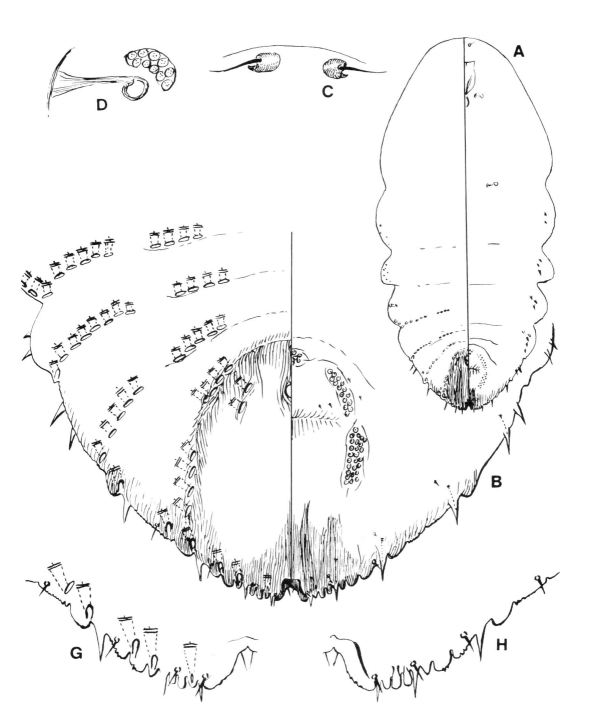

Fig. 109. *Pseudaulacaspis cockerelli* (Cooley).

6   Submedian dorsal ducts present on segment 6 on pygidium. Head and prothorax often expanded laterally .... *cockerelli* (Cooley)
-- Submedian dorsal ducts absent from segment 6 on pygidium. Head and prothorax gradually narrowing towards apex 7

7   Marginal dorsal ducts present as far forward as mesothorax. Single ducts present on lateral angles of head, and a single duct present on midline of head .... *coloisuvae* sp. n.
-- Marginal dorsal ducts absent from mesothorax except for occasional microducts. Ducts absent from head 8

8   Median lobes prominent, not forming a deep notch at apex of pygidium. Submarginal groups of ducts on segment 3 numbering 3-5 .... *samoana* (Doane & Ferris)
-- Median lobes forming deep notch at apex of pygidium. Submarginal groups of ducts on segment 3 numbering 1 or 2 .... *leveri* sp. n.

Pseudaulacaspis cockerelli (Cooley) (Fig. 109)

Chionaspis cockerelli Cooley, 1897: 278.
Chionaspis dilatata Green, 1899: 148. Lectotype female, Sri Lanka (BMNH), here designated [Examined].
Phenacaspis cockerelli (Cooley), Fernald, 1903: 237; Ferris, 1955: 46; Borchsenius 1966: 119.
Chionaspis inday Banks, 1906b: 787. Syn. n.
Phenacaspis inday (Banks), Ferris, 1955: 50; Beardsley 1966: 553.
Pseudaulacaspis cockerelli (Cooley), Takagi, 1970: 43.
Pseudaulacaspis inday (Banks), Takagi, 1985: 46.

Description
An extremely variable species, the body, and especially the head and thorax, often swollen in the South Pacific specimens, but sometimes slender. The characters vary considerably depending on whether specimens are bark or leaf forms. Specimens on the bark often have the median lobes prominent and well developed, and those on the leaves have the median lobes slender, recessed into the pygidium, forming a notch. Despite the variability, the species should be identifiable from the illustration, the most important characters being the submarginal and submedian rows of ducts present on segments 2 to 5, and the submedian group of 1 or 2 ducts on segment 6, this group apparently present in all specimens from the South Pacific region.

Material examined
FIJI. Viti Levu, 1949. On *Cocos nucifera*.
LORD HOWE I. 1911. On Arecaceae.
NEW CALEDONIA. New Caledonia, 1980. On *Carica papaya*.
PAPUA NEW GUINEA. C.P.: 1964. On *Cocos nucifera, Mangifera indica*.
SOLOMON IS. G.P.: Guadalcanal, 1954. M.P.: Malaita, 1950. W.P.: New Georgia Is, Gizo, 1984, Kolombangara, 1954; Baeroko, 1985. On *Cocos nucifera*.
VANUATU. Efate, 1983; Espiritu Santo, 1983. On *Carica papaya, Cocos nucifera*.

Comments
There are many synonyms listed by Borchsenius (1966) and Takagi (1970). The species is common throughout southern Asia, and has been reported from Africa, the Malagasian area and Australia, extending into the temperate areas. In the New World it is present in Florida, and in the North Pacific it is common in Hawaii and Micronesia. It is probably widespread throughout the tropical Pacific area on numerous species of plants, but it is particularly common on coconut.

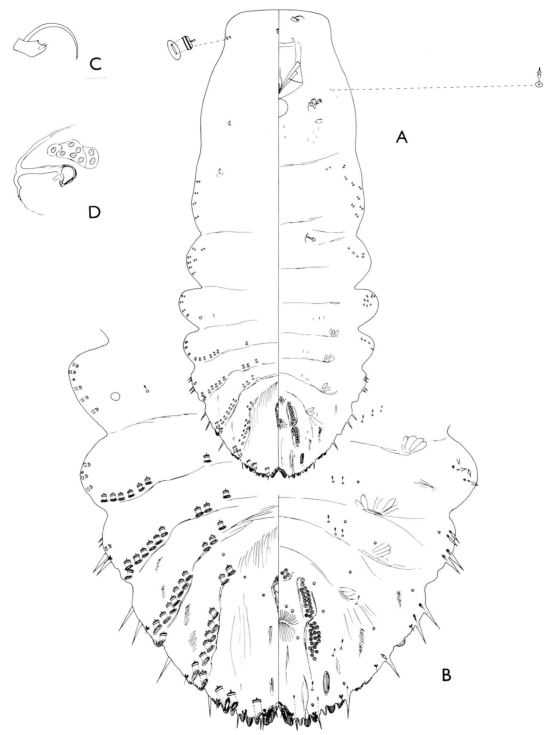

Fig. 110. *Pseudaulacaspis coloisuvae* sp. n.

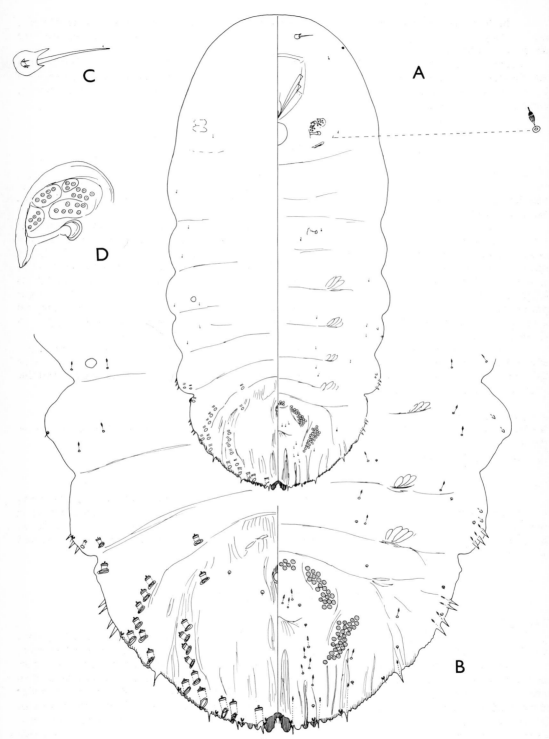

Fig. 111. *Pseudaulacaspis leveri* sp. n.

The lectotype of *Chionaspis dilatata* is selected from four specimens on Green's 'type' slide, labelled Ceylon, Peradeniya, June 97, and is clearly marked. The other three specimens are here designated paralectotypes.

Some specimens are available labelled 'part of type material' of *Chionaspis inday* Banks from the Philippines, on *Cocos* sp., sent by C. Banks to E.E. Green. Some contain dorsal ducts on segment 6, and the characters come well within the range of variation discussed by Takagi (1970). The name, therefore, is here sunk as a synonym of *P. cockerelli*.

Pseudaulacaspis coloisuvae sp. n. (Fig. 110)

Description

Adult female, slide-mounted, up to 1.2 mm long, elongate; pygidium rounded; head straight at anterior margin; lateral lobes of free abdominal segments moderately produced; body membranous except for pygidium.

Pygidium with median lobes well developed, each with the inner margins much longer than outer margins, forming a deep notch at apex of abdomen. Second lobes well developed, inner lobules projecting beyond median lobes, outer lobules shorter. Third lobes moderately developed. Gland spines arranged singly on fifth and posterior segments; decreasing in size and more numerous forward to abdominal segment 1; present also submarginally on mesothorax and metathorax. Marginal ducts longer than wide on pygidial margins, shorter further forward; submarginal and submedian groups present on segments 2 to 5 on each side, absent from these positions on segment 6. Smaller marginal ducts present on either side of abdominal segment 3, and as far forward as mesothorax. A similar-sized duct also present at each lateral angle of head, and on midline of head.

Ventral surface with perivulvar pores in 5 elongate groups. Microducts present on submargins of pygidium, submedially on free abdominal segments and on prothorax. Small ducts, each with thick bars, present on margins of mesothorax and metathorax. Antennae each with 1 long seta. Anterior spiracles each with a group of 6-9 pores.

Material examined

Holotype female. **FIJI.** Viti Levu, Colo-i-Suva, on unidentified plant, 23.iii.1957 (*B.A. O'Connor*) (BMNH).
Paratypes female. **FIJI.** Same data as holotype. 4 (BMNH).

Comments

This species seems to follow the general pattern of the genus, but is peculiar in possessing dorsal ducts at the lateral angles and midline of the head. One specimen has one duct lacking at one side. In its arrangement of the other characters, however, it seems to be related to *P. brideliae* (Takahashi) described from Taiwan, but the latter has a narrower and more rounded head.

Pseudaulacaspis leveri sp. n. (Fig. 111)

Description

An elongate species; adult female about 0.9 mm long; pygidium and head rounded; body membranous except for pygidium; lateral lobes of free abdominal segments only moderately developed.

Median lobes prominent but recessed into pygidium, forming a notch at apex, the inner edges longer than outer edges, each lobe rounded distally. Second lobes bilobed, much smaller than median lobes. Third lobes represented by serrations on margins. Gland spines arranged singly on each side of fifth and posterior segments; more numerous on third and fourth segments, and represented by 1 or 2 that are much smaller on margins of second and third segments. Dorsal ducts wider than long, present on pygidial margins; others shorter, present in submarginal rows on segments 4 and 5 on each side, and as a small marginal group on segment 3. Single submedian ducts present on segments 4 and

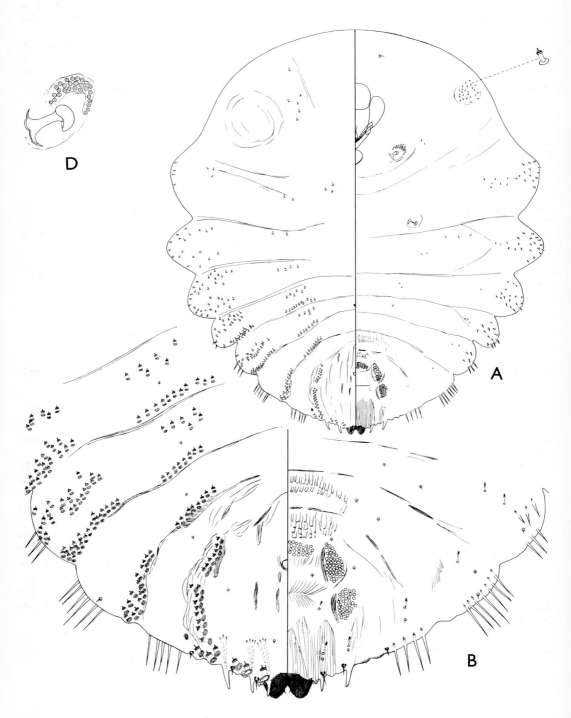

Fig. 112. *Pseudaulacaspis major* (Cockerell).

5 on each side, only rarely 2 ducts present instead of 1. Microducts sparse in submarginal areas forward to prothorax. A submarginal boss present on each side of segment 1. Anus situated towards base of pygidium.

Ventral surface with numerous perivulvar pores present in 5 elongate groups. Microducts on pygidium present in a group anterior to vulva, in a row above each second lobe to vulva, and on submargins. Further forward, microducts sparsely present on free abdominal segments and near spiracles. Antennae each with 1 long seta. Anterior spiracles each with a group of 23-30 disc pores.

Material examined
Holotype female. **FIJI.** Viti Levu, Tomanivi, on Pandanaceae, 6.vii.1944 (*R.A. Lever*) (BMNH).
Paratypes female. **FIJI.** Same data as holotype. 8 (BMNH), 1 (DSIR), 1 (USNM).

Comments
This species resembles some species of *Aulacaspis*, but there are 2 short setae present between the median lobes. They are difficult to see in some specimens, but the species is left in *Pseudaulacaspis* for the present. *P. subcorticalis* (Green) has a similar arrangement of dorsal ducts, and may be a close relative; but *P. leveri* has the median lobes recessed into the apex of the pygidium to form a notch, whereas the median lobes of *P. subcorticalis* are prominent.

Pseudaulacaspis major (Cockerell) (Fig. 112)

Chionaspis major Cockerell, 1894: 43.
Aulacaspis flacourtiae Rutherford, 1914: 259. [Synonymised by Morrison, 1924: 232].
Pseudaulacaspis flacourtiae (Rutherford), MacGillivray, 1921: 316.
Aulacaspis major (Cockerell), Morrison, 1924: 232.
Pseudaulacaspis major (Cockerell), Mamet, 1941: 30; Zimmerman, 1948: 381; Borchsenius, 1966: 175.

Description
Female scale normally white but obscured by plant tissue, and according to Mamet (1941), colour also ranging through reddish brown to black; subcircular to oval, with brown exuviae marginal. Male scale not known.

Adult female broadly oval to turbinate, widest at about first abdominal segment, the prepygidial segments with well-developed lateral lobes; body becoming moderately sclerotized at maturity.

Pygidium with well-developed and prominent median lobes rounded, the edges with numerous notches, yoked deeply at base. Second lobes represented by points. Gland spines 'fleshy' on segments 6-8, arranged singly, each rounded at apex; those on segment 8 overlapping median lobes dorsally, each with a few elongate microducts. Anterior gland spines elongate, each with short microduct, present in marginal groups on segments 3-5, gradually smaller in larger groups as far forward as mesothorax. Marginal ducts on fifth and posterior segments with larger sclerotized rims. Dorsal ducts on pygidium only slightly smaller than marginal ducts, but progressively smaller further forward; present in submarginal and submedian rows on each side of abdominal segments, and in small groups on thoracic segments. A small boss present on submargins of segment 5.

Ventral surface with perivulvar pores present in 5 large groups. Minute ducts present in median areas of abdomen and laterally on thorax, including a large conspicuous group on prothorax. Antennae close together. Anterior spiracles each with a group of numerous disc pores.

Material examined
TONGA. Tongatapu Group, Tongatapu, 1974. On *Elattostachys falcata*.

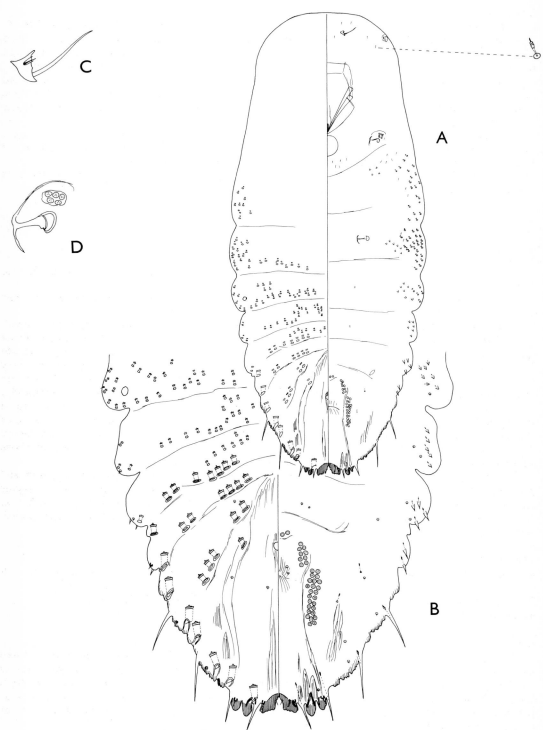

Fig. 113. *Pseudaulacaspis multiducta* sp. n.

Comments

In being broadly oval and almost turbinate, this species can only be confused with *P. pentagona* in the South Pacific area. It differs in possessing fleshy gland spines, rounded distally, at the apex of the pygidium, whereas the gland spines in *P. pentagona* are more robust and branched apically. Other species of *Pseudaulacaspis* in the area are more elongate than *P. major*.

The species was described from Antigua, in the West Indies, but it is known from Central and South America, southern Asia, the Malagasian area and Hawaii on numerous host-plants. This is the first record from the South Pacific region, where it may have a wider distribution. No damage has been reported.

There have been available some specimens from the original material described from Antigua. The illustration has been prepared from material collected in Sri Lanka on *Flacourtia*, the same locality and host-plant from which the material of *A. flacourtiae* was described.

Takagi (1985) has suggested that the species belongs to *Rutherfordia* MacGillivray, but it is left in *Pseudaulacaspis* pending further research.

Pseudaulacaspis multiducta sp. n. (Fig. 113)

Description

Female scale elongate, brilliant white, exuviae yellow-brown; on the edges of the leaves. Male scale not observed.

Adult female narrowly elongate, longest specimens 1.5 mm long, free abdominal segments moderately developed, head rounded.

Pygidium elongate with median lobes well developed, divergent and wide, straight or rounded, giving the lobes a squat appearance; each lobe with distal edge notched; median lobe bases recessed into apex, forming a notch. Second lobes prominent, bilobed, each lobule rounded, inner lobule longer than median lobes. Third lobes represented by notches on margin. Gland spines unusually long on pygidial margin, longer than lobes, arranged singly; further forward, they are shorter and more numerous, and on thorax are replaced by duct tubercles. Marginal macroducts longer than wide. Dorsal ducts shorter than marginal ducts, there being on each side a single submedian duct present on segment 6, and 2 or 3 submarginal ducts and 2-5 submedian ducts on segments 3 to 5. Smaller ducts present, numerous in rows or bands across anterior end of segment 3, and across the metathorax and segments 1 and 2. Marginally, they are present as far forward as mesothorax. A submarginal boss present on each side of segment 1. Anus situated towards base of pygidium.

Ventral surface with 5 elongate groups of perivulvar pores. Microducts sparse on pygidium and free abdominal segments, and present in groups on prothorax, next to labium, and on head margins (but see comments). Small ducts, similar to the small ducts on dorsum, on margins of mesothorax, metathorax and abdominal segment 1. Antennae each with a single long seta. Anterior spiracles each with 3-6 disc pores.

Material examined

Holotype female. **IRIAN JAYA.** Sarmi District, Bodem, on leaves of dicotyledonous plant, 17.vii.1959 (*T. Maa*) (BPBM).
Paratypes female. **IRIAN JAYA.** Same data as holotype. 5 (BPBM), 5 (BMNH).
Non-type material. **IRIAN JAYA.** Sentani, nr Jayapura, 24.vi.1959. (*T. Maa*).

Comments

Important distinguishing characters of this species are the almost continuous rows of small ducts across the metathorax and first three abdominal segments, and the unusually long gland spines on the pygidium. Another species with similar gland spines is *P. centreesa* Ferris, described from China. Both species also share the similar habit of occurring on the leaf margins, but *P. centreesa* possesses numerous spicules on the ventral surface, and lacks the continuous rows of ducts anterior to the pygidium which are present in *P. multiducta*.

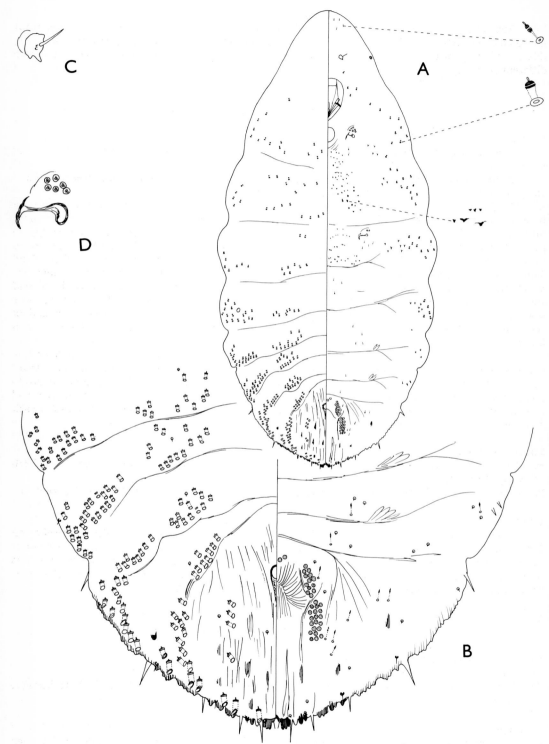

Fig. 114. *Pseudaulacaspis papulosa* sp. n.

Specimens collected at Sentani differ from the type material in having more numerous microducts on the venter, there being a conspicuous row between the posterior spiracles: however, this is probably just a variation, and there is nothing to suggest that the specimens represent a different species.

Pseudaulacaspis papulosa sp. n. (Fig. 114)

Description
Scale of adult female elongate, convex, edges often wrapped around branchlets; white, exuviae yellow-brown. Male scale not observed.

Adult female fusiform, about 1.85 mm long; anterior end narrow but rounded, free abdominal segments with lateral lobes distinct but not well developed.

Pygidium with median lobes separated by a space equal to width of 1 lobe, narrowly yoked at base; each lobe almost quadrate, the distal edge usually with 2 deep notches present. Second lobes much smaller than median lobes, the lobules elongate and rounded, often covered by overlapping pygidial margins, and recognizable by the presence of ventral paraphyses. Third lobes represented by serrations of margin. Gland spines, longer than lobes, arranged singly on pygidium; short gland spines present in small groups on segments 2 and 3, and replaced by duct tubercles on mesothorax, metathorax and abdominal segment 1. Marginal macroducts longer than wide. Dorsal ducts shorter and smaller than marginal ducts, arranged in submedian groups forward to abdominal segment 1, more scattered on thoracic segments to median areas. A boss present laterally on each side of abdominal segment 1, and a scar-like boss present laterally on either side of segment 5. Anus situated towards base of pygidium.

Ventral surface with 5 groups of perivulvar pores present. Microducts present lateral to vulva, in median areas of abdomen, near spiracles, and on head. Small ducts, similar to the small ducts on dorsum, present in lateral areas of abdominal segment 1, thoracic segments, and forward to clypeolabral shield. Spicules of various sizes numerous in median areas of mesothorax and metathorax, the largest on mesothorax. Antennae each with a single seta. Anterior spiracles each with 3-6 disc pores; posterior spiracles each with 1 or 2 pores.

Material examined
Holotype female. **IRIAN JAYA.** Baliem Valley, Wamena, on branchlets of *Casuarina* sp., 24.ii.1960 (*T. Maa*) (BPBM).
Paratypes female. **IRIAN JAYA.** Same data as holotype. 3 (BPBM), 4 (BMNH), 1 (USNM).

Comments
This species is so close to *P. hilli* (Laing) described from New South Wales, Australia, on *Casuarina* sp., that there may be some justification in uniting them. In all the material at hand of *P. hilli*, however, the ventral spicules are few, represented by 2-4 large ones on the mesothorax and 2 on the metathorax, with only one or two that are minute in the same areas. In *P. papulosa* they are abundant in the median areas, reaching as far forward as the labium.

Pseudaulacaspis pentagona (Targioni) (Fig. 115)

Diaspis pentagona Targioni, 1886: 184.
Aspidiotus vitiensis Maskell, 1895: 40.
Pseudaulacaspis pentagona (Targioni), MacGillivray, 1921: 305; Ferris, 1937a: 109; Balachowsky, 1954: 236; Takagi, 1970: 42.

Description
Female scale subcircular, white; exuviae marginal, reddish brown. Male scale smaller than female scale, elongate, white and flat.

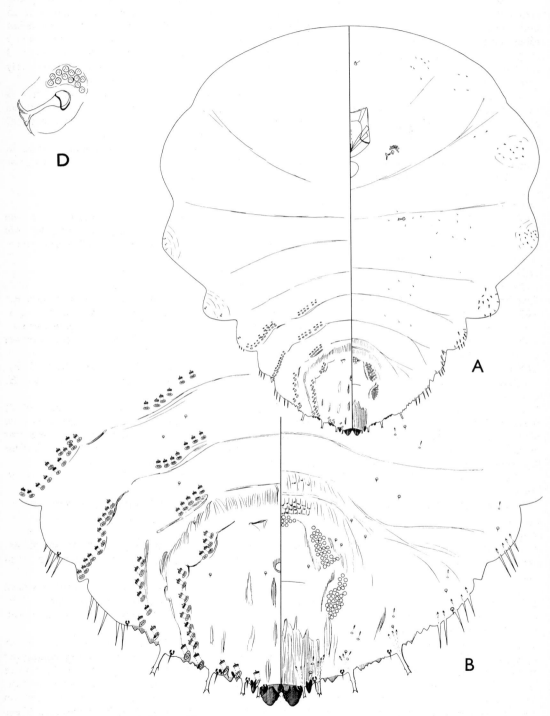

Fig. 115. *Pseudaulacaspis pentagona* (Targioni).

Adult female, on slide, broadly turbinate, widest at about mesothorax, the prepygidial segments with lateral lobes well developed. Median lobes prominent, each notched more than once on both sides; apex rounded; ventral paraphyses well developed. Second lobes much smaller than median lobes, pointed or rounded, the inner lobule either minute or lacking. Marginal gland spines usually single on either side of fifth and posterior segments, branched apically, each with 2 or 3 microducts; present in groups as far forward as abdominal segment 1; pointed and progressively shorter. Minute gland spines or duct tubercles present on metathorax and prothorax.

Pygidium with macroducts present on margin, and in well-defined submarginal and submedian rows on each side of abdominal segment 5 and decreasing in size forward to segment 2. A few minute ducts also present on margins of abdominal segment 1 and metathorax. Anus situated just above centre of pygidium. Base of pygidium with a pair of preanal scleroses present.

Ventral surface with numerous perivulvar pores present in 5 groups. Microducts present in submedian areas of abdomen and on head and prothorax. Antennae set close together. Anterior spiracles each with as many as 18 disc pores.

Material examined
FIJI. Viti Levu, 1899. On *Allemanda* sp., *Calophyllum* sp., *Capsicum* sp., *Cassia alata*, *Erythrina* sp., *Hibiscus* sp., *H. esculentus*, *H. manihot*, *Ipomoea* sp., *Lagerstroemia* sp., *Ligustrum* sp., *Manihot esculenta*, *Morus* sp., *Nerium oleander*, *Passiflora* sp., *Platanocephalus indicus*, *Plumeria rubra*, *Stachytarpheta* sp., *Verbena* sp.
NEW CALEDONIA. New Caledonia, 1899. On *Acacia* sp.
NORFOLK I. 1947. On *Citrus reticulata*, ?*Nicotiana tabacum*, *Pelargonium* sp.,
PAPUA NEW GUINEA. C.P.: 1959. E.H.P.: 1959. E.N.B.P.: 1941. Madang P.: 1960. Manus P.: 1959. M.B.P.: 1985. Morobe P.: 1956. W.N.B.P.: 1961. On *Albizia stipulata*, *Allemanda* sp., *A. cathartica*, *Capsicum* sp., *Carica papaya*, *Codiaeum* sp., *Crotolaria* sp., *Erythrina* sp., *Fuchsia* sp., *Hibiscus manihot*, *Hibiscus mutabilis*, *H. rosa-sinensis*, *Kalanchoe* sp., *Macaranga* sp., *Manihot esculenta*, Orchidaceae, *Pedilanthus tithymaloides*, *Stachytarpheta* sp.
SOLOMON IS. G.P.: Guadalcanal, 1970. T.P.: Santa Cruz Is, Ndende (=Santa Cruz), 1984. W.P.: Shortland Is, Maleai, 1984. On 'aibika cabbage', *Capsicum* sp., *C. grossum*, *H. esculentus*, *H. manihot*, 'leguminous tree', *Manihot esculenta*.
TONGA. Ha'apai Group, Foa, 1977, Lifuka, 1977; Tongatapu Group, 'Eua, 1977, Tongatapu, 1973; Vava'u Group, Vava'u 1974. On *Allemanda cathartica*, *Capsicum* sp., *Cliffortia polygonifolia*, *Ervatamia orientalis*, 'fagamimi', *Glycine max*, *Gossypium brasiliense*, *Helianthus* sp., *Hibiscus manihot*, *Lycopersicon esculentum*, *Nerium oleander*, *Phaseolus vulgaris*, *Plumeria rubra*, *Ricinus communis*, *Solanum uporo*, *S. verbascifolium*, *Stachytarpheta* sp., *Vigna* sp.
VANUATU. Espiritu Santo, 1945; Tanna, 1983. On *Citrus maxima*, *Erythrina* sp., *H. manihot*, *R. communis*.
WESTERN SAMOA. Upolu, 1978. On *Cedrela toona*, *Morinda citrifolia*, *Passiflora edulis*.

Comments
This is a cosmopolitan and destructive species recorded from a long series of plants, occurring on bark, leaves and fruit; it is recognizable by the branched or bifid apices to the gland spines on the fifth and posterior segments.

Maskell (1895) described the species as *Aspidiotus vitiensis* from Fiji, from material collected by A. Koebele on different forest trees. Specimens are at hand sent by Koebele to E.E. Green, which are probably part of the original material. The species must have become well established in Fiji by 1895.

In addition to the above list, *P. pentagona* has been recorded from Irian Jaya on *Manihot esculenta* by Thomas (1962a); from New Caledonia as a serious pest of most introduced plants, and from Wallis Is as a pest of *Broussonetia papyrifera* by Cohic (1958a), although its numbers were later reduced by the fungi *Fusarium coccidicola* and *Septobasidium bogoriense* (Cohic, 1959); in Fiji, Lever (1946) described it as a pest on *Passiflora quadrangularis*, and it was also recorded on *Verbena bonariensis* (Greenwood, 1940), *Malvastrum tricuspidatum*, *Triumfetta bartramia* and *Urena lobata* (Greenwood, 1977). Further host records from the South Pacific literature, too numerous to

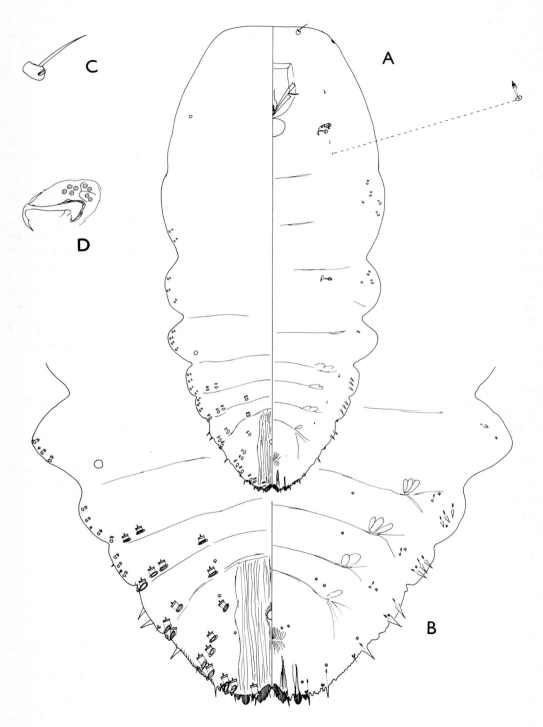

Fig. 116. *Pseudaulacaspis ponticula* sp. n.

cite individually, include *Althaea* sp., *Angelonia salicariaefolia*, *Argyreia nervosa*, *Broussonetia* sp., *Cassia* sp., *Catharanthus roseus*, *Cocos nucifera*, *Delphinium* sp., Euphorbiaceae, *Helianthus annuus*, *Hibiscus diversifolius*, *Lagerstroemia flos-reginae*, *Ligustrum japonicum*, *Mangifera indica*, *Morus alba*, *M. ?indica*, *Passiflora quadrangularis*, *Plumeria acutifolia*, *Prunus amygdalus*, *Stachytarpheta dichotoma*, *S. indica*, *S. mutabilis*, *S. urticifolia* and *Tournefortia argentea*.

Pseudaulacaspis ponticula sp. n. (Fig. 116)

Description
    Adult female on slide elongate-oval, widest at about mesothorax; length about 0.95 mm; head gently rounded to flat; body membranous except for sclerotized pygidium, median areas of pygidium heavily sclerotized.
    Pygidium with median lobes diverging, recessed into apex, forming a notch; each lobe with inner margins longer than outer margins, notched a few times and tending to be pointed apically; the ventral paraphyses well developed and divergent. Second lobes bilobed, well developed, each inner lobule rounded and extending beyond median lobes; outer lobule smaller than inner lobule. Third lobes represented by short points. Gland spines single to segment 3, double on segment 4. Shorter gland spines present further forward, about 3 laterally on each of segments 2 and 3, and present singly on mesothorax, metathorax and abdominal segment 1. Dorsal pygidial ducts all about same size, present in marginal groups as far forward as segment 2; single submedian ducts present on segments 3 to 5, but sometimes absent from segment 3. Smaller ducts present on margins of segment 3 and forward to mesothorax. Submarginal bosses present on abdominal segment 1 and prothorax. Anus situated just above centre of pygidium.
    Ventral surface with perivulvar pores absent. Microducts sparse, present on submargins of free abdominal segments and prothorax. Small ducts present laterally on mesothorax and metathorax. Antennae each with a single seta. Anterior spiracles each with a group of disc pores.

Material examined
Holotype female. **PAPUA NEW GUINEA.** Morobe P., Buso, on fern, 13.x.1979 (*J.H. Martin*) (BMNH).
Paratypes female. **PAPUA NEW GUINEA.** Same data as holotype. 4 (BMNH).

Comments
    This should be easily distinguishable from other species of *Pseudaulacaspis* in the southern Pacific area in lacking perivulvar pores. In this respect it resembles *P. atalantiae* (Takahashi), described from Taiwan; but this species has the head and thorax sclerotized, whereas in *P. ponticula* it is membranous.

Pseudaulacaspis samoana (Doane & Ferris) (Fig. 117)

Chionaspis samoana Doane & Ferris, 1916: 399.
Phenacaspis samoana (Doane & Ferris), MacGillivray, 1921: 346; Ferris, 1955: 52.
Pseudaulacaspis samoana (Doane & Ferris), Takagi, 1985: 49.

Description
    Female scale fusiform, about 1.5 mm long; white; yellow exuviae terminal. Male scale not known.
    Adult female on slide elongate-oval, widest at about first abdominal segment; body membranous except for pygidium; free abdominal segments moderately developed.
    Pygidium with median lobes prominent, rounded, the mesal margins slightly longer than outer margins, with a few notches present; yoked at base, ventral paraphyses well developed. Second lobes much smaller than the median lobes, bilobed. Third lobes represented by notches on body margin.

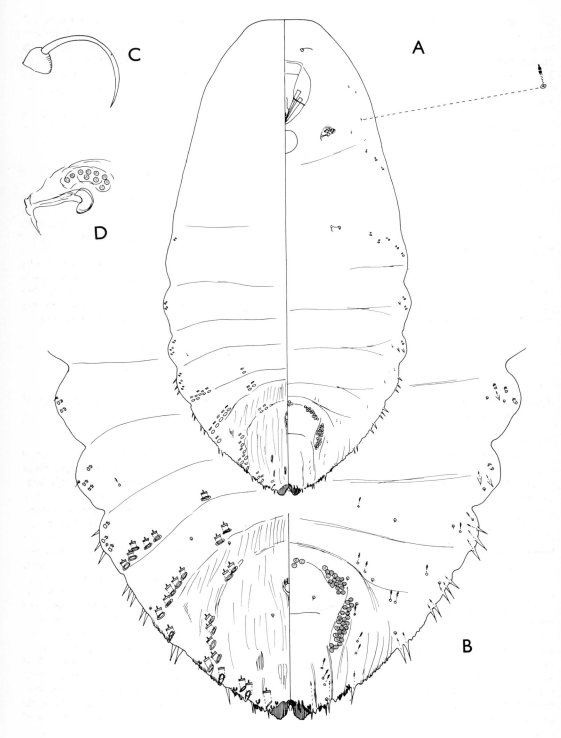

Fig. 117. *Pseudaulacaspis samoana* (Doane & Ferris).

Gland spines about same length as lobes or longer, 1 present between median and second lobes, 2 between second and third lobes, 1 lateral to third lobes; present in groups of 2 or 3 further forward, but represented by only 1 or 2 laterally on segments 1 and 2. Marginal macroducts numbering 6 on each side. Dorsal ducts slightly shorter than marginal ducts, absent from segment 6; 3-5 present in rows submarginally on segments 3 to 5; usually 2 submedially on each side of segments 4 and 5, and 0-1 in these positions on segment 3. Small ducts few, present on margins forward to metathorax.

Ventral surface with perivulvar pores present in 5 elongate groups. Microducts present in submarginal rows on pygidium, sparse on free abdominal segments and prothorax. Small ducts situated laterally on mesothorax, metathorax and abdominal segment 1.

Material examined
SAMOA. ?1916. On Arecaceae.

Comments
The paucity of dorsal ducts and the prominent median lobes are the main distinguishing characters of this species. There have been available for study a few specimens collected by R.W. Doane labelled 'cotype', kindly made available by R.O. Schuster (UCD).

Lever (1945a) recorded this species from Fiji as *Phenacaspis samoana*, on palm; Hinckley (1965) records it on ?*Cocos nucifera* there. O'Connor (1949) mentions it on *C. nucifera* in Tonga, where Dumbleton (1954) describes it as a pest on coconut; he remarks, however, that the species is also present in Western Samoa on coconut, but is not a pest there. There is no indication in the original description whether the species was described from Western Samoa or American Samoa.

Genus **SCHIZENTASPIDUS** Mamet

Schizentaspidus Mamet, 1958: 421. Type-species *Schizentaspidus loranthi* Mamet by original designation.

Description
Adult female sclerotized; prepygidium broad, with constrictions and articulations present between mesothorax and metathorax, and between metathorax and abdominal segment 1; pygidium produced, quite broad. Ducts 1-barred. 3 Pairs of well-developed lobes present, never bilobed. Median lobes at least as long as wide, well sclerotized, each notched on either side, with a sclerosis at the base. Second and third lobes narrower than median lobes, with more or less rounded tips; second lobes each notched subapically on either side, third lobes notched on lateral margins. Plates as long as lobes, 2 present between median lobes, 2 between median and second lobes, 3 between second and third lobes, and 8-9 present lateral to third lobe. Plates fringed between lobes, and as far laterally as dorsal marginal seta of fifth segment; plates beyond this seta broad, each with dentate margins and a few long processes at mesal angle. Dorsal marginal seta at outer corner of each median lobe short and slender; setae on second and third lobes shorter than lobes, slender or lanceolate (broadened and flattened); dorsal marginal seta of fifth segment less so; marginal setae further forward long and slender, especially those on ventral surface. Anus long, narrow oval, situated at about posterior third of pygidium. Dorsal macroducts elongate, not slender; one marginal duct present between each pair of lobes, and one on fifth segment; other macroducts forming a submarginal band on pygidium; marginal macroducts present on abdominal segments. Dorsal submarginal bosses situated on abdominal segments 1 and 3; thoracic tubercle present on posterior lateral angles of mesothorax. Perivulvar pores present or absent. Vulva well developed, shaped like a wide rounded V. An area of dermal granulation and a smooth tubercle associated with each anterior spiracle. Antennae each with one or more small sclerotized spurs and 1 seta.

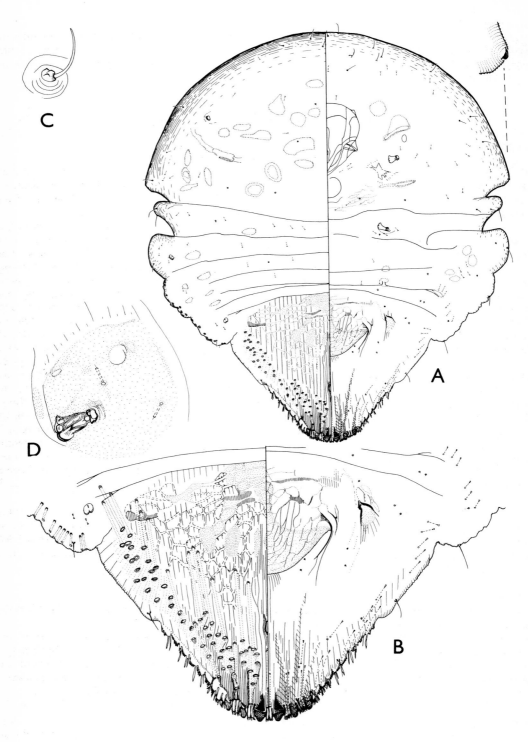

Fig. 118. *Schizentaspidus silvicola* sp. n.

Comments
Mamet (1958) based this genus on 2 species, *S. loranthi* from the Molucca Is, and an unnamed species from New Guinea on *Loranthus rigidiflorus*. This latter species has not been seen for this work. More species probably remain to be discovered in this genus, which apparently originates in southern Asia.

Schizentaspidus silvicola sp. n. (Fig. 118)

Description
Scale of adult female circular, with central exuviae. Male scale not known.
Slide-mounted female 0.8-1.6 mm long; prepygidium broad, sclerotized, with constrictions and articulations present between mesothorax and metathorax, and metathorax and abdominal segment 1; pygidium narrower; becoming more sclerotized with maturity. Central area of pygidium with faint mozaic pattern; venter of pygidium anterior to vulva also with slight mozaic.
Pygidium produced, sometimes quite broad at base, with heavier sclerotization dorsally at the base of second and third lobes. Median lobes well developed, each at least as wide as long, notched on either side; both lobes separated by more than one third lobe width; with a sclerosis present at the base of each lobe, separated from lobe by a very narrow membranous strip of cuticle. Second and third lobes as long as median lobes, about two-thirds as wide, with rounded or slightly squared tips; second lobes notched on either side; third lobes notched on lateral margins. Plates as long as lobes; 2 present between median lobes, 2 between median and second lobes, 3 between second and third lobes; 8-9 plates present lateral to third lobe, reaching almost to dorsal marginal seta of fourth segment. Plates between lobes finely fringed; those present between third lobe and dorsal marginal seta of fifth segment more coarsely fringed; plates beyond this seta broad, with sclerotized dentate margins and 1 or a few long fleshy processes, sometimes furcate at mesal angles. Dorsal marginal seta situated at outer basal corner of each median lobe short and slender; those on second and third lobes shorter than lobes, lanceolate (broadened and flattened), dorsal marginal seta of fifth segment less so. Anus long, narrow oval, situated at posterior third of pygidium. Dorsal macroducts 20-30 times as long as wide; 1 marginal macroduct present between each pair of lobes, and 1 near dorsal marginal seta on either side of fifth segment. Submarginal macroducts in 3 sclerotized poriferous furrows running roughly parallel to margin on each side, 48-106 ducts present each side. Marginal macroducts present from anterior edge of segment 4 to segment 1 or 2, total 14-30. Dorsal microducts few; occasional short submarginal ducts present around margin, and some longer ducts scattered on abdominal segments. Dorsal submarginal bosses on segments 1 and 3 sclerotized.
Venter without perivulvar pores. Vulva large, wide, forming a shallow rounded V. Microducts moderate length, submarginal, most numerous on pygidium and abdominal segments 2 and 3; a few scattered on prosoma. An area of dermal granulation, a smooth tubercle and 1 microduct associated with each moderately sclerotized anterior spiracle. Antennal bases small, sclerotized, each with 3 slight spurs and 1 seta.

Material examined
Holotype female. **PAPUA NEW GUINEA.** Morobe P. coast, Buso, lower garden area, on *Myristica* sp. gall, 5.xi.1979 (*J. H. Martin*) (BMNH).
Paratypes female. **PAPUA NEW GUINEA.** Same data as holotype, 1 (BMNH); Morobe P. coast, Buso, on *Salacea* sp. 19.x.1979 (*J.H. Martin*), 2 (BMNH); same locality data, on ?*Schefflera* sp., 3.xi.1979 (*J. H. Martin*), 1 (BMNH); E.H.P., Kratke Mts, Kassam, 1350m., on undetermined Sapotaceae, 5.xi.1959 (*I. Maa*) 1 (BPBM). **SOLOMON IS.** G.P., Guadalcanal, Rua Vatu, on *Pandanus* sp. fruits, 5.iv.1955 (*E.S. Brown*), 8 (BMNH), 1 (USNM), 1 (DSIR), 1 (ARSDC).

Comments
The prosomal constrictions, thoracic tubercle, submarginal band of pygidial macroducts, sclerotization and large vulva place this species near *Schizentaspidus loranthi* Mamet. *S. silvicola*

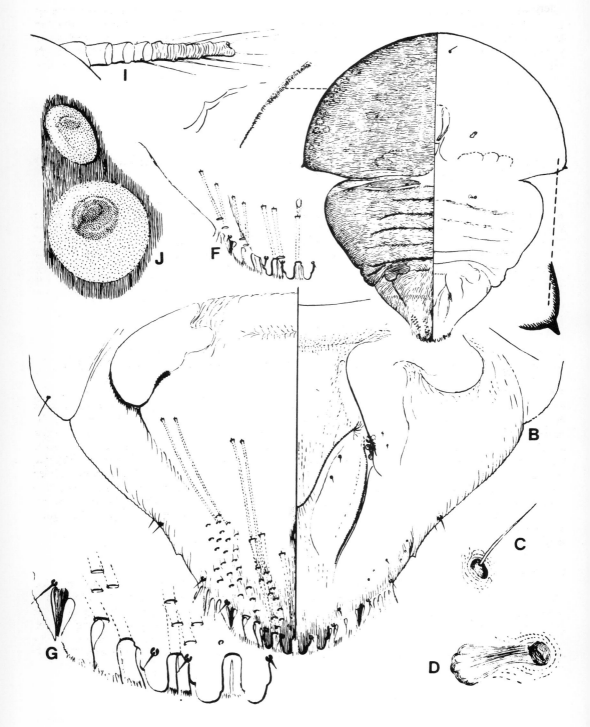

Fig. 119. *Selenaspidus articulatus* (Morgan).

shows some similarities to *Octaspidiotus*, because there is a clear strip of cuticle at the base of the median lobes, and broad sclerotized dentate plates with fleshy processes reaching from the dorsal marginal seta of the fifth segment to the edge of the fourth segment. *S. silvicola* is probably endemic and occurs on plants in the rain forest, particularly in the canopy, in small numbers; it clearly has the ability to occasionally form larger colonies on cultivated plants like *Pandanus*, however.

Genus **SELENASPIDUS** Cockerell

Selenaspidus Cockerell, 1897a: 14; Ferris, 1938: 264; Mamet, 1958: 362. Type-species *Aspidiotus articulatus* Morgan, by original designation.

All species in this genus have elongate 1-barred ducts, fringed plates and 3 pairs of lobes, the third pair conspicuously spur-like. Characteristic of the genus is the constriction between the mesothorax and metathorax. *Schizentaspidus*, a closely related genus, differs in having all 3 lobes the same shape, and in possessing a constriction between the metathorax and first abdominal segment.

Nearly 30 species are recognized, mostly from Africa, and the genus was revised by Mamet (1958). The only species known from the South Pacific area is *S. articulatus* (Morgan), which can be identified from the key to genera.

Selenaspidus articulatus (Morgan) (Fig. 119)

Aspidiotus articulatus Morgan, 1889: 352.
Selenaspidus articulatus (Morgan), Fernald, 1903: 284; Ferris, 1938: 265; Mamet, 1958: 370.

Material examined
FIJI. Vanua Levu, 1977; Viti Levu, 1956. On *Acalypha tricolor, Anacardium occidentale, Annona muricata, A. squamosa, Artocarpus heterophyllus, Arundina bambusifolia, Averrhoa bilimbi, A. carambola, Barringtonia asiatica, Calathea* sp., *Calophyllum inophyllum, Camellia sinensis, Cassia alata, Citrus* sp., *C. aurantifolia, C. aurantium, C. limon, C. maxima, C. reticulata, C. sinensis, Cocos nucifera, Codiaeum* sp., *C. variegatum, Coffea arabica, C. canephora, Decaspermum* sp., *Eugenia* sp., *Excoecaria* sp., *Fortunella japonica, Gardenia* sp., *G. scandens, Gliricidia sepium, Hevea brasiliensis, Hibiscus syriacus, Ixora coccinea, Lagerstroemia indica, Litchi chinensis, Mangifera indica, Nephelium* sp., *Passiflora edulis, Persea americana, Phoenix dactylifera, Plumeria rubra, Saccharum officinarum, Spondias dulcis, Tamarindus* sp., *T. indica, Thespesia propulnea*, 'uiniglu', *Vitis vinifera, Xanthosoma sagittifolium.*
SOLOMON IS. G.P.: Guadalcanal, 1956. On *Annona muricata, Cocos nucifera.*

Comments
This is a tropicopolitan species, and the only species in the genus to have a wide distribution. It has been recorded on numerous species of plants. Dumbleton (1954) recorded it from Fiji on *Citrus* sp; Swaine (1971) described it as a minor pest on many hosts including *Camellia sinensis*, but reported that it thrives best on citrus leaves and fruit, rapidly drying the tissues and sometimes killing the tree; Maddison (1976) classed it as a major pest of citrus there. It has also been recorded from Fiji on *Bauhinia variegata, Lantana camara* and *Musa* spp. (Hinckley, 1965).

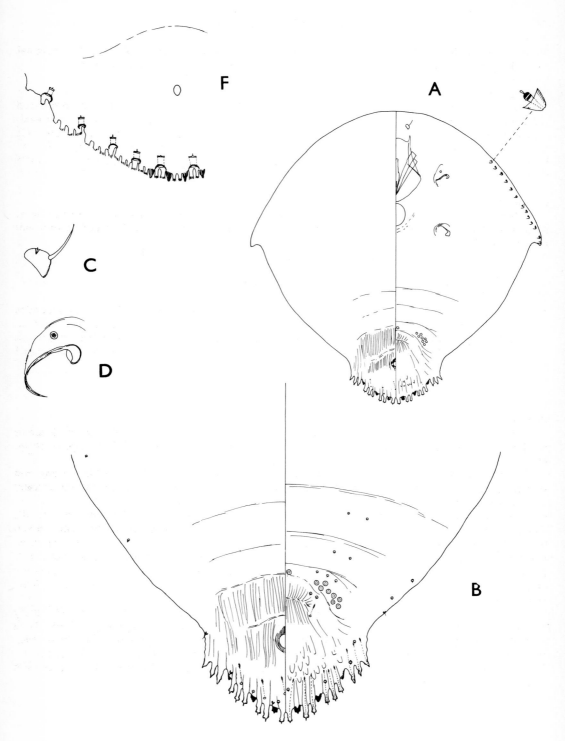

Fig. 120. *Silvestraspis ficaria* sp. n.

Genus **SILVESTRASPIS** Bellio

Silvestraspis Bellio, 1929: 159. Type-species *Silvestraspis sinensis* Bellio, by original designation. Later found to be a synonym of *Cryptoparlatorea uberifera* Lindinger.

Description

A pupillarial genus. Adult female often wider than long, with thorax prolonged postero-laterally into lobes. Pygidium with 3 pairs of lobes, small and single, well separated. Plates longer than lobes, swollen apically. Dorsal pygidial ducts represented by a few microducts on margins. Anus situated at centre of pygidium. Perivulvar pores present in 3-5 groups. Second instars with fringed plates and 2-barred ducts, the duct orifices surrounded by sclerotized lunate rims.

Comments

This genus belongs to the *Parlatoria* group of genera and is probably of southern Asian origin. Apart from the type-species, another species is described here as new from the Solomon Islands and Papua New Guinea.

Key to species of *Silvestraspis*

1 Lateral lobes of thorax poorly developed, body usually longer than wide. Duct tubercles absent from lateral to anterior spiracles .... *ficaria* sp. n.
   Lateral lobes of thorax well developed, body often wider than long. Duct tubercles present in groups lateral to anterior spiracles .... *uberifera* (Lindinger)

Silvestraspis ficaria sp. n. (Fig. 120)

Description

Scale of female oval, about 0.75 mm long, convex, yellow-brown, shiny except for dark centre patch; exuviae of first instar terminal. Male scale smaller and flatter than female scale, elongate.

Adult female, on slide, small, about 0.45 mm long, but usually about as wide as long and sometimes wider; membranous except for lightly sclerotized pygidium. Thorax prolonged laterally into weakly developed lobes.

Pygidium with 3 pairs of lobes. Median lobes well separated, each pointed at apex, small, about as wide as long. Second and third lobes small, triangular. Plates swollen apically, much longer than lobes, present in pairs between median lobes and between each median and second lobe, but 3 present between second and third lobes; more pointed lateral to third lobes, sometimes short. Dorsal ducts slender, there being 1 present between median plates, 1 at base of plates between median and second lobes, and 1 at base of centre plate between second and third lobes. Anus situated in centre of pygidium.

Ventral surface with 3 groups of perivulvar pores present, a median group of 1-2, rarely absent, and lateral groups of 7-10 pores. Microducts reduced in number to 1 or 2 on pygidium. Duct tubercles present in a more or less single row on thorax, terminating posteriorly at tip of each lateral lobe. Antennae each with 1 long seta. Anterior spiracles each usually with 1 disc pore.

Second instar with 2 pairs of recognizable lobes. Plates fringed. Marginal ducts, each with lunate orifice, around margins apparently forward to thorax.

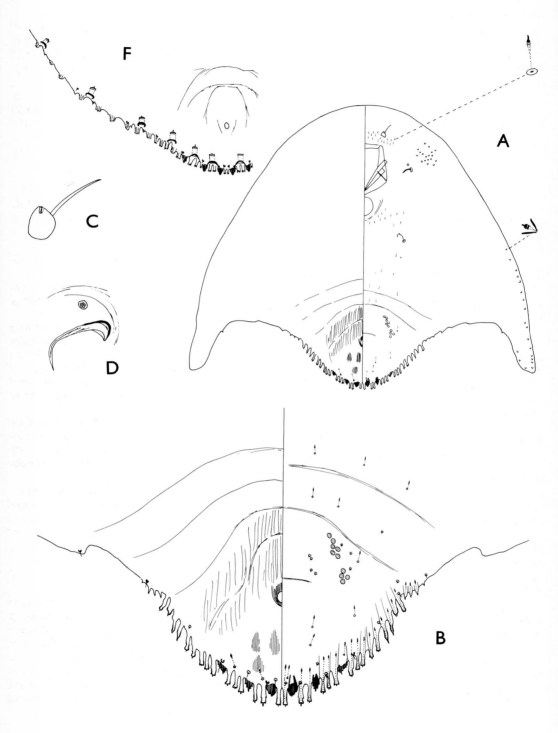

Fig. 121. *Silvestraspis uberifera* (Lindinger).

Material examined
Holotype female. **SOLOMON IS.** Guadalcanal, Tenaru, on leaves *Ficus glandifera*, 21.ii.1984 (*M. Bigger*) (BMNH).
Paratypes female. **SOLOMON IS.** Same data as holotype. 2 (BMNH), 1 (ARSDC). **PAPUA NEW GUINEA.** C.P., Port Moresby, on *Ficus* sp., 3.ix.1959 (*T. Maa*). 8 (BPBM), 4 (BMNH), 2 (DSIR).

Comments
The main differences between this species and *S. uberifera* are the poorly developed lateral lobes on the thorax, and the absence of gland spines lateral to the anterior spiracles.

Silvestraspis uberifera (Lindinger) (Fig. 121)

Cryptoparlatorea uberifera Lindinger, 1911: 126.
Silvestraspis sinensis Bellio, 1929: 160; Ferris, 1941a: 20. [Synonymized by Takahashi 1942: 46].
Silvestraspis uberifera (Lindinger), Takahashi, 1942: 46; Takagi, 1969: 47.

Description
Scale of pupillarial female about 0.75 mm long, pale yellow; exuviae of first instar terminal. Male scale smaller than female scale, elongate.

Adult female, slide-mounted, about 0.6 mm wide and 0.5 mm long; wider than long, with thorax produced postero-laterally into long lobe-like processes; membranous except for lightly sclerotized pygidium.

Pygidium with 3 pairs of lobes. Median pair pointed, triangular, longer than wide, each with 2-3 notches on each side. Second lobes narrower than median lobes, elongate. Third lobes wider than long. Plates longer than lobes, swollen apically, there being 2 present between median lobes; 2 between median and second lobes; 3 between second and third lobes, and a series lateral to third lobes, gradually shorter and more pointed further forward. Dorsal ducts slender, represented usually by 2 pairs present on margins between median and third lobes. Anus situated at about centre of pygidium.

Ventral surface with perivulvar pores present in 4 groups. Microducts few, located on submargins of pygidium; present also in median areas of free abdominal segments, between anterior spiracles and between antennae. Duct tubercles present in a conspicuous group lateral to each anterior spiracle, and in a marginal row on each thoracic process. Anterior spiracles each usually with a single disc pore.

Second instar becoming progressively sclerotized before moulting, with fringed plates and 3 pairs of lobes. Marginal macroducts numbering 6 on each side of pygidium, each with sclerotized lunate orifice.

Material examined
IRIAN JAYA. BIAK, 1959. On leaves of herbaceous vine.

Comments
This species was described originally from Celebes (now Sulawesi) and the Philippines. It is also known from China, including Taiwan and Hong Kong, and from Cambodia. Known host-plants are *Artocarpus* sp., *Mallotus philippinensis*, *Cinnamomum zeylanicum* and *Eugenia jambos*. Its distribution is probably much wider, and the record from Irian Jaya above suggests that it will be found in other areas of the tropical South Pacific area.

The specimens from Irian Jaya differ from the descriptions and illustrations given by Ferris (1941a) and Takagi (1969) in possessing more numerous plates beyond the third lobes, but for the present they are regarded as the same species. There have been available a few specimens of *Silvestraspis sinensis* labelled 'co-type', collected in China.

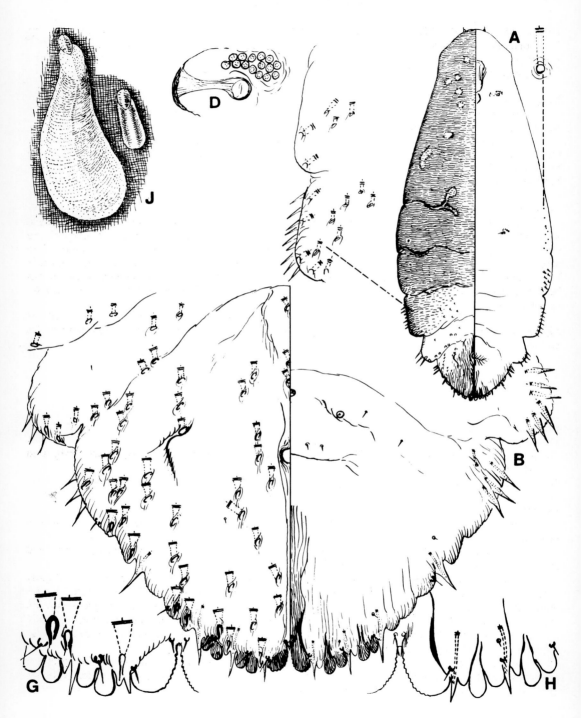

Fig. 122. *Unaspis citri* (Comstock).

Genus **UNASPIS** MacGillivray

Unaspis MacGillivray, 1921: 308; Rao, 1949: 59. Type-species *Chionaspis acuminata* Green, by original designation.
Prontaspis MacGillivray, 1921: 311. Type-species *Chionaspis citri* Comstock by original designation. [Synonymised by Ferris, 1936: 26].

Description
Female scale brownish, elongate, usually with a median longitudinal ridge; exuviae terminal. Male scale white, felted, elongate.

The important characters are 2-barred ducts; dorsal ducts numerous on pygidium posteriorly to eighth (median lobe) segment; 3 pairs of lobes present, the median lobes not zygotic, the second and third lobes bilobed. Gland spines absent from between median lobes but present laterally. Perivulvar pores present or absent.

Comments
About 15 species are now recognized in this genus, and most have been described from southern Asia, China and Japan. *U. citri* has become widely distributed on citrus throughout many of the citrus-growing areas of the world, and is the only one recorded from the South Pacific area. The arrowhead scale, *U. yannonensis* (Kuwana), causes considerable damage to citrus, and although it is known to occur in China and Japan, it has also become established in the Mediterranean area. It has been intercepted at quarantine inspection in Hawaii from the South Pacific region, but there are no definite records.

The genus differs from *Chionaspis* in having the median lobes separated at the base, and not zygotic; furthermore, the dorsal ducts of the pygidium are numerous, and occupy all segments to the eighth.

Unaspis citri (Comstock) (Fig. 122)

Chionaspis citri Comstock, 1883: 100.
Prontaspis citri (Comstock), MacGillivray, 1921: 311.
Dinaspis veitchi Green & Laing, 1923: 123. Lectotype female, Fiji (BMNH), here designated [examined].
Unaspis citri (Comstock), Ferris, 1936: 26; Ferris 1937a: 129; Balachowsky, 1954: 290.

Description
Female scale elongate, brown, flattened, but with a median ridge. Male scale white, with a median ridge.

Adult female, slide-mounted, elongate, fusiform; mature specimens with head, thorax and anterior abdominal segments heavily sclerotized.

Pygidium with 3 pairs of prominent lobes. Median lobes close together at base, slightly recessed into apex of pygidium, well developed; appearing divergent because inner margins longer than outer margins. Second and third lobes bilobed, rounded. Gland spines arranged singly on pygidial margins, but more numerous forward to segment 2. Marginal macroducts numbering 6 or 7 pairs, only a little larger than other pygidial dorsal ducts; smaller forward to segment 1. Pygidial ducts totalling 60-75.

Ventral surface with perivulvar pores present or absent; when present, usually numbering 1-3 at each side. Duct tubercles present on metathorax and abdominal segment 1. Anterior spiracles each with 8-15 disc pores; posterior spiracles each with 3-5 pores.

Material examined
COOK IS. Aitutaki, 1975; Atiu, 1970; Mangaia, 1975; Rarotonga, 1937. On *Citrus* sp., *C. aurantifolia, C. limon, C. maxima, C. paradisi, C. reticulata, C. sinensis, Musa* sp.
FIJI. Lau Is, Lakeba, 1945; Taveuni, 1944; Vanua Levu, 1975; Viti Levu, 1914. On *Citrus* sp., *C. limon, C. maxima, C. paradisi, C. reticulata, C. sinensis, Psidium guajava.*
KIRIBATI. Abemama, 1976. On *C. aurantifolia.*
NEW CALEDONIA. New Caledonia, 1928. On *Citrus* sp.
NIUE. 1954. On *Annona muricata, Citrus* sp., *C. aurantifolia, C. aurantium, C. limon, C. paradisi, C. reticulata, C. sinensis.*
PAPUA NEW GUINEA. C.P.: 1960. E.N.B.P.: 1940. M.B.P.: 1985; Louisiade Archipelago, Misima, 1985; Trobriand Is, Kiriwina, 1985. Morobe P.: 1959. S.H.P.: 1983. W.N.B.P.: 1984. W.P.: 1985. On *Citrus* sp., *C. aurantifolia, C. aurantium.*
SOLOMON IS. C.I.P.: Bellona, 1974; Russell Is, Mbanika, 1935. G.P.: Guadalcanal, 1932. W.P.: New Georgia Is, Kolombangara, 1980. On *Citrus* sp., *C. aurantifolia, C. paradisi.*
TONGA. Tongatapu Group, 'Eua, 1974, Tongatapu, 1973. On *C. aurantium, C. limon, C. maxima, C. reticulata.*
VANUATU. Efate, 1925; Espiritu Santo, 1925; Malekula, 1983. On *Citrus* sp., *C. aurantifolia, C. limon, C. maxima,* causing yellow spots on leaves; *C. paradisi.*
WESTERN SAMOA. Manono, 1977; Savai'i, 1975; Upolu, 1970. On *C. aurantifolia, C. grandis, C. limon, C. maxima, C. paradisi, C. reticulata, C. sinensis.*

Comments
Although this species has been recorded from a few plant families, it seems to prefer the family Rutaceae, especially citrus. It is usually common throughout citrus plantations, mainly on the leaves and fruit, wherever it is introduced.

Balachowsky (1954) mentions *U. citri* being injurious to citrus in Fiji, Samoa and New Caledonia. Lever (1933) records it from Solomon Is causing enough damage to citrus to be worth spraying, while Stapley (1976) describes it as a major pest of *Citrus* spp. and *Capsicum* spp. there. Maddison (1976) classes it as a major pest in Tonga on *C. aurantifolia, C. grandis* and *C. sinensis,* and on citrus in Wallis Is. In addition to the above material, *U. citri* has been recorded from Cook Is on *C. grandis* (Maddison, 1976) and *Artocarpus heterophyllus* (Valentine, 1975); from Fiji on *C. aurantifolia* and *C. grandis* (Lever, 1945a, 1945b); from New Caledonia on *Citrus aurantium, C. grandis, C. limon, C. paradisi, C. reticulata* by Cohic (1956), who noted it was often associated with the fungi *Fusarium coccidicola, Septobasidium alni* and *S. bogoriense* there; from New Guinea on *Cocos nucifera* (Froggatt, 1940); and from Vanuatu, Cohic (1950b) and Maddison (1976) record it as the only pest of citrus of any importance, and Cohic (1953) mentions it on *Citrus reticulata.* Bark splitting of citrus, as discussed in the introduction, has been reported from Fiji by Swaine (1971), from Niue by Meister (1975) and from Papua New Guinea (Anon, 1969; Brough, 1986). *U. citri* has been intercepted at quarantine inspection in New Zealand on *Citrus* fruit from the South Pacific area.

Systematic list of host-plants of the tropical South Pacific Diaspididae

The plant classification follows the system of Parker (1982), with reference to Cronquist (1981), Willis (1973) and Fosberg *et al.* (1979).

FILICOPHYTA

ADIANTACEAE

ADIANTUM
  Pinnaspis aspidistrae

ADIANTUM AETHIOPICUM
  Pinnaspis buxi

ASPLENIACEAE

ASPLENIUM NIDUS
  Aspidiotus maddisoni
  Aspidiotus nerii
  Pinnaspis strachani

TMESIPTERIDACEAE

TMESIPTERIS TANNENSIS
  Aspidiotus nerii

PINOPHYTA (Gymnospermae)

ARAUCARIACEAE

AGATHIS LANCEOLATA
  Chrysomphalus aonidum

AGATHIS MOOREI
  Chrysomphalus aonidum

ARAUCARIA
  Carulaspis giffardi
  Genaparlatoria araucariae
  Lepidosaphes rubrovittata
  Oceanaspidiotus araucariae

ARAUCARIA BRASILIANA
  Lindingaspis rossi

ARAUCARIA COOKII
  Carulaspis giffardi
  Oceanaspidiotus araucariae

ARAUCARIA CUNNINGHAMII
  Carulaspis giffardi

ARAUCARIA HETEROPHYLLA
  Carulaspis giffardi
  Oceanaspidiotus araucariae

ARAUCARIA HUNSTEINII
  Lepidosaphes rubrovittata
  Octaspidiotus australiensis

CUPRESSACEAE

CUPRESSUS MACROCARPA
  Carulaspis minima

CYCADACEAE

CYCAS
  Aonidiella aurantii
  Aspidiotus destructor
  Chrysomphalus aonidum
  Chrysomphalus dictyospermi
  Fiorinia fioriniae
  Hemiberlesia lataniae
  Hemiberlesia palmae
  Pinnaspis strachani

CYCAS CIRCINALIS
  Hemiberlesia palmae

CYCAS RUMPHII
  Aonidiella aurantii

PINACEAE

PINUS
  Chrysomphalus aonidum

PINUS CARIBAEA
  Chrysomphalus aonidum
  Chrysomphalus dictyospermi
  Hemiberlesia palmae
  Lopholeucaspis cockerelli

PODOCARPACEAE

PODOCARPUS GNIDIOIDES
  Aonidia longa

MAGNOLIOPHYTA (Angiospermae)

MAGNOLIOPSIDA (Dicotyledonae)

ACANTHACEAE

GRAPTOPHYLLUM PICTUM
  Hemiberlesia palmae

JUSTICIA
  Ischnaspis longirostris
  Pinnaspis strachani

AMARANTHACEAE

GOMPHRENA GLOBOSA
  Hemiberlesia lataniae

ANACARDIACEAE

ANACARDIUM OCCIDENTALE
  Chrysomphalus dictyospermi
  Hemiberlesia palmae
  Pinnaspis strachani
  Pseudaonidia trilobitiformis
  Selenaspidus articulatus

MANGIFERA INDICA
  Abgrallaspis cyanophylli
  Aspidiotus destructor
  Chrysomphalus dictyospermi
  Clavaspis herculeana
  Fiorinia proboscidaria
  Hemiberlesia palmae
  Ischnaspis longirostris
  Morganella longispina
  Pseudaonidia trilobitiformis
  Pseudaulacaspis cockerelli
  Pseudaulacaspis pentagona
  Selenaspidus articulatus

SCHINUS TEREBINTHIFOLIUS
  Lindingaspis rossi
  Pseudaonidia trilobitiformis

SPONDIAS DULCIS
  Aspidiotus destructor
  Chrysomphalus dictyospermi
  Hemiberlesia palmae
  Selenaspidus articulatus

ANNONACEAE

ANNONA
  Abgrallaspis cyanophylli
  Hemiberlesia palmae
  Pseudaonidia trilobitiformis
  Unaspis citri

ANNONA CHERIMOLIA
  Aspidiotus destructor

ANNONA MURICATA
  Abgrallaspis cyanophylli
  Aspidiotus destructor
  Chrysomphalus aonidum
  Hemiberlesia palmae
  Howardia biclavis
  Parlatoria cinerea
  Pinnaspis strachani
  Selenaspidus articulatus
  Unaspis citri

ANNONA RETICULATA
  Aspidiotus destructor
  Chrysomphalus aonidum
  Hemiberlesia palmae
  Pinnaspis strachani

ANNONA SQUAMOSA
  Abgrallaspis cyanophylli
  Aspidiotus destructor
  Chrysomphalus aonidum
  Hemiberlesia palmae
  Howardia biclavis
  Pseudaonidia trilobitiformis
  Selenaspidus articulatus

CANANGA ODORATA
  Pinnaspis strachani

APIACEAE

DAUCUS CAROTA
  Fiorinia proboscidaria

APOCYNACEAE

ALLEMANDA
  Aonidiella inornata
  Pseudaulacaspis pentagona

ALLEMANDA CATHARTICA
　Aonidiella inornata
　Howardia biclavis
　Pseudaulacaspis pentagona

ALLEMANDA HENDERSONI
　Aspidiotus destructor

ALLEMANDA SCHOTTII
　Howardia biclavis

CATHARANTHUS ROSEUS
　Pseudaonidia trilobitiformis
　Pseudaulacaspis pentagona

CERBERA OPPOSITIFOLIA
　Pseudaonidia trilobitiformis

ERVATAMIA CORONARIA
　Parlatoria proteus

ERVATAMIA ORIENTALIS
　Howardia biclavis
　Pseudaulacaspis pentagona

NERIUM OLEANDER
　Chrysomphalus aonidum
　Hemiberlesia lataniae
　Pseudaonidia trilobitiformis
　Pseudaulacaspis pentagona

PLUMERIA
　Chrysomphalus aonidum
　Chrysomphalus dictyospermi
　Hemiberlesia palmae
　Pinnaspis strachani

PLUMERIA ACUTIFOLIA
　Abgrallaspis cyanophylli
　Aspidiotus destructor
　Aspidiotus nerii
　Howardia biclavis
　Pinnaspis aspidistrae
　Pinnaspis strachani
　Pseudaonidia trilobitiformis
　Pseudaulacaspis pentagona

PLUMERIA RUBRA
　Aonidiella inornata
　Aspidiotus destructor
　Chrysomphalus aonidum
　Hemiberlesia lataniae
　Hemiberlesia palmae
　Ischnaspis longirostris
　Lepidosaphes tapleyi
　Pinnaspis strachani
　Pseudaonidia trilobitiformis
　Pseudaulacaspis pentagona
　Selenaspidus articulatus

ARALIACEAE

CHEIRODENDRON MARQUESENSE
　Lepidosaphes marginata

MERYTA MACROPHYLLA
　Chrysomphalus dictyospermi

REYNOLDSIA MARCHIONENSIS
　Lepidosaphes marginata

SCHEFFLERA
　Chrysomphalus dictyospermi
　Hemiberlesia lataniae
　Lopholeucaspis cockerelli
　Pinnaspis strachani
　Schizentaspidus silvicola

SCHEFFLERA ACTINOPHYLLA
　Hemiberlesia lataniae

ASTERACEAE

GENUS INDET.
　Lepidosaphes rubrovittata
　Lepidosaphes tapleyi

CHRYSANTHEMUM
　Hemiberlesia lataniae

ELEPHANTOPUS SCABER
　Hemiberlesia lataniae

FITCHIA
　Hemiberlesia lataniae
　Pinnaspis strachani

GERBERA
　Chrysomphalus aonidum

GERBERA JAMESONI
　Chrysomphalus aonidum

HELIANTHUS
　Pseudaulacaspis pentagona

HELIANTHUS ANNUUS
　Pseudaulacaspis pentagona

PLUCHEA ODORATA
Aonidiella comperei

BEGONIACEAE

BEGONIA
Hemiberlesia lataniae

BIGNONIACEAE

CRESCENTIA CUJETE
Pseudaonidia trilobitiformis

KIGELIA PINNATA
Hemiberlesia palmae

PYROSTEGIA VENUSTA
Pseudaonidia trilobitiformis

TECOMA STANS
Morganella longispina

BIXACEAE

BIXA ORELLANA
Hemiberlesia palmae
Howardia biclavis
Parlatoria proteus

BOMBACACEAE

CEIBA PENTANDRA
Abgrallaspis cyanophylli
Aspidiotus destructor
Hemiberlesia lataniae
Hemiberlesia palmae

DURIO ZIBETHINUS
Aulacaspis vitis

BORAGINACEAE

CORDIA
Hemiberlesia lataniae

CORDIA SUBCORDATA
Oceanaspidiotus pangoensis

TOURNEFORTIA
Pinnaspis strachani

TOURNEFORTIA ARGENTEA
Parlatoria pergandii

Pinnaspis strachani
Pseudaulacaspis pentagona

BRASSICACEAE

BRASSICA CHINENSIS
Aspidiotus destructor

BRASSICA NAPUS
Aspidiotus destructor

BRASSICA OLERACEA
Aspidiotus destructor

RAPHANUS SATIVUS
Aspidiotus destructor

BURSERACEAE

CANARIUM COMMUNE
Pinnaspis buxi

CACTACEAE

OPUNTIA
Diaspis echinocacti

CAESALPINIACEAE

BAUHINIA
Abgrallaspis cyanophylli
Hemiberlesia palmae
Morganella longispina
Pinnaspis strachani

BAUHINIA MONANDRA
Hemiberlesia palmae

BAUHINIA VARIEGATA
Chrysomphalus aonidum
Morganella longispina
Pseudaonidia trilobitiformis
Selenaspidus articulatus

CAESALPINIA PULCHERRIMA
Pinnaspis strachani

CASSIA
Aonidiella inornata
Aspidiotus destructor
Hemiberlesia lataniae
Pseudaulacaspis pentagona

CASSIA ALATA
  Pinnaspis strachani
  Pseudaulacaspis pentagona
  Selenaspidus articulatus

CASSIA NODOSA
  Aspidiotus destructor

CASSIA OCCIDENTALIS
  Aspidiotus destructor
  Pinnaspis strachani

CASSIA TORA
  Aspidiotus destructor

TAMARINDUS
  Selenaspidus articulatus

TAMARINDUS INDICA
  Chrysomphalus aonidum
  Selenaspidus articulatus

CARICACEAE

CARICA PAPAYA
  Aonidiella orientalis
  Aspidiotus destructor
  Aspidiotus excisus
  Aspidiotus macfarlanei
  Chrysomphalus dictyospermi
  Howardia biclavis
  Morganella longispina
  Pseudaonidia trilobitiformis
  Pseudaulacaspis cockerelli
  Pseudaulacaspis pentagona

CASUARINACEAE

CASUARINA
  Chrysomphalus dictyospermi
  Diaspis casuarinae
  Hemiberlesia lataniae
  Hemiberlesia palmae
  Pseudaulacaspis papulosa

CASUARINA COLLINA
  Diaspis casuarinae

CHRYSOBALANACEAE

PARINARI LAURINUM
  Hemiberlesia lataniae
  Howardia biclavis

CLUSIACEAE

CALOPHYLLUM
  Hemiberlesia palmae
  Howardia biclavis
  Pseudaulacaspis pentagona

CALOPHYLLUM INOPHYLLUM
  Aspidiotus destructor
  Chrysomphalus aonidum
  Chrysomphalus dictyospermi
  Hemiberlesia palmae
  Lopholeucaspis cockerelli
  Pinnaspis strachani
  Pseudaonidia trilobitiformis
  Selenaspidus articulatus

COMBRETACEAE

TERMINALIA
  Pinnaspis strachani

TERMINALIA CALAMANSANAY
  Pinnaspis strachani

TERMINALIA CATAPPA
  Chrysomphalus dictyospermi
  Howardia biclavis
  Lopholeucaspis cockerelli
  Pinnaspis aspidistrae
  Pinnaspis strachani

TERMINALIA COMPLANATA
  Pinnaspis strachani

CONVOLVULACEAE

ARGYREIA NERVOSA
  Pseudaulacaspis pentagona

IPOMOEA
  Pseudaulacaspis pentagona

IPOMOEA BATATAS
  Aspidiella hartii
  Oceanaspidiotus nendeanus

CRASSULACEAE

KALANCHOE
  Pseudaulacaspis pentagona

CUCURBITACEAE

GENUS INDET.
Aspidiotus destructor
Pinnaspis strachani

CUCUMIS SATIVUS
Aspidiotus destructor
Pinnaspis strachani

CUCURBITA MAXIMA
Pinnaspis strachani

CUCURBITA PEPO
Aonidiella aurantii

MOMORDICA
Pinnaspis buxi

MOMORDICA CHARANTIA
Pinnaspis buxi

DILLENIACEAE

DILLENIA BIFLORA
Aspidiotus destructor

EBENACEAE

DIOSPYROS
Aonidiella comperei
Aspidiotus nerii
Oceanaspidiotus spinosus

DIOSPYROS KAKI
Ischnaspis longirostris
Pinnaspis strachani

ELAEOCARPACEAE

ELAEOCARPUS TONGANUS
Hemiberlesia lataniae

ELAEODENDRON CURTIPENDULUM
Lindingaspis rossi

EPACRIDACEAE

DRACOPHYLLUM
Aspidiotus cochereaui

DRACOPHYLLUM RAMOSUM
Aspidiotus cochereaui

LEUCOPOGON
Chrysomphalus aonidum

EUPHORBIACEAE

GENUS INDET.
Pseudaulacaspis pentagona

ACALYPHA GODSEFFIANA
Pinnaspis aspidistrae

ACALYPHA HISPIDA
Abgrallaspis cyanophylli
Hemiberlesia palmae

ACALYPHA TRICOLOR
Selenaspidus articulatus

ALEURITES FORDII
Hemiberlesia lataniae

ALEURITES MOLUCCANA
Aspidiotus destructor
Chrysomphalus aonidum
Hemiberlesia lataniae
Hemiberlesia palmae
Lepidosaphes stepta
Lopholeucaspis cockerelli
Pinnaspis strachani
Pseudaonidia trilobitiformis

ALEURITES MONTANA
Pseudaonidia trilobitiformis

ALEURITES TRILOBA
Aspidiotus destructor

ANNESIJOA
Aonidiella comperei
Aonidiella inornata

BISCHOFIA JAVANICA
Aonidiella eremocitri
Aonidiella inornata

BREYNIA DISTICHA
Aspidiotus destructor

CODIAEUM
Cryptophyllaspis ruebsaameni
Lepidosaphes gloverii
Lepidosaphes tokionis
Parlatoria crotonis
Pseudaonidia trilobitiformis

Pseudaulacaspis pentagona
Selenaspidus articulatus

CODIAEUM VARIEGATUM
[?]*Lepidosaphes beckii*
Lepidosaphes tokionis
Parlatoria crotonis
Selenaspidus articulatus

CROTON
Lepidosaphes stepta
Parlatoria crotonis

EUPHORBIA
Aonidiella inornata
Aspidiotus destructor
Aspidiotus excisus
Chrysomphalus aonidum

EUPHORBIA HETEROPHYLLA
Hemiberlesia palmae
Odonaspis ruthae
Pinnaspis strachani

EUPHORBIA PULCHERRIMA
Aspidiotus destructor

EXCOECARIA
Selenaspidus articulatus

EXCOECARIA AGALLOCHA
Pinnaspis strachani

HEVEA BRASILIENSIS
Abgrallaspis cyanophylli
Aspidiotus destructor
Hemiberlesia palmae
Pinnaspis buxi
Selenaspidus articulatus

JATROPHA CURCAS
Abgrallaspis cyanophylli

MACARANGA
Chrysomphalus dictyospermi
Howardia biclavis
Pseudaulacaspis pentagona

MACARANGA SEEMANNII
Aspidiotus destructor

MANIHOT ESCULENTA
Abgrallaspis cyanophylli
Chrysomphalus dictyospermi

Hemiberlesia lataniae
Pinnaspis strachani
Pseudaulacaspis pentagona

PEDILANTHUS
Pinnaspis strachani

PEDILANTHUS TITHYMALOIDES
Pseudaulacaspis pentagona

RICINUS COMMUNIS
Aonidiella aurantii
Chrysomphalus aonidum
Pinnaspis strachani
Pseudaulacaspis pentagona

FABACEAE

CANAVALIA
Hemiberlesia lataniae
Hemiberlesia palmae

CLITORIA TERNEATA
Pseudaonidia trilobitiformis

CROTALARIA
Pinnaspis strachani
Pseudaulacaspis pentagona

CROTALARIA MUCRONATA
Aspidiotus destructor

CROTALARIA SALTIANA
Aspidiotus destructor

CROTALARIA USARAMOENSIS
Pinnaspis strachani

DERRIS ELLIPTICA
Parlatoria proteus

ERYTHRINA
Lepidosaphes gloverii
Pseudaulacaspis pentagona

ERYTHRINA FASTIGIATA
Oceanaspidiotus caledonicus

ERYTHRINA FUSCA
Pinnaspis strachani

ERYTHRINA INDICA
Aonidiella aurantii
[?]*Aonidiella citrina*

Clavaspis herculeana
Pinnaspis aspidistrae
Pinnaspis strachani

ERYTHRINA LITHOSPERMA
Aonidiella aurantii
Fijifiorinia oconnori
Pinnaspis strachani

ERYTHRINA SUBUMBRANS
Pinnaspis strachani

GLIRICIDIA SEPIUM
Hemiberlesia palmae
Selenaspidus articulatus

GLYCINE MAX
Pseudaulacaspis pentagona

INDIGOFERA ANIL
Howardia biclavis

INOCARPUS
Lepidosaphes rubrovittata

INOCARPUS FAGIFER
Aspidiotus destructor
Ischnaspis longirostris
Lepidosaphes rubrovittata
Lopholeucaspis cockerelli
Parlatoria crotonis
Parlatoria pergandii
Pinnaspis aspidistrae
Pinnaspis buxi
Pinnaspis strachani

MUCUNA BENNETTII
Ischnaspis longirostris
Pseudaonidia trilobitiformis

MUNDULEA SUBEROSA
Howardia biclavis

PHASEOLUS VULGARIS
Pinnaspis buxi
Pinnaspis strachani
Pseudaulacaspis pentagona

SOPHORA TOMENTOSA
Pinnaspis strachani

VIGNA
Pseudaulacaspis pentagona

VIGNA UNGUICULATA
Aspidiotus destructor

FAGACEAE

NOTHOFAGUS AEQUILATERALIS
Chrysomphalus aonidum

NOTHOFAGUS CODONANDRA
Aspidiotus cochereaui

GERANIACEAE

PELARGONIUM
Pseudaulacaspis pentagona

GESNERIACEAE

SAINTPAULIA
Pinnaspis strachani

HERNANDIACEAE

HERNANDIA OVIGERA
Pinnaspis strachani

HERNANDIA PELTATA
Oceanaspidiotus pangoensis
Pinnaspis strachani

HERNANDIA SONORA
Lindingaspis similis

HIPPOCRATEACEAE

SALACEA
Aonidiella inornata
Schizentaspidus silvicola

LAMIACEAE

COLEUS
Abgrallaspis cyanophylli

OCIMUM GRATISSIMUM
Pinnaspis strachani

LAUREACEAE

CASSYTHA FILIFORMIS
Pinnaspis strachani

CINNAMOMUM ZEYLANICUM
Abgrallaspis cyanophylli
Aspidiotus destructor

LAURUS NOBILIS
Chrysomphalus aonidum
Pseudaonidia trilobitiformis

LITSEA VITIENSIS
Aspidiotus destructor

PERSEA
Pseudaonidia trilobitiformis

PERSEA AMERICANA
Abgrallaspis cyanophylli
Aspidiotus destructor
Chrysomphalus dictyospermi
Fiorinia fioriniae
Hemiberlesia lataniae
Hemiberlesia palmae
Lopholeucaspis cockerelli
Morganella longispina
Oceanaspidiotus spinosus
Pinnaspis strachani
Pseudaonidia trilobitiformis
Selenaspidus articulatus

LECYTHIDACEAE

BARRINGTONIA
Abgrallaspis cyanophylli
Aonidiella eremocitri
Aonidiella inornata
Aspidiotus destructor
Chrysomphalus dictyospermi
Hemiberlesia lataniae
Hemiberlesia palmae
Lepidosaphes rubrovittata
Lopholeucaspis cockerelli
Pinnaspis strachani

BARRINGTONIA ASIATICA
Aspidiotus destructor
Chrysomphalus aonidum
Hemiberlesia lataniae
Hemiberlesia palmae
Pinnaspis strachani
Pseudaonidia trilobitiformis
Selenaspidus articulatus

BARRINGTONIA BUTONICA
Pinnaspis strachani

BARRINGTONIA EDULIS
Pinnaspis buxi

BARRINGTONIA RACEMOSA
Chrysomphalus dictyospermi
Hemiberlesia palmae
Lopholeucaspis cockerelli

BARRINGTONIA SPECIOSA
Chrysomphalus aonidum

LOGANIACEAE

FAGRAEA
Chrysomphalus aonidum

GENIOSTOMA
Lepidosaphes geniostomae

STRYCHNOS NUX-VOMICA
Ischnaspis longirostris

LORANTHACEAE

LORANTHUS RIGIDIFLORUS
Schizentaspidus

LYTHRACEAE

LAGERSTROEMIA
Pseudaulacaspis pentagona

LAGERSTROEMIA FLOS-REGINAE
Chrysomphalus dictyospermi
Morganella longispina
Pseudaulacaspis pentagona

LAGERSTROEMIA INDICA
Selenaspidus articulatus

PEMPHIS ACIDULA
Hemiberlesia lataniae

MAGNOLIACEAE

ELMERRILLEA PAPUANA
Lepidosaphes elmerrilleae

MALVACEAE

GENUS INDET.
Howardia biclavis

ABUTILON
  Hemiberlesia lataniae
  Pinnaspis strachani

ABUTILON GRAVEOLENS
  Hemiberlesia lataniae

ABUTILON HYBRIDUM
  Pinnaspis strachani

ALTHAEA
  Pseudaulacaspis pentagona

GOSSYPIUM
  Clavaspis herculeana
  Hemiberlesia lataniae
  Lepidosaphes tokionis
  Pinnaspis strachani

GOSSYPIUM BRASILIENSE
  Pseudaulacaspis pentagona

HIBISCUS
  Aspidiotus destructor
  Hemiberlesia palmae
  Lepidosaphes beckii
  Lepidosaphes eurychlidonis
  Lepidosaphes tapleyi
  Odonaspis ruthae
  Pinnaspis strachani
  Pseudaulacaspis pentagona

HIBISCUS DIVERSIFOLIUS
  Pseudaulacaspis pentagona

HIBISCUS ESCULENTUS
  Pseudaulacaspis pentagona

HIBISCUS MANIHOT
  Pinnaspis strachani
  Pseudaulacaspis pentagona

HIBISCUS MUTABILIS
  Pseudaulacaspis pentagona

HIBISCUS ROSA-SINENSIS
  Andaspis numerata
  Howardia biclavis
  Morganella longispina
  Pinnaspis aspidistrae
  Pinnaspis strachani
  Pseudaulacaspis pentagona

HIBISCUS SYRIACUS
  Abgrallaspis cyanophylli
  Howardia biclavis
  Pinnaspis strachani
  Selenaspidus articulatus

HIBISCUS TILIACEUS
  Hemiberlesia palmae
  Pinnaspis strachani

MALVASTRUM TRICUSPIDATUM
  Pseudaulacaspis pentagona

MALVAVISCUS ARBOREUS
  Pinnaspis strachani

THESPESIA PROPULNEA
  Pinnaspis strachani
  Selenaspidus articulatus

URENA LOBATA
  Howardia biclavis
  Pinnaspis strachani
  Pseudaulacaspis pentagona

MELASTOMATACEAE

ASTRONIDIUM
  Fijifiorinia astronidii

MELIACEAE

CEDRELA TOONA
  Abgrallaspis cyanophylli
  Pseudaulacaspis pentagona

KHAYA GRANDIFOLIA
  Howardia biclavis
  Howardia stricklandi

MELIA AZEDERACH
  Aspidiotus nerii
  Hemiberlesia lataniae

SWIETENIA MACROPHYLLA
  Abgrallaspis cyanophylli
  Chrysomphalus dictyospermi
  Howardia biclavis
  Ischnaspis longirostris

MIMOSACEAE

ACACIA
  Hemiberlesia lataniae
  Pinnaspis strachani
  Pseudaulacaspis pentagona

ACACIA MELLIFERA
  Howardia biclavis

ACACIA SIMPLICIFOLIA
  Pseudaonidia trilobitiformis

ACACIA SPIRORBIS
  Howardia biclavis
  Pseudaonidia trilobitiformis

ALBIZIA FALCATARIA
  Andaspis hawaiiensis
  Hemiberlesia lataniae

ALBIZIA LEBBEK
  Aspidiotus destructor
  Pinnaspis strachani

ALBIZIA STIPULATA
  Howardia biclavis
  Pseudaulacaspis pentagona

LEUCAENA GLAUCA
  Hemiberlesia lataniae

LEUCAENA LEUCOCEPHALA
  Pinnaspis strachani

MIMOSA PUDICA
  Clavaspis herculeana
  Hemiberlesia lataniae
  Pinnaspis strachani

PITHECELLOBIUM DULCE
  Hemiberlesia lataniae

PITHECELLOBIUM SAMAN
  Clavaspis herculeana
  Hemiberlesia lataniae
  Hemiberlesia palmae

PROSOPIS INSULARUM
  Hemiberlesia lataniae

MORACEAE

ARTOCARPUS
  Hemiberlesia lataniae
  Pseudaonidia trilobitiformis

ARTOCARPUS ALTILIS
  Abgrallaspis cyanophylli
  Aonidiella aurantii
  Aspidiotus destructor
  Chrysomphalus aonidum
  Chrysomphalus dictyospermi
  Hemiberlesia palmae
  Pinnaspis strachani
  Pseudaonidia trilobitiformis

ARTOCARPUS HETEROPHYLLUS
  Chrysomphalus dictyospermi
  Hemiberlesia lataniae
  Hemiberlesia palmae
  Pinnaspis buxi
  Pinnaspis strachani
  Pseudaonidia trilobitiformis
  Selenaspidus articulatus
  Unaspis citri

ARTOCARPUS INCISA
  Aonidiella aurantii
  Aspidiotus destructor
  Chrysomphalus dictyospermi
  Pseudaonidia trilobitiformis

BROUSSONETIA
  Pseudaulacaspis pentagona

BROUSSONETIA PAPYRIFERA
  Aonidiella aurantii
  Oceanaspidiotus pangoensis
  Pseudaulacaspis pentagona

FICUS
  Abgrallaspis cyanophylli
  Andaspis spinosa
  Aonidiella comperei
  Aspidiotus destructor
  Chrysomphalus aonidum
  Chrysomphalus dictyospermi
  Duplaspidiotus claviger
  Hemiberlesia lataniae
  Ischnaspis longirostris
  Lepidosaphes rubrovittata
  Leucaspis bugnicourti
  Morganella longispina

Octaspidiotus australiensis
  Silvestraspis ficaria
  Pseudaonidia trilobitiformis

FICUS CARICA
  Hemiberlesia lataniae
  Morganella longispina

FICUS GLANDIFERA
  Hemiberlesia palmae
  Silvestraspis ficaria

FICUS GLANDULIFERA
  Hemiberlesia lataniae
  Hemiberlesia palmae

FICUS PUMILA
  Chrysomphalus aonidum
  Pseudaonidia trilobitiformis

FICUS TINCTORIA
  Chrysomphalus dictyospermi

MORUS
  Pseudaulacaspis pentagona

MORUS ALBA
  Howardia biclavis
  Pseudaulacaspis pentagona

MORUS INDICA
  Pseudaulacaspis pentagona

MYRISTICACEAE

MYRISTICA
  Chrysomphalus dictyospermi
  Schizentaspidus silvicola

MYRISTICA HYPARGYRAEA
  Clavaspis herculeana
  Hemiberlesia lataniae
  Lindingaspis samoana

MYRISTICA MACRANTHA
  Fijifiorinia oconnori
  Howardia biclavis

MYRTACEAE

BAECKEA PINIFOLIA
  Fiorinia neocaledonica

DECASPERMUM
  Hemiberlesia palmae
  Selenaspidus articulatus

EUCALYPTUS DEGLUPTA
  Aspidiotus destructor
  Chionaspis keravatana

EUCALYPTUS GRANDIS
  Hemiberlesia lataniae
  Pinnaspis strachani

EUGENIA
  Abgrallaspis cyanophylli
  Aspidiotus destructor
  Chrysomphalus aonidum
  Morganella longispina
  Selenaspidus articulatus

EUGENIA CUMINI
  Chrysomphalus aonidum

EUGENIA MALACCENSIS
  Aspidiotus destructor
  Chrysomphalus dictyospermi
  Hemiberlesia lataniae
  Hemiberlesia palmae
  Lepidosaphes rubrovittata
  Parlatoria proteus

MELALEUCA
  Aonidiella inornata
  Chrysomphalus aonidum

MELALEUCA LEUCADENDRON
  Chrysomphalus aonidum

MELALEUCA QUINQUENERVIA
  Chrysomphalus aonidum

PSIDIUM
  Abgrallaspis cyanophylli
  Pseudaonidia trilobitiformis

PSIDIUM CATTLEIANUM
  Hemiberlesia lataniae
  Hemiberlesia palmae
  Morganella longispina
  Pseudaonidia trilobitiformis

PSIDIUM GUAJAVA
  Abgrallaspis cyanophylli
  Aspidiotus destructor
  Chrysomphalus aonidum

Chrysomphalus dictyospermi
Hemiberlesia lataniae
Hemiberlesia palmae
Morganella longispina
Pseudaonidia trilobitiformis
Unaspis citri

SPERMOLEPIS GUMMIFERA
Chrysomphalus dictyospermi

NYCTAGINACEAE

BOUGAINVILLEA
Chrysomphalus aonidum

OCHNACEAE

SCHUURMANSIA
Pinnaspis buxi

SCHUURMANSIA HENNINGSII
Chrysomphalus dictyospermi
Hemiberlesia palmae

OLEACEAE

JASMINUM
Aspidiotus destructor
Chrysomphalus aonidum
Ischnaspis longirostris
Morganella longispina

JASMINUM OFFICINALE
Aspidiotus destructor
Ischnaspis longirostris

JASMINUM SAMBAC
Aspidiotus destructor
Chrysomphalus aonidum
Howardia biclavis
Ischnaspis longirostris
Morganella longispina

LIGUSTRUM
Pseudaulacaspis pentagona

LIGUSTRUM JAPONICUM
Ischnaspis longirostris
Pseudaulacaspis pentagona

OLEA APETALA
Lindingaspis rossi

OLEA VERRUCOSA
Lindingaspis rossi

ONAGRACEAE

FUCHSIA
Pseudaulacaspis pentagona

LUDWIGIA OCTOVALVIS
Chrysomphalus aonidum

OXALIDACEAE

AVERRHOA BILIMBI
Selenaspidus articulatus

AVERRHOA CARAMBOLA
Aspidiotus destructor
Hemiberlesia lataniae
Morganella longispina
Selenaspidus articulatus

PASSIFLORACEAE

PASSIFLORA
Pseudaonidia trilobitiformis
Pseudaulacaspis pentagona

PASSIFLORA EDULIS
Hemiberlesia lataniae
Hemiberlesia palmae
Lopholeucaspis cockerelli
Pseudaonidia trilobitiformis
Pseudaulacaspis pentagona
Selenaspidus articulatus

PASSIFLORA LAURIFOLIA
Pseudaonidia trilobitiformis

PASSIFLORA QUADRANGULARIS
Aspidiotus destructor
Pseudaonidia trilobitiformis
Pseudaulacaspis pentagona

PIPERACEAE

PIPER
Aspidiotus destructor
Hemiberlesia palmae
Pinnaspis buxi

PIPER ADUNCUM
   Aonidiella inornata
   Hemiberlesia palmae
   Lopholeucaspis cockerelli

PIPER MACGILLIVRAYI
   Aspidiotus destructor

PIPER METHYSTICUM
   Abgrallaspis cyanophylli
   Aonidiella inornata
   Aspidiotus destructor
   Hemiberlesia lataniae
   Howardia biclavis

PIPER PUBERULUM
   Aspidiotus destructor

PITTOSPORACEAE

PITTOSPORUM
   Oceanaspidiotus caledonicus

POLYGONACEAE

COCCOLOBA
   Pinnaspis strachani

COCCOLOBA UVIFERA
   Abgrallaspis cyanophylli

POLYGONUM
   Aonidiella inornata

PORTULACACEAE

PORTULACA
   Pinnaspis strachani

PORTULACA LUTEA
   Pinnaspis aspidistrae

PROTEACEAE

GREVILLEA ROBUSTA
   Howardia biclavis

HELICIA
   Pinnaspis buxi

MACADAMIA TERNIFOLIA
   Lindingaspis rossi

MACADAMIA TETRAPHYLLA
   Abgrallaspis cyanophylli
   Chrysomphalus dictyospermi
   Duplaspidiotus claviger
   Hemiberlesia palmae
   Howardia biclavis
   Parlatoria proteus

PUNICACEAE

PUNICA GRANATUM
   Howardia biclavis

RANUNCULACEAE

DELPHINIUM
   Pseudaulacaspis pentagona

RHIZOPHORACEAE

BRUGUIERA GYMNORHIZA
   Hemiberlesia palmae
   Oceanaspidiotus pangoensis
   Pinnaspis strachani

RHIZOPHORA
   Aulacaspis martini

RHIZOPHORA KANDELIA
   Aulacaspis martini

RHIZOPHORA MANGLE
   Pinnaspis strachani

RHIZOPHORA MUCRONATA
   Hemiberlesia lataniae

ROSACEAE

CLIFFORTIA POLYGONIFOLIA
   Pseudaulacaspis pentagona

ERIOBOTRYA JAPONICA
   Abgrallaspis cyanophylli
   Chrysomphalus aonidum
   Hemiberlesia lataniae
   Pseudaonidia trilobitiformis

MALUS
   Hemiberlesia palmae ·

MALUS PUMILA
   Parlatoria pergandii

MALUS SYLVESTRIS
   Chrysomphalus dictyospermi
   Hemiberlesia lataniae
   Parlatoria cinerea

PRUNUS AMYGDALUS
   Pseudaulacaspis pentagona

PRUNUS PERSICA
   Hemiberlesia lataniae

PYRUS COMMUNIS
   Hemiberlesia palmae

ROSA
   Aonidiella aurantii
   Aulacaspis rosarum
   Chrysomphalus aonidum
   Diaspis boisduvalii
   Fiorinia proboscidaria

ROSA INDICA
   Aonidiella aurantii
   Aulacaspis rosarum
   Chrysomphalus dictyospermi
   Clavaspis herculeana
   Hemiberlesia lataniae

RUBUS OCCIDENTALIS
   Aulacaspis rosarum

RUBIACEAE

COFFEA
   Abgrallaspis cyanophylli
   Ischnaspis longirostris

COFFEA ARABICA
   Abgrallaspis cyanophylli
   Ischnaspis longirostris
   Selenaspidus articulatus

COFFEA CANEPHORA
(COFFEA ROBUSTA)
   Hemiberlesia lataniae
   Hemiberlesia palmae
   Ischnaspis longirostris
   Parlatoria proteus
   Selenaspidus articulatus

COPROSMA LAEVIGATA
   Pinnaspis buxi

GARDENIA
   Chrysomphalus aonidum
   Duplaspidiotus claviger
   Howardia biclavis
   Parlatoria proteus
   Selenaspidus articulatus

GARDENIA SCANDENS
   Parlatoria proteus
   Selenaspidus articulatus

GUETTARDA SPECIOSA
   Abgrallaspis cyanophylli
   Hemiberlesia palmae
   Pinnaspis aspidistrae

IXORA
   Ischnaspis longirostris

IXORA COCCINEA
   Howardia biclavis
   Selenaspidus articulatus

MORINDA CITRIFOLIA
   Aonidiella comperei
   Chrysomphalus aonidum
   Hemiberlesia lataniae
   Hemiberlesia palmae
   Pinnaspis buxi
   Pinnaspis strachani
   Pseudaulacaspis pentagona

NEONAUCLEA
   Howardia biclavis

PLATANOCEPHALUS INDICUS
   Pseudaulacaspis pentagona

PLATANOCEPHALUS MORINDAEFOLIUS
   Aonidiella inornata
   Aspidiotus destructor

RANDIA TAHITENSIS
   Howardia biclavis
   Parlatoria proteus
   Selenaspidus articulatus

TIMONIUS
   Hemiberlesia lataniae

RUTACEAE

CITRUS
   Andaspis hawaiiensis

Aonidiella aurantii
Aonidiella eremocitri
Aspidiotus excisus
Aspidiotus nerii
[?]*Aulacaspis sumatrensis*
Chrysomphalus aonidum
Chrysomphalus dictyospermi
Chrysomphalus pinnulifer
Fiorinia proboscidaria
Fiorinia theae
Hemiberlesia lataniae
Hemiberlesia palmae
Howardia biclavis
Ischnaspis longirostris
Lepidosaphes beckii
Lepidosaphes gloverii
Lopholeucaspis baluanensis
Lopholeucaspis cockerelli
Morganella longispina
Parlatoria cinerea
Parlatoria pergandii
Parlatoria proteus
Parlatoria ziziphi
Pinnaspis aspidistrae
Pinnaspis buxi
Pinnaspis strachani
Pseudaonidia trilobitiformis
Selenaspidus articulatus
Unaspis citri

CITRUS AURANTIFOLIA
Aonidiella aurantii
Aspidiotus excisus
Chrysomphalus aonidum
Chrysomphalus dictyospermi
Fiorinia proboscidaria
Hemiberlesia lataniae
Lepidosaphes beckii
Lepidosaphes gloverii
Lopholeucaspis cockerelli
Parlatoria cinerea
Parlatoria pergandii
Pinnaspis strachani
Selenaspidus articulatus
Unaspis citri

CITRUS AURANTIUM
Aonidiella aurantii
Chrysomphalus aonidum
Chrysomphalus dictyospermi
Fiorinia proboscidaria
Howardia biclavis
Lepidosaphes beckii
Lepidosaphes gloverii
Lopholeucaspis cockerelli
Morganella longispina
Parlatoria cinerea
Pseudaonidia trilobitiformis
Selenaspidus articulatus
Unaspis citri

CITRUS GRANDIS
Aonidiella aurantii
Aspidiotus destructor
Chrysomphalus aonidum
Chrysomphalus dictyospermi
Lepidosaphes beckii
Lepidosaphes gloverii
Morganella longispina
Parlatoria cinerea
Pseudaonidia trilobitiformis
Unaspis citri

CITRUS LIMON
Aonidiella aurantii
Chrysomphalus aonidum
Chrysomphalus dictyospermi
Duplaspidiotus claviger
Fiorinia proboscidaria
Hemiberlesia palmae
Lepidosaphes beckii
Lepidosaphes gloverii
Lopholeucaspis cockerelli
Morganella longispina
[?]*Odonaspis morrisoni*
Parlatoria cinerea
Parlatoria pergandii
Pinnaspis strachani
Pseudaonidia trilobitiformis
Selenaspidus articulatus
Unaspis citri

CITRUS MAXIMA
Aonidiella aurantii
Aspidiotus destructor
Chrysomphalus dictyospermi
Fiorinia proboscidaria
Hemiberlesia lataniae
Lepidosaphes beckii
Lepidosaphes gloverii
Lepidosaphes tokionis
Lopholeucaspis cockerelli
Morganella longispina
Parlatoria cinerea
Parlatoria pergandii
Pinnaspis strachani
Pseudaonidia trilobitiformis
Pseudaulacaspis pentagona

   Selenaspidus articulatus
   Unaspis citri

CITRUS PARADISI
   Aonidiella aurantii
   Chrysomphalus aonidum
   Chrysomphalus dictyospermi
   Fiorinia proboscidaria
   Hemiberlesia lataniae
   Lepidosaphes beckii
   Morganella longispina
   Parlatoria cinerea
   Parlatoria pergandii
   Pseudaonidia trilobitiformis
   Unaspis citri

CITRUS RETICULATA
   Aonidiella aurantii
   Chrysomphalus aonidum
   Fiorinia fioriniae
   Fiorinia proboscidaria
   Hemiberlesia lataniae
   Howardia biclavis
   Lepidosaphes beckii
   Lepidosaphes gloverii
   Morganella longispina
   Parlatoria cinerea
   Parlatoria pergandii
   Pseudaonidia trilobitiformis
   Pseudaulacaspis pentagona
   Selenaspidus articulatus
   Unaspis citri

CITRUS SINENSIS
   Aonidiella aurantii
   Chrysomphalus aonidum
   Chrysomphalus dictyospermi
   [?]*Diaspis boisduvalii*
   Duplaspidiotus claviger
   Fiorinia proboscidaria
   Hemiberlesia lataniae
   Howardia biclavis
   Lepidosaphes beckii
   Lepidosaphes gloverii
   Parlatoria cinerea
   Parlatoria citri
   Parlatoria pergandii
   Pinnaspis strachani
   Selenaspidus articulatus
   Unaspis citri

FORTUNELLA JAPONICA
   Fiorinia proboscidaria

   Lepidosaphes gloverii
   Selenaspidus articulatus

MICROMELUM MINUTUM
   Pinnaspis strachani

MURRAYA EXOTICA
   Lepidosaphes beckii
   Pseudaonidia trilobitiformis

SANTALACEAE

SANTALUM AUSTRO-CALEDONICUM
   Chrysomphalus aonidum
   Fiorinia fioriniae
   Pseudaonidia trilobitiformis

SAPINDACEAE

DODONAEA VISCOSA
   Chrysomphalus aonidum
   Duplaspidiotus claviger
   Hemiberlesia lataniae
   Pinnaspis strachani
   Pseudaonidia trilobitiformis

ELATTOSTACHYS FALCATA
   Pseudaulacaspis major

LITCHI CHINENSIS
   Howardia biclavis
   Ischnaspis longirostris
   Pinnaspis strachani
   Selenaspidus articulatus

NEPHELIUM
   Selenaspidus articulatus

NEPHELIUM LAPPACEUM
   Howardia biclavis

SAPOTACEAE

GENUS INDET.
   Hemiberlesia lataniae
   Lepidosaphes rubrovittata
   Lepidosaphes securicula
   Octaspidiotus australiensis
   Parlatoria crotonis
   Schizentaspidus silvicola

CHRYSOPHYLLUM CAINITO
   Hemiberlesia lataniae

Hemiberlesia palmae
  Howardia biclavis

MANILKARA ZAPOTA
  Howardia biclavis

POMETIA PINNATA
  Lepidosaphes pometiae

SCROPHULARIACEAE

  ANGELONIA SALICARIAEFOLIA
    Pseudaulacaspis pentagona

SOLANACEAE

  GENUS INDET.
    Aspidiotus destructor
    Chrysomphalus dictyospermi
    Pinnaspis aspidistrae

  CAPSICUM
    Aspidiotus destructor
    Pseudaonidia trilobitiformis
    Pseudaulacaspis pentagona
    Unaspis citri

  CAPSICUM ANNUUM
    Aspidiotus destructor
    Pinnaspis strachani
    Pseudaonidia trilobitiformis

  CAPSICUM FRUTESCENS
    Aonidiella aurantii
    Aspidiotus destructor
    Hemiberlesia lataniae
    Lepidosaphes tapleyi
    Pinnaspis strachani
    Pseudaonidia trilobitiformis

  CAPSICUM GROSSUM
    Pseudaulacaspis pentagona

  CAPSICUM MINIMUM
    Aspidiotus destructor

  CAPSICUM OVATUM
    Abgrallaspis cyanophylli

  DATURA METEL
    Pinnaspis strachani

  LYCOPERSICON ESCULENTUM
    Aspidiotus destructor
    Howardia biclavis
    Lepidosaphes tapleyi
    Pinnaspis aspidistrae
    Pinnaspis buxi
    Pinnaspis strachani
    Pseudaulacaspis pentagona

  NICOTIANA TABACUM
    Pseudaulacaspis pentagona

  PHYSALIS LANCEOLATA
    Aspidiotus destructor

  PHYSALIS PERUVIANA
    Aspidiotus destructor

  SOLANUM
    Hemiberlesia lataniae

  SOLANUM MELONGENA
    Aspidiotus destructor
    Chrysomphalus dictyospermi
    Pinnaspis aspidistrae
    Pinnaspis strachani

  SOLANUM TORVUM
    Pinnaspis strachani

  SOLANUM UPORO
    Pseudaulacaspis pentagona

  SOLANUM VERBASCIFOLIUM
    Pseudaulacaspis pentagona

STERCULIACEAE

  THEOBROMA CACAO
    Abgrallaspis cyanophylli
    Aspidiotus destructor
    Hemiberlesia palmae
    Lopholeucaspis cockerelli
    Parlatoria proteus
    Pseudaonidia trilobitiformis

THEACEAE

  CAMELLIA
    Fiorinia fioriniae

  CAMELLIA SINENSIS
    Abgrallaspis cyanophylli
    Andaspis numerata

Chrysomphalus dictyospermi
Hemiberlesia palmae
Howardia biclavis
Selenaspidus articulatus

TILIACEAE

TRIUMFETTA BARTRAMIA
Pseudaulacaspis pentagona

TRIUMFETTA RHOMBOIDEA
Pinnaspis strachani

URTICACEAE

LAPORTEA
Pinnaspis strachani

LAPORTEA PHOTINIPHYLLA
Aspidiotus destructor

VERBENACEAE

AVICENNIA NITIDA
Hemiberlesia lataniae
Lindingaspis rossi

AVICENNIA OFFICINALIS
Lindingaspis rossi

CITHAREXYLUM SPINOSUM
Howardia biclavis

CLERODENDRUM
Abgrallaspis cyanophylli

GMELINA ARBOREA
Pinnaspis strachani

LANTANA CAMARA
Aspidiotus destructor
Selenaspidus articulatus

PREMNA
Chrysomphalus aonidum
Pseudaonidia trilobitiformis

STACHYTARPHETA
Lepidosaphes stepta
Pinnaspis strachani
Pseudaulacaspis pentagona

STACHYTARPHETA DICHOTOMA
Pseudaulacaspis pentagona

STACHYTARPHETA INDICA
Pseudaulacaspis pentagona

STACHYTARPHETA MUTABILIS
Pseudaulacaspis pentagona

STACHYTARPHETA URTICIFOLIA
Pseudaulacaspis pentagona

VERBENA
Pseudaulacaspis pentagona

VERBENA BONARIENSIS
Pseudaulacaspis pentagona

VITACEAE

VITIS VINIFERA
Aonidiella inornata
Aspidiotus nerii
Hemiberlesia lataniae
Oceanaspidiotus spinosus
Parlatoria cinerea
Pinnaspis strachani
Selenaspidus articulatus

WINTERACEAE

DRIMYS PAUCIFLORA
Hemiberlesia rapax

ZYGOGYNUM
Hemiberlesia lataniae

LILIOPSIDA (Monocotyledonae)

AGAVACEAE

AGAVE AMERICANA
Pinnaspis strachani

CORDYLINE
Hemiberlesia lataniae
Hemiberlesia palmae
Pinnaspis aspidistrae
Pinnaspis strachani
Pseudaonidia trilobitiformis

CORDYLINE NEO-CALEDONICA
Hemiberlesia lataniae
Pinnaspis aspidistrae
Pseudaonidia trilobitiformis

CORDYLINE TERMINALIS
Abgrallaspis cyanophylli
Pinnaspis aspidistrae
Pinnaspis buxi
Pinnaspis strachani

SANSEVIERIA
Chrysomphalus aonidum

YUCCA ALOIFOLIA
Hemiberlesia lataniae

ALOEACEAE

ALOE
Ischnaspis longirostris
Pinnaspis strachani

ARACEAE

ALOCASIA MACRORHIZA
Aspidiella sacchari
Lepidosaphes gloverii

ANTHURIUM
Fiorinia fioriniae

COLOCASIA
Hemiberlesia palmae

COLOCASIA ESCULENTA
Aspidiotus destructor
Hemiberlesia palmae
Pinnaspis buxi
Pinnaspis strachani

CYRTOSPERMA CHAMISSONIS
Hemiberlesia lataniae

EPIPREMNUM PINNATUM
Aspidiotus [?]excisus
Chrysomphalus dictyospermi
Fiorinia proboscidaria
Hemiberlesia palmae
Parlatoria proteus

POTHOS AUREUS
Pseudaonidia trilobitiformis

RHAPHIDOPHORA
Chionaspis rhaphidophorae
Pinnaspis strachani

RHAPHIDOPHORA ACUMINATA
Pinnaspis aspidistrae

RHAPHIDOPHORA VITIENSIS
Fiorinia proboscidaria

XANTHOSOMA
Chrysomphalus dictyospermi

XANTHOSOMA SAGITTIFOLIUM
Aspidiotus destructor
Hemiberlesia palmae
Selenaspidus articulatus

ARECACEAE

GENUS INDET.
Aspidiotus destructor
Aulacaspis sumatrensis
Chrysomphalus aonidum
Chrysomphalus dictyospermi
Ischnaspis longirostris
Oceanaspidiotus spinosus
Parlatoria crotonis
Parlatoria proteus
Pinnaspis aspidistrae
Pinnaspis buxi
Pinnaspis strachani
Pseudaulacaspis cockerelli
Pseudaulacaspis samoana

ARECA
Chrysomphalus dictyospermi
Pinnaspis buxi

ARECA CATECHU
Chrysomphalus dictyospermi
Fiorinia proboscidaria
Pinnaspis buxi

BALAKA
Hemiberlesia lataniae

COCOS
Parlatoria crotonis

COCOS NUCIFERA
Abgrallaspis cyanophylli
Aonidiella aurantii
Aonidiella eremocitri
Aspidiotus destructor
Aspidiotus macfarlanei
Aspidiotus nerii
Aspidiotus pacificus

Aulacaspis sumatrensis
Chrysomphalus aonidum
Chrysomphalus dictyospermi
Chrysomphalus propsimus
Fiorinia coronata
Fiorinia fioriniae
Hemiberlesia lataniae
Hemiberlesia palmae
Ischnaspis longirostris
Lepidosaphes gloverii
Lepidosaphes karkarica
Lepidosaphes tapleyi
Lindingaspis similis
Oceanaspidiotus pangoensis
Pinnaspis aspidistrae
Pinnaspis buxi
Pinnaspis strachani
Pseudaulacaspis cockerelli
Pseudaulacaspis pentagona
Pseudaulacaspis samoana
Selenaspidus articulatus
Unaspis citri

ELAEIS GUINEENSIS
   Aspidiotus destructor
   Chrysomphalus dictyospermi
   Ischnaspis longirostris
   Pinnaspis aspidistrae
   Pinnaspis strachani

HOWEIA FORSTERIANA
   Chrysomphalus dictyospermi
   Fiorinia fioriniae

LATANIA COMMERSONII
   Chrysomphalus aonidum

MONSTERA DELICIOSA
   Ischnaspis longirostris
   Pinnaspis buxi

NYPA FRUTICANS
   Fiorinia coronata
   Odonaspis ruthae

PHOENIX
   Aspidiotus destructor
   Aspidiotus nerii
   Hemiberlesia lataniae

PHOENIX DACTYLIFERA
   Selenaspidus articulatus

ROYSTONEA REGIA
   Chrysomphalus dictyospermi
   Hemiberlesia palmae

VEITCHIA JOANNIS
   Aonidiella inornata
   Chrysomphalus dictyospermi
   Hemiberlesia lataniae
   Pinnaspis buxi
   Pinnaspis strachani

BROMELIACEAE

ANANAS
   Diaspis boisduvalii
   Diaspis bromeliae

ANANAS COMOSUS
   Diaspis boisduvalii
   Diaspis bromeliae

ANANAS SATIVUS
   Aspidiotus nerii
   Diaspis bromeliae

CANNACEAE

CANNA INDICA
   Aspidiotus destructor
   Pinnaspis strachani

CYPERACEAE

GENUS INDET.
   Agrophaspis buxtoni

CAREX
   Agrophaspis buxtoni

DIOSCOREACEAE

DIOSCOREA
   Abgrallaspis cyanophylli
   Aspidiella hartii
   Aspidiotus destructor
   Hemiberlesia lataniae
   Hemiberlesia palmae
   Pinnaspis strachani

DIOSCOREA ALATA
   Abgrallaspis cyanophylli
   Aspidiella hartii
   Pinnaspis strachani

DIOSCOREA BULBIFERA
  Hemiberlesia palmae
  Pinnaspis strachani

DIOSCOREA ESCULENTA
  Aspidiella hartii
  Pinnaspis strachani

DIOSCOREA NUMMULARIA
  Aspidiotus destructor

DIOSCOREA SATIVA
  Aspidiella hartii

FLAGELLARIACEAE

FLAGELLARIA
  Galleraspis

HELICONIACEAE

HELICONIA
  Chrysomphalus aonidum
  Fiorinia fijiensis
  Hemiberlesia lataniae
  Hemiberlesia palmae
  Lopholeucaspis cockerelli
  Pinnaspis buxi
  Pinnaspis strachani

HELICONIA BIHAI
  Aspidiotus destructor
  Fiorinia fijiensis
  Hemiberlesia palmae

HELICONIA BRASILIENSIS
  Pinnaspis aspidistrae

IRIDACEAE

GLADIOLUS
  Chrysomphalus aonidum
  Hemiberlesia lataniae

LILIACEAE

ASPARAGUS PLUMOSUS
  Chrysomphalus dictyospermi
  Pinnaspis strachani

ASPIDISTRA
  Chrysomphalus aonidum

CRINUM
  Pinnaspis aspidistrae
  Pinnaspis strachani

CRINUM ASIATICUM
  Pinnaspis aspidistrae
  Pinnaspis strachani

CRINUM PEDUNCULATUM
  Pinnaspis aspidistrae

DIANELLA INTERMEDIA
  Chrysomphalus dictyospermi

EUCHARIS
  Pinnaspis strachani

EUCHARIS GRANDIFLORA
  Pinnaspis strachani

HIPPEASTRUM
  Pinnaspis aspidistrae

HIPPEASTRUM EQUESTRE
  Pinnaspis aspidistrae

MARANTACEAE

CALATHEA
  Chrysomphalus dictyospermi
  Selenaspidus articulatus

MARANTA
  Pinnaspis buxi
  Pinnaspis strachani
  Pseudaonidia trilobitiformis

MUSACEAE

MUSA
  Abgrallaspis cyanophylli
  Aonidiella aurantii
  Aonidiella comperei
  Aonidiella inornata
  Aspidiotus destructor
  Aspidiotus [?]excisus
  Aspidiotus musae
  Chrysomphalus aonidum
  Chrysomphalus dictyospermi
  Hemiberlesia lataniae
  Hemiberlesia palmae
  Ischnaspis longirostris
  Pinnaspis aspidistrae
  Pinnaspis buxi

    Pinnaspis strachani
    Selenaspidus articulatus
    Unaspis citri

MUSA FEHI
    Pinnaspis strachani

MUSA NANA
    Pinnaspis strachani

MUSA PARADISIACA
    Aspidiotus destructor
    Hemiberlesia lataniae
    Hemiberlesia palmae
    Pinnaspis aspidistrae

MUSA SAPIENTUM
    Abgrallaspis cyanophylli
    Aonidiella aurantii
    Aspidiotus destructor
    Aspidiotus nerii
    Chrysomphalus aonidum
    Chrysomphalus dictyospermi
    Hemiberlesia lataniae
    Hemiberlesia palmae
    Pinnaspis aspidistrae
    Pinnaspis strachani

ORCHIDACEAE

GENUS INDET.
    Aspidiotus destructor
    Furcaspis biformis
    Genaparlatoria pseudaspidiotus
    Lepidosaphes rubrovittata
    Octaspidiotus australiensis
    Pinnaspis strachani
    Pseudaulacaspis pentagona

ARUNDINA BAMBUSIFOLIA
    Selenaspidus articulatus

CYMBIDIUM
    Genaparlatoria pseudaspidiotus

DENDROBIUM
    Parlatoria proteus

GRAMMATOPHYLLUM PAPUANUM
    Hemiberlesia palmae
    Ischnaspis longirostris

MILTONIA REGNELLI
    Furcaspis biformis

ONCIDIUM
    Furcaspis biformis

ORCHIS
    Pinnaspis strachani

VANDA
    Genaparlatoria pseudaspidiotus

VANILLA
    Chrysomphalus aonidum
    Chrysomphalus dictyospermi
    Hemiberlesia palmae

VANILLA FRAGRANS
    Chrysomphalus dictyospermi

VANILLA PLANIFOLIA
    Hemiberlesia palmae

PANDANACEAE

GENUS INDET.
    Parlatoria crotonis
    Pseudaulacaspis leveri

FREYCINETIA
    Chionaspis freycinetiae
    Parlatoria proteus

PANDANUS
    Aspidiotus nerii
    Aulacaspis madiunensis
    Chionaspis pandanicola
    Chrysomphalus aonidum
    Fiorinia coronata
    Hemiberlesia lataniae
    Lindingaspis similis
    Pinnaspis buxi
    Pinnaspis strachani
    Schizentaspidus silvicola

PANDANUS ODORATISSIMUS
    Aonidiella inornata
    Chrysomphalus aonidum
    Chrysomphalus propsimus
    Hemiberlesia palmae
    Lepidosaphes esakii
    Parlatoria crotonis
    Pinnaspis aspidistrae
    Pinnaspis strachani

PANDANUS UPOLUENSIS
    Pinnaspis buxi

POACEAE

GENUS INDET.
Aspidiella sacchari
Odonaspis morrisoni
Odonaspis ruthae

BAMBUSA
[?]Odonaspis secreta

BAMBUSA VULGARIS
[?]Aonidiella aurantii
Odonaspis greenii

BRACHIARIA MUTICA
Aspidiella sacchari

CHLORIS
Hemiberlesia lataniae
Odonaspis ruthae

CYMBOPOGON CITRATUS
Odonaspis ruthae

CYNODON DACTYLON
Odonaspis ruthae

ISCHAEMUM
Aspidiella sacchari

LEPTURUS REPENS
Odonaspis ruthae

MISCANTHUS
Hemiberlesia palmae

PANICUM
Odonaspis ruthae

SACCHARUM
Odonaspis saccharicaulis

SACCHARUM OFFICINARUM
Aspidiella sacchari
Aspidiotus destructor
Aulacaspis tegalensis
Selenaspidus articulatus

SCHIZOSTACHYUM GLAUCIFOLIUM
Froggattiella penicillata
[?]Odonaspis secreta

TRIPSACUM LAXUM
Hemiberlesia palmae

SMILACACEAE

SMILAX
Oceanaspidiotus pangoensis

STRELITZIACEAE

STRELITZIA
Pinnaspis strachani

ZINGIBERACEAE

ALPINIA BOIA
Fiorinia fijiensis

ALPINIA NUTANS
Aspidiotus destructor
Chrysomphalus dictyospermi

ALPINIA PURPURATA
Hemiberlesia palmae
Pinnaspis strachani

CURCUMA LONGA
Aspidiella hartii
Hemiberlesia palmae

ELETTARIA CARDAMOMUM
Abgrallaspis cyanophylli
Aonidiella inornata
Hemiberlesia palmae

PHAEOMERIA SPECIOSA
Chrysomphalus dictyospermi

ZINGIBER
Chrysomphalus dictyospermi

ZINGIBER OFFICINALE
Aspidiella hartii
Aspidiotus destructor
Pinnaspis strachani

Index to plant genera in plant families

Abutilon - Malvaceae
Acacia - Mimosaceae
Acalypha - Euphorbiaceae
Adiantum - Adiantaceae
Agathis - Araucariaceae
Agave - Agavaceae
Albizia - Mimosaceae
Aleurites - Euphorbiaceae
Allemanda - Apocynaceae
Alocasia - Araceae
Aloe - Aloeaceae
Alpinia - Zingiberaceae
Althaea - Malvaceae
Anacardium - Anacardiaceae
Ananas - Bromeliaceae
Angelonia - Scrophulariaceae
Annesijoa - Euphorbiaceae
Annona - Annonaceae
Anthurium - Araceae
Araucaria - Araucariaceae
Areca - Arecaceae
Argyreia - Convolvulaceae
Artocarpus - Moraceae
Arundina - Orchidaceae
Asparagus - Liliaceae
Aspidistra - Liliaceae
Asplenium - Aspleniaceae
Astronidium - Melastomataceae
Averrhoa - Oxalidaceae
Avicennia - Verbenaceae

Baeckea - Myrtaceae
Balaka - Arecaceae
Bambusa - Poaceae
Barringtonia - Lecythidaceae
Bauhinia - Caesalpiniaceae
Begonia - Begoniaceae
Bischofia - Euphorbiaceae
Bixa - Bixaceae
Bougainvillea - Nyctaginaceae
Brachiaria - Poaceae
Brassica - Brassicaceae
Breynia - Euphorbiaceae
Broussonetia - Moraceae
Bruguiera - Rhizophoraceae

Caesalpinia - Caesalpiniaceae
Calathea - Marantaceae
Calophyllum - Clusiaceae
Camellia - Theaceae

Cananga - Annonaceae
Canarium - Burseraceae
Canavalia - Fabaceae
Canna - Cannaceae
Capsicum - Solanaceae
Carex - Cyperaceae
Carica - Caricaceae
Cassia - Caesalpiniaceae
Cassytha - Laureaceae
Casuarina - Casuarinaceae
Catharanthus - Apocynaceae
Cedrela - Meliaceae
Ceiba - Bombacaceae
Cerbera - Apocynaceae
Cheirodendron - Araliaceae
Chloris - Poaceae
Chrysanthemum - Asteraceae
Chrysophyllum - Sapotaceae
Cinnamomum - Laureaceae
Citharexylum - Verbenaceae
Citrus - Rutaceae
Clerodendrum - Verbenaceae
Cliffortia - Rosaceae
Clitoria - Fabaceae
Coccoloba - Polygonaceae
Cocos - Arecaceae
Codiaeum - Euphorbiaceae
Coffea - Rubiaceae
Coleus - Lamiaceae
Colocasia - Araceae
Coprosma - Rubiaceae
Cordia - Boraginaceae
Cordyline - Agavaceae
Crescentia - Bignoniaceae
Crinum - Liliaceae
Crotalaria - Fabaceae
Croton - Euphorbiaceae
Cucumis - Cucurbitaceae
Cucurbita - Cucurbitaceae
Curcuma - Zingiberaceae
Cupressus - Cupressaceae
Cycas - Cycadaceae
Cymbidium - Orchidaceae
Cymbopogon - Poaceae
Cynodon - Poaceae
Cyrtosperma - Araceae

Datura - Solanaceae
Daucus - Apiaceae
Decaspermum - Myrtaceae

Delphinium - Ranunculaceae
Dendrobium - Orchidaceae
Derris - Fabaceae
Dianella - Liliaceae
Dillenia - Dilleniaceae
Dioscorea - Dioscoreaceae
Diospyros - Ebenaceae
Dodonaea - Sapindaceae
Dracophyllum - Epacridaceae
Drimys - Winteraceae
Durio - Bombacaceae

Elaeis - Arecaceae
Elaeocarpus - Elaeocarpaceae
Elaeodendron - Elaeocarpaceae
Elattostachys - Sapindaceae
Elephantopus - Asteraceae
Elettaria - Zingiberaceae
Elmerrillea - Magnoliaceae
Epipremnum - Araceae
Eriobotrya - Rosaceae
Ervatamia - Apocynaceae
Erythrina - Fabaceae
Eucalyptus - Myrtaceae
Eucharis - Liliaceae
Eugenia - Myrtaceae
Euphorbia - Euphorbiaceae
Excoecaria - Euphorbiaceae

Fagraea - Loganiaceae
Ficus - Moraceae
Fitchia - Asteraceae
Flagellaria - Flagellariaceae
Fortunella - Rutaceae
Freycinetia - Pandanaceae
Fuchsia - Onagraceae

Gardenia - Rubiaceae
Geniostoma - Loganiaceae
Gerbera - Asteraceae
Gladiolus - Iridaceae
Gliricidia - Fabaceae
Glycine - Fabaceae
Gmelina - Verbenaceae
Gomphrena - Amaranthaceae
Gossypium - Malvaceae
Grammatophyllum - Orchidaceae
Graptophyllum - Acanthaceae
Grevillea - Proteaceae
Guettarda - Rubiaceae

Helianthus - Asteraceae
Helicia - Proteaceae
Heliconia - Heliconiaceae

Hernandia - Hernandiaceae
Hevea - Euphorbiaceae
Hibiscus - Malvaceae
Hippeastrum - Liliaceae
Howeia - Arecaceae

Indigofera - Fabaceae
Inocarpus - Fabaceae
Ipomoea - Convolvulaceae
Ischaemum - Poaceae
Ixora - Rubiaceae
Jasminum - Oleaceae
Jatropha - Euphorbiaceae
Justicia - Acanthaceae
Kalanchoe - Crassulaceae
Khaya - Meliaceae
Kigelia - Bignoniaceae

Lagerstroemia - Lythraceae
Lantana - Verbenaceae
Laportea - Urticaceae
Latania - Arecaceae
Laurus - Laureaceae
Lepturus - Poaceae
Leucaena - Mimosaceae
Leucopogon - Epacridaceae
Ligustrum - Oleaceae
Litchi - Sapindaceae
Litsea - Laureaceae
Loranthus - Loranthaceae
Ludwigia - Onagraceae
Lycopersicon - Solanaceae

Macadamia - Proteaceae
Macaranga - Euphorbiaceae
Malus - Rosaceae
Malvastrum - Malvaceae
Malvaviscus - Malvaceae
Mangifera - Anacardiaceae
Manihot - Euphorbiaceae
Manilkara - Sapotaceae
Maranta - Marantaceae
Melaleuca - Myrtaceae
Melia - Meliaceae
Meryta - Araliaceae
Micromelum - Rutaceae
Miltonia - Orchidaceae
Mimosa - Mimosaceae
Miscanthus - Poaceae
Momordica - Cucurbitaceae
Monstera - Arecaceae
Morinda - Rubiaceae
Morus - Moraceae
Mucuna - Fabaceae

Mundulea - Fabaceae
Murraya - Rutaceae
Musa - Musaceae
Myristica - Myristicaceae

Neonauclea - Rubiaceae
Nephelium - Sapindaceae
Nerium - Apocynaceae
Nicotiana - Solanaceae
Nothofagus - Fagaceae
Nypa - Arecaceae

Ocimum - Lamiaceae
Olea - Oleaceae
Oncidium - Orchidaceae
Opuntia - Cactaceae
Orchis - Orchidaceae

Pandanus - Pandanaceae
Panicum - Poaceae
Parinari - Chrysobalanaceae
Passiflora - Passifloraceae
Pedilanthus - Euphorbiaceae
Pelargonium - Geraniaceae
Pemphis - Lythraceae
Persea - Laureaceae
Phaeomeria - Zingiberaceae
Phaseolus - Fabaceae
Phoenix - Arecaceae
Phyllanthus - Euphorbiaceae
Physalis - Solanaceae
Pinus - Pinaceae
Piper - Piperaceae
Pithecellobium - Mimosaceae
Pittosporum - Pittosporaceae
Platanocephalus - Rubiaceae
Pluchea - Asteraceae
Plumeria - Apocynaceae
Podocarpus - Podocarpaceae
Polygonum - Polygonaceae
Pometia - Sapotaceae
Portulaca - Portulacaceae
Pothos - Araceae
Premna - Verbenaceae
Prosopis - Mimosaceae
Prunus - Rosaceae
Psidium - Myrtaceae
Punica - Punicaceae
Pyrostegia - Bignoniaceae
Pyrus - Rosaceae

Randia - Rubiaceae
Raphanus - Brassicaceae

Reynoldsia - Araliaceae
Rhaphidophora - Araceae
Rhizophora - Rhizophoraceae
Ricinus - Euphorbiaceae
Rosa - Rosaceae
Roystonea - Arecaceae
Rubus - Rosaceae
Saccharum - Poaceae
Saintpaulia - Gesneriaceae
Salacea - Hippocrateaceae
Sansevieria - Agavaceae
Santalum - Santalaceae
Schefflera - Araliaceae
Schinus - Anacardiaceae
Schizostachyum - Poaceae
Schuurmansia - Ochnaceae
Smilax - Smilacaceae
Solanum - Solanaceae
Sophora - Fabaceae
Spermolepis - Myrtaceae
Spondias - Anacardiaceae
Stachytarpheta - Verbenaceae
Strelitzia - Strelitziaceae
Strychnos - Loganiaceae
Swietenia - Meliaceae

Tamarindus - Caesalpiniaceae
Tecoma - Bignoniaceae
Terminalia - Combretaceae
Theobroma - Sterculiaceae
Thespesia - Malvaceae
Timonius - Rubiaceae
Tmesipteris - Tmesipteridaceae
Tournefortia - Boraginaceae
Tripsacum - Poaceae
Triumphetta - Tiliaceae

Urena - Malvaceae

Vanda - Orchidaceae
Vanilla - Orchidaceae
Veitchia - Arecaceae
Verbena - Verbenaceae
Vigna - Fabaceae
Vitis - Vitaceae

Xanthosoma - Araceae

Yucca - Agavaceae

Zingiber - Zingiberaceae
Zygogynum - Winteraceae

References

ADACHI, M.S. & FULLAWAY, D.T. (1953). Two new Diaspidid scales on *Araucaria*. *Proc. Hawaii. ent. Soc.* **15**, 87-91.

ANON (1969). Insect pest survey for the year ending 30th June 1967. Phytophagous insect pests in Papua and New Guinea. *Papua New Guin. agric. J.* **21**, 49-75.

ASHMEAD, W.H. (1880). On the red or circular scale of the orange (*Chrysomphalus ficus* Riley ms.) *Am. Entomologist* **3**, 267-269.

BALACHOWSKY, A.S. (1948). Les cochenilles de France, d'Europe, du nord de l'Afrique et du bassin méditerranéan. IV. Monographie des Coccoidea, classification - Diaspidinae (Première partie). *Actualités sci. industr.* **1054**, 1-154.

BALACHOWSKY, A.S. (1953). Les cochenilles de France, d'Europe, du nord de l'Afrique et du bassin méditerranéan. VII. Monographie des Coccoidea, Diaspidinae - IV. Odonaspidini - Parlatorini. *Actualités sci. industr.* **1202**, 1-207.

BALACHOWSKY, A.S. (1954). Les cochenilles paléarctiques de la tribu des Diaspidini. 450 pp. Paris, Institut Pasteur.

BALACHOWSKY, A.S. (1956). Les cochenilles du Continent Africain Noir Vol. I - Aspidiotini (1ère partie). *Ann. Mus. Congo belge*, 4to N.S. **3**, 7-142.

BALACHOWSKY, A.S. (1958). Les cochenilles du Continent Africain Noir Vol. II. Aspidiotini (2me partie), Odonaspidini et Parlatorini. *Ann. Mus. Congo. belge*, 4to N.S. **4**, 149-356.

BANKS, C.S. (1906a). New Philippine insects. *Philipp. J. Sci.* **1**, 229-236.

BANKS, C.S. (1906b). A change of name in Coccidae. *Philipp. J. Sci.* **1**, 787.

BARANYOVITS, F. (1953). Some aspects of the biology of armoured scale insects. *Endeavour* **12**, 202-209.

BEARDSLEY, J.W. (1966). Insects of Micronesia, Homoptera: Coccoidea. *Insects Micronesia* **6**, 377-562.

BEARDSLEY, J.W. (1975). Insects of Micronesia, Homoptera: Coccoidea, Supplement. *Insects Micronesia* **6**, 657-662.

BEARDSLEY, J.W. & GONZALEZ, R.H. (1975). The biology and ecology of armoured scales. *A. Rev. Ent.* **20**, 47-73.

BELLIO, G. (1929). Descrizione di un nuovo genre di Diaspinae (Hemiptera: Coccidae) dell'estremo oriente. *Boll. Lab. Zool. gen. agr. R. Scuola Agric. Portici* **22**, 159-165.

BEN-DOV, Y. (1974). A revision of *Ischnaspis* Douglas with a description of a new allied genus (Homoptera: Diaspididae). *J. ent.* (B) **43**, 19-32.

BEN-DOV, Y. (?1987). A taxonomic analysis of the armored scale tribe Odonaspidini of the world (Homoptera: Coccoidea: Diaspididae). *Tech. Bull. U.S. Dep. Agric.* 1723 (in press).

BEN-DOV, Y. & MATILE-FERRERO, D. (1984). On the association of ants, genus *Melissotarsus* (Formicidae), with armoured scale insects (Diaspididae) in Africa. *Verh. X int. Symp. Entomofaunistik Mitteleur.* pp. 378-380.

BERLESE, A. (1895). Le cocciniglie Italiane viventi sugli agrumi. Parte III. I. Diaspiti. *Riv. Patol. veg.*, Padova **4**, 74-170.

BERLESE, A. & LEONARDI, G. (1896). Diagnosi di cocciniglie nuove. (Cont.). *Riv. Patol. veg.*, Padova **4**, 345-352.

BIGGER, M. (1985). Forest entomology in the Solomon Islands. Solomon Islands forest insect reference collection. Section C - Appendix. 89pp. Tropical Development Research Institute, Unpublished Report. London.

BODENHEIMER, F.S. (1951). Citrus entomology in the Middle East. pp. xii, 663. 'S. Gravenhage, W. Junk.

BORATYNSKI, K. (1957). On the two species of the genus *Carulaspis* MacGillivray (Homoptera: Coccoidea, Diaspidini) in Britain. *Entomologist's mon. Mag.* **93**, 246-251.

BORCHSENIUS, N.S. (1949). Identification of the soft and armoured scales (Coccoidea) of Armenia [In Russian]. 271 pp. Izd. Akad. Nauk Armianskoi SSR. Erevan.
BORCHSENIUS, N.S. (1958). Notes on the Coccoidea of China. 2. Descriptions of some new species of Pseudococcidae, Aclerdidae and Diaspididae (Homoptera, Coccoidea). *Ent. Obozr.* **37,** 156-173.
BORCHSENIUS, N.S. (1963). On the revision of the genus *Lepidosaphes* Shimer (Coccoidea, Homoptera, Insecta). *Zool. Zh.* **42,** 1161-1174.
BORCHSENIUS, N.S. (1966). A catalogue of the armoured scale insects (Diaspidoidea) of the world. 449 pp. 'Nauka', Moscow, Leningrad.
BORCHSENIUS, N.S. & WILLIAMS, D.J. (1963). A study of the types of some little-known genera of Diaspididae with descriptions of new genera (Hemiptera: Coccoidea). *Bull. Br. Mus. nat. Hist. Ent* **13**: 353-394.
BOUCHÉ, P.F. (1833). Naturgeschichte der Schädlichen und nutzlichen Garteninsekten. 176 pp. Berlin, Nicolai.
BOUCHÉ, P.F. (1851). Neue Arten der Schildlaus-Familie. *Stettin. ent. Ztg* **12,** 110-112.
BRIMBLECOMBE, A.R. (1959). Studies of the Coccoidea. 10. New species of Diaspididae. *Qd J. agric. Sci.* **16,** 381-407.
BROUGH, E.J. (1986). Citrus entomology in Papua New Guinea. *News Bull. ent. Soc. Qd* **14,** 43-47.
BRUN, L.O. & CHAZEAU, J. (1980). Catalogue des ravageurs d'interêt agricole de Nouvelle-Calédonie. 125 pp. Laboratoire de Zoologie Appliquée, O.R.S.T.O.M., Noumea.
BRUN, L.O. & CHAZEAU, J. (1984). Complément au catalogue des ravageurs d'interêt agricole de Nouvelle-Calédonie. Quatrième conférence régionale de la Protection des végétaux. 8pp. SPC/Plant Protection 4/WP.3.
CHARLÍN, R. (1973). Coccoidea de Isla de Pascua. *Revta chil. Ent.* **7,** 111-114.
CHEN FANG-G. (1983). The Chionaspidini (Diaspididae, Coccoidea, Homoptera) from China. 174 pp. Sichuan, China.
CHOU, I. (1986). Monographia Diaspididarum Sinensium. Vol. 3. pp. 435-771. Shaanxi.
COCKERELL, T.D.A. (1892a). Scale Insects of the Cocoanut. *Jamaica Post* (Jan. 29), 5.
COCKERELL, T.D.A. (1892b). Museum notes. *J. Inst. Jamaica* **1,** 134-137.
COCKERELL, T.D.A. (1893a). Museum notes. Coccidae. *J. Inst. Jamaica* **1,** 180.
COCKERELL, T.D.A. (1893b). A list of West Indian Coccidae. *J. Inst. Jamaica* **1,** 252-256.
COCKERELL, T.D.A. (1893c). West Indian Coccidae. *Entomologist's mon. Mag.* **29,** 38-41.
COCKERELL, T.D.A. (1893d). Coccidae, or scale insects, which live on orchids. *Gdnrs' Chron.* **13,** 548.
COCKERELL, T.D.A. (1894). *Diaspis lanatus. Ent. News.* **5,** 43.
COCKERELL, T.D.A. (1895). New species of Coccidae. *Psyche* **7,** 7-8.
COCKERELL, T.D.A. (1896). A check-list of the Coccidae. *Bull. Ill. St. Lab. nat. Hist.* **4,** 318-339.
COCKERELL, T.D.A. (1897a). The San Jose scale and its nearest allies. *Tech. ser. Bur. Ent. U.S.* **6,** 1-31.
COCKERELL, T.D.A. (1897b). Contributions to coccidology - II. *Am. Nat.* **31,** 588-592.
COCKERELL, T.D.A. (1899a). Some notes on Coccidae. *Proc. Acad. nat. Sci. Philad.*, 259-275.
COCKERELL, T.D.A. (1899b). First supplement to the check-list of the Coccidae. *Bull. Ill. nat. Hist. Surv.* **5,** 389-398.
COCKERELL, T.D.A. (1900). Some Coccidae quarantined at San Francisco. *Psyche* **9,** 70-72.
COCKERELL, T.D.A. (1901). South African Coccidae. *Entomologist* **34,** 223-227.
COCKERELL, T.D.A. (1902a). The Coccid genus *Aulacaspis. Entomologist* **35,** 58-59.
COCKERELL, T.D.A. (1902b). A new gall-making Coccid. *Ann. Mag. nat. Hist.* (7) **9,** 20-26.
COCKERELL, T.D.A. (1905a). Tables for the identification of Rocky Mountain Coccidae (scale insects and mealybugs). *Univ. Colo. Stud. gen. Ser.* **2,** 189-203.

COCKERELL, T.D.A. (1905b). Some Coccidae from the Philippine Islands. *Proc. Davenport Acad. Sci.* **10**, 127-136.
COHIC, F. (1950a). Activité saissonière de la faune économique Néo-Calédonienne. *Rev. fr. Ent.*, **17**, 6-87.
COHIC, F. (1950b). Les insectes nuisibles aux plantes cultivées dans les Wallis et Futuna. *Agron. trop., Nogent* **5**, 563-581.
COHIC, F. (1953). Enquête phytosanitaire sur les plantations aux Nouvelles-Hébrides. *Revue agric. Nouv. Caléd.* **4**, 11-21.
COHIC, F. (1955). Rapport d'une mission aux établissements français d'l'Océanie. Fascicle III. Enquête sur les parasites animaux des cultures. 68 pp. Institut Français d'Océanie, Nouméa, O.R.S.T.O.M.
COHIC, F. (1956). Parasites animaux des plantes cultivees en Nouvelle-Caledonie et dependances. pp. 1-91. Noumea, Inst. franc. d'Oceanie. O.R.S.T.O.M.
COHIC, F. (1958a). Contribution a l'etude des cochenilles d'interêt économique de Nouvelle-Calédonie et dépendances. *Tech. Pap. S. Pacif. Commn* **116**, 1-35.
COHIC, F. (1958b). Contribution a l'étude des cochenilles de Nouvelle-Calédonie [Hom.] [Description d'une nouvelle espèce de *Leucaspis*.] *Bull. Soc. ent. Fr.* **63**, 49-54.
COHIC, F. (1959). Enquête sur les parasites animaux d'interêt agricole à Wallis. 69 pp. Institut Français d'Océanie. Nouméa, O.R.S.T.O.M.
COMSTOCK, J.H. (1881). Report of the entomologist. Part II. Report on scale insects. *Rep. U.S. Dep. Agric.* (1880) pp. 276-349.
COMSTOCK, J.H. (1883). Second report on scale insects. Including a monograph of the subfamily Diaspinae of the family Coccidae and a list, with notes, of the other species of scale insects found in North America. *Rep. Cornell Univ. Coll. Agric. Exp. Stn* (1882-83) **2**, 47-142.
COOLEY, R.A. (1897). New species of *Chionaspis*. *Can. Ent.* **29**, 278-282.
COOLEY, R.A. (1899). The Coccid genera *Chionaspis* and *Hemichionaspis*. *Bull. Hatch agric. Exp. Stn.* (Special Bull.) 57 pp.
COSTA, O.G. (1835). Fauna del regno di Napoli, famiglia de' coccinigliferi, o de' gallinsetti. 23pp. Napoli.
CRONQUIST, A. (1981). An integrated system of classification of flowering plants. 1262 pp. New York.
CURTIS, J. (1843). The small brown scale, *Aspidiotus proteus*, nobis. *Gdnrs' Chron.* **39**, 676.
DALE, P.S. (1959). Pest control in Samoa. *Dep. Bull. Agric. For. Fish. West. Samoa* **3**, 1-15.
DICKSON, R.C. (1951). Construction of the scale covering of *Aonidiella aurantii* (Mask.). *Ann. ent. Soc. Am.* **44**, 596-602.
DOANE, R.W. (1908). Notes on *Aspidiotus destructor* (Sig.) and its chalcid parasite in Tahiti. *J. econ. Ent.* **1**, 341-2.
DOANE, R.W. (1909). Notes on insects affecting the cocoanut trees in the Society Islands. *J. econ. Ent.* **2**, 220-223.
DOANE, R.W. & FERRIS, G.F. (1916). Notes on Samoan Coccidae with descriptions of three new species. *Bull. ent. Res.* **6**, 399-402.
DOANE, R.W. & HADDEN, E. (1909). Coccidae from the Society Islands. *Can. Ent.* **41**, 296-300.
DOUGLAS, J.W. (1887a). Note on some British Coccidae (No. 6). *Entomologist's mon. Mag.* **23**, 239-243.
DOUGLAS, J.W. (1887b). Notes on some British Coccidae (No.7) *Entomologist's mon. Mag.* **24**, 21-28.
DUMBLETON, L.J. (1954). A list of insect pests recorded in South Pacific Territories. *Tech. Pap. S. Pacif. Commn.* **79**, 1-202.
FERNALD, M.E. (1903). A catalogue of the Coccidae of the world. *Bull. Hatch agric. Exp. Stn* **88**, 360 pp.
FERRIS, G.F. (1921). Some Coccidae from Eastern Asia. *Bull. ent. Res.* **12**, 211-220.
FERRIS, G.F. (1936). Contributions to the knowledge of the Coccoidea (Homoptera). II. (Contribution No. 2). *Microentomology* **1**: 17-92.

FERRIS, G.F. (1937a). Atlas of the scale insects of North America (Series 1) [vol. 1]. Serial Nos. SI-1 to SI-136. California, Stanford Univ. Press.
FERRIS, G.F. (1937b). Contributions to the knowledge of the Coccoidea (Homoptera) V. *Microentomology* **2**, 47-101.
FERRIS, G.F. (1938). Atlas of the scale insects of North America (Series 2) [vol.2]. Serial Nos SII-1a, SII-2a, and SII-137 to SII-268. California, Stanford Univ. Press.
FERRIS, G.F. (1939). Scale insects (Hemiptera: Coccoidea) from the Marquesas. *Bull. Bernice P. Bishop Mus.* **142**, 125-131.
FERRIS, G.F. (1941a). Atlas of the scale insects of North America (Series 3) [vol. 3]. Serial Nos SIII-2b and SIII-269 to SIII-384. California, Stanford Univ. Press.
FERRIS, G.F. (1941b). The genus *Aspidiotus* (Homoptera: Coccoidea: Diaspididae). *Microentomology* **6**, 33-69.
FERRIS, G.F. (1942). Atlas of the scale insects of North America, (Series 4) [vol. 4]. Serial Nos SIV-2c and SIV-385 to SIV-448 (SIV-445, 11 pp.; SIV-446, 70pp.; SIV-447, 4pp.; SIV-448, 7pp.). California, Stanford Univ. Press.
FERRIS, G.F. (1955). The genus *Phenacaspis* Cooley and Cockerell. Part I. (Insecta: Homoptera: Coccoidea). *Microentomology* **20**, 41-54.
FERRIS, G.F. & RAO, V.P. (1947). The genus *Pinnaspis* Cockerell (Homoptera: Coccoidea: Diaspididae). *Microentomology* **12**, 25-58.
FIRMAN, I.D. (1982). California red scale in Fiji. *Inf. Circ. S. Pacif. Commn* No. 90.
FOSBERG, F.R., SACHET, M. & OLIVER, R. (1979). A geographical checklist of the Micronesian Dicotyledonae. *Micronesica* **15**, 41-295.
FROGGATT, J.L. (1936). Some insects recorded from the Mandated Territory of New Guinea. *New Guinea agric. Gaz.* **2**, 15-18.
FROGGATT, J.L. (1940). Annual report of the Department of Agriculture for the year ending 30th June, 1939. Entomologist's Report. *New Guinea agric. Gaz.* **6**, 9-13.
GHAURI, M.S.K. (1962). The morphology and taxonomy of male scale insects (Homoptera: Coccoidea), 221 pp. London, Br. Mus. Nat. Hist.
GRANDPRÉ, A.D. de & CHARMOY, D. d'E. de (1899). Liste raisonnee des cochenilles de l'Ile de Maurice. pp. 20-49. The Planters and Commercial Gazette. Soc. Amic. Scient.
GREATHEAD, D.J. (1971). A review of biological control in the Ethiopian Region. *Tech. Commun. Commonw. Inst. biol. Control* **5**, 1-162.
GREATHEAD, D.J. (1975). The ecology of a scale insect, *Aulacaspis tegalensis*, on sugar cane in East Africa. *Trans. R. ent. Soc. Lond.* **127**, 101-114.
GREEN, E.E. (1890). Insect pests of the tea plant. Part I. 104 pp. Colombo, Independent Press.
GREEN, E.E. (1896a). The Coccidae of Ceylon. Part I. p. 1-103. London, Dulau.
GREEN, E.E. (1896b). Catalogue of Coccidae collected in Ceylon. *Indian Mus. Notes* **4**, 2-10.
GREEN, E.E. (1899). The Coccidae of Ceylon, Part II. p. 105-169. London, Dulau.
GREEN, E.E. (1900). Supplementary notes on the Coccidae of Ceylon. *J. Bombay nat. Hist. Soc.* **13**, 66-76, 252-257.
GREEN, E.E. (1905). Supplementary notes on the Coccidae of Ceylon. *J. Bombay nat. Hist. Soc.* **16**, 340-357.
GREEN, E.E. (1907). Notes on the Coccidae collected by the Percy Sladen Trust Expedition to the Indian Ocean: supplemented by a collection received from Mr. R. Dupont, Director of Agriculture, Seychelles. *Trans. Linn. Soc. Lond.* Ser. Zool. **12**, 197-207.
GREEN, E.E. (1911). On some Coccidae affecting rubber trees in Ceylon, with descriptions of new species. *J. econ. Biol.* **6**, 27-37.
GREEN, E.E. (1915). Notes on Coccidae collected by F.P. Jepson, Government Entomologist, Fiji. *Bull. ent. Res.* **6**, 44.
GREEN, E.E. (1916). Remarks on Coccidae from Northern Australia - II. *Bull. ent. Res.* **7**, 53-65.
GREEN, E.E. (1930). Fauna Sumatrensis (Bijdrad nr. 65) Coccidae. *Tijdschr. Ent.* **73**, 279-297.

GREEN, E.E. & LAING, F. (1923). Descriptions of some species and some new records of Coccidae. - I. Diaspidinae. *Bull. ent. Res* **14**, 123-131.

GREENWOOD, W. (1929). The food plants or hosts of some Fijian insects. *Proc. Linn. Soc. N.S.W.* **54**, 344-352.

GREENWOOD, W. (1940). The food plants of some Fijian insects. IV. *Proc. Linn. Soc. N.S.W.* **65**, 211-218.

GREENWOOD, W. (1977). The food plants or hosts of some Fijian insects. V. *Proc. Linn. Soc. N.S.W.* **101**, 237-241.

GRESSITT, J.L. (1954). Insects of Micronesia, Introduction. *Insects Micronesia* **1**, i-ix, 1-257.

GREVE, J.E. van S. & ISMAY, J.W. [Eds] (1983). Crop insect survey of Papua New Guinea from July 1st 1969 to December 31st 1978. *Papua New Guin. agric. J.* **32**, i - iv, 1-120.

GUTIERREZ, J. (1981). Actualisation des données sur l'entomologie économique à Wallis et à Futuna. 23 pp. O.R.S.T.O.M.

HAMON, A.B. (1985). *Oceanaspidiotus araucariae* (Adachi & Fullaway) (Homoptera: Coccoidea: Diaspididae). *Entomology Circ. Fla Dep. Agric.* No. 276.

HINCKLEY, A.D. (1965). Trophic records of some insects, mites, and ticks in Fiji. *Bull. Dep. Agric. Fiji* **45**, 1-116.

HOWELL, J.O. (1980). The value of second-stage males in armoured scale insects (Diaspididae) phyletics. *Israel J. Ent.* **14**, 87-96.

IKIN, R. (1984). Plant Protection News. *Inf. Circ. S. Pacif. Commn* No. 93.

JEPSON, F.P. (1913). Some peliminary notes on a scale insect infesting the banana in Fiji. *Bull. Dep. Agric. Fiji* **5**, 1-7.

JEPSON, F.P. (1915). III. Division of Entomology. pp. 17-27. *In* Agriculture (Annual report for the year 1914). *Rep. Dep. Agric. Fiji* Council Paper **24**, 1-37.

KERNER, J.S. (1778). Naturgeschichte des Coccus Bromeliae oder des Ananaschildes. 8,56 pp. Stuttgart, Erhard.

KIRKALDY, G.W. (1902). Hemiptera. *Fauna hawaii.* **3**, 93-174.

KOEBELE, A. (1893). Studies of parasitic and predaceous insects in New Zealand, Australia, and adjacent islands. 39 pp. Washington, U.S.D.A.

KOMOSINSKA, H. (1969). Studies on the genus *Abgrallaspis* Balachowsky, 1948, (Homoptera, Coccoidea, Diaspididae). *Acta zool. cracov.* **14**, 43-85.

KUWANA, I. (1902). Coccidae (Scale Insects) of Japan. *Proc. Calif. Acad. Sci.* (ser. S, Zool). **3**, 43-98.

KUWANA, I. & MURAMATSU, K. (1931). New scale insects and white fly found upon plants entering Japanese ports. *Zool. Mag. Tokyo* **43**, 647-660.

LAING, F. (1927). Coccidae, Aphididae and Aleyrodidae. *Insects Samoa* **2**, 35-45.

LAING, F. (1933). The Coccidae of New Caledonia. *Ann. Mag. nat. Hist.* **11**, 675-678.

LEONARDI, G. (1897). Intorno al genere *Aspidiotus*. *Riv. Patol. veg., Padova* **5**, 375.

LEONARDI, G. (1898a). Monographia del genere *Mytilaspis*. Nota Preventiva. *Riv. Patol. veg., Padova* **6**, 45-47.

LEONARDI, G. (1898b). Generi e specie di Diaspiti. Saggio di Sistematica degli *Aspidiotus*. *Riv. Patol. veg., Padova* (1897) **6**, 48-78 (208-236).

LEONARDI, G. (1899). Generi e specie di Diaspiti. Saggio di Sistematica degli *Aspidiotus*. *Riv. Patol. veg., Padova* **7**, 173-225.

LEVER, R.J.A.W. (1933). Entomologist's annual report for the year 1931-32. *Agric. Gaz. Br. Solomon Isl.* **1**, 3-6.

LEVER, R.J.A.W. (1940). Insect pests of citrus, pineapple and tobacco. *Agric. J. Dep. Agric. Fiji* **11**, 1-3.

LEVER, R.J.A.W. (1945a). An annotated check list of the mealybugs and scale insects of Fiji. *Agric. J. Dep. Agric. Fiji* **16**, 41-44.

LEVER, R.J.A.W. (1945b). Insect pests of some economic crops in Fiji. *Bull. ent. Res.* **35**, 367-377.

LEVER, R.J.A.W. (1946). Insect pests in Fiji. *Bull. Dep. Agric. Fiji* **23**, 1-36.

LEVER, R.J.A.W. (1947). Insect pests of some economic crops in Fiji. No 2. *Bull. ent. Res.* **38**, 137-143.
LEVER, R.J.A.W. (1968). A check list of economic insects and mites of crops in the British Solomon Islands. *Tech. Docum. F.A.O. Plant Prot. Comm. S. E. Asia Pac. Reg.* **65**, 1-12.
LINDINGER, L. (1905). Zwei neue Schildläuse aus Asien. *Insektenbörse* **22**, 131-132.
LINDINGER, L. (1908). Coccidenstudien. I. Zur Systematik der Diaspinen. II. Kritische Notizen (1907). *Berl. ent. Z.* **52**, 96-106.
LINDINGER, L. (1911). Beitrage zur Kenntnis der Schildläuse und ihrer Verbreitung. II. *Z. wiss. InsektBiol.* **7**, 126-130, 172-77.
LINDINGER, L. (1912). Die Schildläuse (Coccidae) Europas, Nordafrikas und Vorderasiens, einschliesslich der Azoren, der Kanaren und Madeiras. 388 pp. Stuttgart, Ulmer.
LINNAEUS, C. (1758). Systema Naturae, &c. Editio decima, reformata. Tomus I. Pars II. Regnum Animale. 824 pp. Holmiae.
MACFARLANE, R. (1986). Plant Protection News. *Inf. Circ. S. Pacif. Commn* No. 95.
MACGILLIVRAY, A.D. (1921). The Coccidae. 502 pp. Urbana, Ill., Scarab Co.
MADDISON, P.A. (1976). Interim report to the South Pacific Bureau of Economic Co-operation, on pests of a limited range of crops. *In* U.N.D.P./F.A.O. survey of agricultural pests and diseases. Interim report March 1976. Part 2 Nematology Entomology. *Sth Pac. Bur. Econ. Devel.* **4**, 1-2, 1-57.
MAMET, R. (1941). On some Coccidae (Hemipt. Homopt.) described from Mauritius by de Charmoy. *Bull. Maurit. Inst.* **2**, 23-37.
MAMET, R. (1958). The *Selenaspidus* complex (Homoptera Coccoidea). *Ann. Mus. Congo belge*, 4to N.S. **4**, 359-429.
MANSER, P.D. (1974). Report to the Government of the Gilbert and Ellice Islands Colony on a survey of insect pests of crops. 35pp. U.N.D.P./F.A.O., No. TA 3246, Rome.
MASKELL, W.M. (1879). On some Coccidae in New Zealand *Trans. Proc. N.Z. Inst.* (1878) **11**, 187-228.
MASKELL, W.M. (1891). Further Coccid notes: with descriptions of new species from New Zealand, Australia, and Fiji. *Trans. Proc. N.Z. Inst.* (1890) **23**, 1-36.
MASKELL, W.M. (1895). Further coccid notes, with descriptions of new species from New Zealand, Australia, Sandwich Islands and elsewhere, and remarks on many species already reported. *Trans. Proc. N.Z. Inst.* **27**, 36-75.
MATILE-FERRERO, D. (1982). Notes sur les *Ischnaspis* Douglas et *Trischnaspis* Ben-Dov afrotropicaux et description de trois espèces nouvelles [Homoptera, Coccoidea, Diaspididae] *Revue fr. Ent.*, (NS) **4**, 63-71.
MATILE-FERRERO, D. & BALACHOWSKY, A. (1973). Deux Aspidiotini (Homoptera Coccoidea - Diaspididae) nouveaux de Nouvelle-Caledonie. *Cah. pacif.* **17**, 239-243.
McKENZIE, H.L. (1937a) Morphological differences distinguishing California red scale, yellow scale, and related species (Homoptera, Diaspididae). *Univ. Calif. Publs Ent.* **6**, 323-336.
McKENZIE, H.L. (1937b). General characteristics of *Aonidiella* Berlese and Leonardi, and a description of a new species from Australia (Homoptera - Diaspididae). *Pan-Pacif. Ent.* **13**, 176-180.
McKENZIE, H.L. (1938). The genus *Aonidiella* (Homoptera: Coccoidea: Diaspididae). *Microentomology* **3**, 1-36.
McKENZIE, H.L. (1939). A revision of the genus *Chrysomphalus* and supplementary notes on the genus *Aonidiella* (Homoptera: Coccoidea: Diaspididae). *Microentomology* **4**, 51-77.
McKENZIE, H.L. (1943). Miscellaneous Diaspid studies including notes on *Chrysomphalus* (Homoptera; Coccoidea; Diaspididae). *Bull. Calif. Dep. Agric.* **32**, 148-162.
McKENZIE, H.L. (1945). A revision of *Parlatoria* and closely allied genera. (Homoptera: Coccoidea: Diaspididae). *Microentomology* **10**, 47-121.
McKENZIE, H.L. (1950). The genera *Lindingaspis* MacGillivray and *Marginaspis* Hall (Homoptera: Coccoidea: Diaspididae). *Microentomology* **15**, 98-124.

McKENZIE, H.L. (1956). The armoured scale insects of California. *Bull. Calif. Insect Surv.* **5**, 1-209.
MEISTER, C.W. (1975). A preliminary survey of citrus on the islands of Tonga, Niue and Western Samoa. 5 pp. Unpublished report, Koronivia Research Station, Fiji.
MILLER, D.R. & KOSZTARAB, M. (1979). Recent advances in the study of scale insects. *A. Rev. Ent.* **24**, 1-27.
MORGAN, A.C.F. (1889). Observations on Coccidae (No. 5). *Entomologist's mon. Mag.* **25**, 349-353.
MORRISON, H. (1924). Synonymical notes on two species of *Aulacaspis* (Hemiptera: Coccidae). *Proc. ent. Soc. Wash.* **26**, 231-232.
MORRISON, H. & MORRISON, E.R. (1966). An annotated list of generic names of the scale insects (Homoptera: Coccoidea). *Misc. Publs U.S. Dep. Agric.* **1015**, 1-206.
MUMFORD, E.P. & ADAMSON, A.M. (193?). Entomological researches in the Marquesas Islands. 17 + iv pp. Unpublished Report [in BMNH].
MUNTING, J. (1977). On the genera *Aulacaspis, Duplachionaspis* and *Ledaspis* from Southern Africa (Homoptera: Diaspididae). *Entomology Mem. Dep. agric. tech. Servs Repub. S. Afr.* **46**, 1-34.
NEWMAN, E. (1869). *Coccus beckii*, a new British Hemipteron of the family Coccidae. *Entomologist* **4**, 217-218.
NEWSTEAD, R. (1893). Observations of Coccidae (No. 5). *Entomologist's mon. Mag.* **29**, 185-188.
NEWSTEAD, R. (1894). Scale insects of Madras. *Indian Mus. Notes* **3**, 21-32.
NEWSTEAD, R. (1901). Monograph of the Coccidae of the British Isles. Vol. 1. 220 pp. London, Ray Soc.
O'CONNOR, B.A. (1949). Some insect pests of Tonga. *Agric. J. Dep. Agric. Fiji* **20**, 47-57.
O'CONNOR, B.A. (1969). Exotic plant pests and diseases. A handbook of plant pests and diseases to be excluded from or prevented from spreading within the area of the South Pacific Commission. pp. viii, 23, [424]. Noumea, South Pacific Commission.
PACKARD, A.S. (1869). Guide to the study of insects and a treatise on those injurious and beneficial to crops: for the use of colleges, farm-schools, and agriculturalists. Massachussetts, Salem, [not seen]. (1870). Second ed. [i]-viii, viii, 702 pp. Salem Naturalist's Book Agency; London, Trubner (seen).
PAINE, R.W. (1935). Economic importance of *A. destructor* in Fiji prior to 1928. pp. 8-11. *In* TAYLOR T.H.C. The campaign against *Aspidiotus destructor*, Sign., in Fiji. *Bull. ent. Res.* **26**, 1-102.
RAO, V.P. (1949). The genus *Unaspis* MacGillivray (Homoptera: Coccoidea: Diaspididae). *Microentomology* **14**, 59-72.
RAO, V.P. & FERRIS, G.F. (1952). The genus *Andaspis* MacGillivray (Insecta: Homoptera: Coccoidea). *Microentomology* **17**, 17-32.
RAO, V.P., GHANI, M.A., SANKARAN, T. & MATHUR, K.C. (1971). A review of the biological control of insects and other pests in South-East Asia and the Pacific Region. *Tech. Commun. Commonw. Inst. biol. Control* **6**, 1-149.
REBOUL, J.L. (1976). Principaux parasites et maladies des plantes cultivees en Polynesie Francaise. Service de Economie Rurale, Recherche Agronomique. No. 129/ER/RA. 58 pp. Pirae.
REDDY, D.B. (1970). A preliminary list of pests and diseases of plants in Western Samoa. *Tech. Docum. FAO Plant Prot. Comm. S.E. Asia Pac. Reg.* **77**, 1-15.
REYNE, A. (1948). Studies on a serious outbreak of *Aspidiotus destructor rigidus* in the cocoanut-palms of Sangi (North Celebes). *Tijdschr. Ent.* **89**, 83-123.
REYNE, A. (1961). Scale insects from Dutch New Guinea. *Beaufortia*, **9**, 121-167.
RISBEC, J. (1937). Observations sur les parasites des plantes cultivées aux Nouvelles-Hébrides. *Faune colon. fr.* **6**, 1-214.
RISBEC, J. (1942). Observations sur les insectes des plantations en Nouvelle-Calédonie. 128 pp. Secrétariat d'état aux Colonies, Paris.

RUTHERFORD, A. (1914). Some Ceylon Coccidae. *Bull. ent. Res.* **5**, 259-268.
RUTHERFORD, A. (1915). Notes on Ceylon Coccidae. *Spolia zeylan.* 10, 103-115.
SANDERS, J.G. (1909). Catalogue of recently described Coccidae. II. *Tech. Ser. Bur. Ent. U.S.* **16**, 33-60.
SASSCER, E.R. (1912). Catalogue of recently described Coccidae. IV. *Tech. Ser. Bur. Ent. U.S.* **16**, 75-82.
SCOTT, C.L. (1952). The scale insect genus *Aulacaspis* in eastern Asia (Homoptera: Coccoidea: Diaspididae). *Microentomology* **17**, 33-60.
SHIMER, H. (1868). Notes on the 'Apple Bark-Louse' (*Lepidosaphes conchiformis*, Gmelin sp.) with a description of a supposed new *Acarus*. *Trans. Am. ent. Soc.*, 361-374.
SIGNORET, V. (1869a) Essai sur les cochenilles (Homoptères-Coccides). *Annls Soc. ent. Fr.* (sér. 4) **8**, 829-876.
SIGNORET, V. (1869b). Essai sur les cochenilles ou gallinsectes (Homoptères - Coccides). *Annls Soc. ent. Fr.* (sér. 4.) **9**, 109-138; 431-452.
SIGNORET, V. (1882). [Observations on various coccids]. *Annls Soc. ent. Fr.* **2**, xxxv-xxxvi.
SIMMONDS, H.W. (1921). The transparent coconut scale *Aspidiotus destructor* and its enemies in southern Pacific. *Agric. Circ. Fiji* **2**, 14-17.
SIMMONDS, H.W. (1925). Pests and diseases of the coconut palm in the islands of the southern Pacific. *Bull. Dep. Agric. Fiji* **16**, 1-31.
STAPLEY, J.H. (1976). Check list of insect pests in the British Solomon Islands. *Tech. Docum. F.A.O. Plant Prot. Comm. S.E. Asia Pac. Reg.* **102**, 1-6.
STOETZEL, M.B. (1976). Scale-cover formation in the Diaspididae (Homoptera: Coccoidea). *Proc. ent. Soc. Wash.* **78**, 323-332.
SWAINE, G. (1971). Agricultural Zoology in Fiji. xix, 424 pp. Overseas Devel. Admin. Overseas Res. Publ. No 18, London.
SWEZEY, O.H. (1924). Notes on insect pests in Samoa. *Hawaii. Plrs' Rec.* **28**, 214-219.
SZENT-IVANY, J.J.H. (1956). New insect pest and host plant records in the territory of Papua and New Guinea. *Papua New Guin. agric. J.* **11**, 1-6.
SZENT-IVANY, J.J.H., WORMSLEY, J.G. & ARDLEY, J.H. (1956). Some insects of *Cycas* in New Guinea. *Papua New Guin. agric. J.* **11**, 1-4.
SZENT-IVANY, J.J.H. & STEVENS, R.M. (1966). Insects associated with *Coffea arabica* and some other crops in the Wau-Bulolo area of New Guinea. *Papua New Guin. agric. J.* **18**, 101-119.
TAKAGI, S. (1969). Diaspididae of Taiwan based on material collected in connection with the Japan - U.S. co-operative science programme, 1965 (Homoptera: Coccoidea) Part I. *Insecta matsum.* **32**, 1-110.
TAKAGI, S. (1970). Diaspididae of Taiwan based on material collected in connection with the Japan - U.S. co-operative science programme, 1965 (Homoptera: Coccoidea) Part II. *Insecta matsum.* **33**, 1-146.
TAKAGI, S. (1980). An interpretation of second instar male characters in the systematics of the Diaspididae. *Israel J. Ent.* **14**, 99-104.
TAKAGI, S. (1984). Some Aspidiotine scale insects with enlarged setae on the pygidial lobes (Homoptera: Coccoidea: Diaspididae). *Insecta matsum.* N.S. 28, 1-69.
TAKAGI, S. (1985). The scale insect genus *Chionaspis*: a revised concept (Homoptera: Coccoidea: Diaspididae). *Insecta matsum.* N.S. **33**, 1-50.
TAKAGI, S. & KAWAI, S. (1967). The genera *Chionaspis* and *Pseudaulacaspis* with a criticism on *Phenacaspis* (Homoptera: Coccoidea). *Insecta matsum.* **30**, 29-43.
TAKAHASHI, R. (1935). Observations on the Coccidae of Formosa, Part V. *Rep. Dep. Agric. Govt. res. Inst. Formosa* **66**, 1-37.
TAKAHASHI, R. (1939). Some Aleyrodidae, Aphididae, and Coccidae from Micronesia (Homoptera). *Tenthredo* **2**, 234-272.
TAKAHASHI, R. (1940). Some Coccidae from Formosa and Japan (Homoptera) V. *Mushi* **13**, 18-28.

TAKAHASHI, R. (1942). Some injurious insects of agricultural plants and forest trees in Thailand and Indo-China, II Coccide. *Rep. Govt. Res. Inst. Dep. Agric. Formosa* **81**, 1-56.

TARGIONI, A. (1867). Studii sulle Cocciniglie. *Memorie Soc. ital. Sci. nat.* **3**, 1-87.

TARGIONI, A. (1868). Introduzione alla seconda memoria per gli studj sulle Cocciniglie; e catalogo dei generi e delle specie della famiglie dei Coccidi. *Atti Soc. ital. Sci. nat.* **11**, 1-45, (1869) 694-738.

TARGIONI, A. (1886). Sull'insetto che danneggia i gelsi. *Boll. Soc. ent. ital.* **19**, 184-186.

TAYLOR, T.H.C. (1935). The campaign against *Aspidiotus destructor* Sign., in Fiji. *Bull. ent. Res.* **26**, 1-102.

THOMAS, R.T.S. (1962a). Host plants of some Sternorrhyncha (Phytophthires) in Netherlands New Guinea (Homoptera). *Pacif. Insects* **4**, 119-120.

THOMAS, R.T.S. (1962b). De plagen van enkele cultuurgewassen in West Nieuw Guinea. *Meded. econ. Zaken, Landb. Serie* **1**, 1-126.

VALENTINE, E.W. (1975). Report on a visit to Rarotonga. Entomology Division, D.S.I.R. Unpublished report.

VEITCH, R. & GREENWOOD, W. (1921). The food plants or hosts of some Fijian insects. *Proc. Linn. Soc. N.S.W.* **46**, 505-517.

VEITCH, R. & GREENWOOD, W. (1924). The food plants or hosts of some Fijian insects. Part ii. *Proc. Linn. Soc. N.S.W.* **49**, 153-161.

WALKER, A.K. & DEITZ, L.L. (1979). A review of entomophagous insects in the Cook Islands. *N.Z. Ent.* **7**, 70-82.

WILLIAMS, D.J. (1960). Some new Diaspidini (Coccoidea: Homoptera) from Africa. *Bull. Br. Mus. nat. Hist. Ent.* **9**, 389-399.

WILLIAMS, D.J. (1963). Synoptic revisions of I. *Lindingaspis* and II. *Andaspis* with two new allied genera (Hemiptera: Coccoidea). *Bull. Br. Mus. nat. Hist.* Ent. **15**, 1-31.

WILLIAMS, D.J. (1970). Two of T.D.A. Cockerell's early articles on scale insects. *Entomologist's mon. Mag.* **106**, 33-34.

WILLIAMS, D.J., (1971). Synoptic discussion of *Lepidosaphes* Shimer and its allies with a key to genera (Homoptera, Coccoidea, Diaspididae) *Bull. ent. Res.* **61**, 7-11.

WILLIAMS, D.J. (1974). The type-species of *Lindingaspis* MacGillivray (Homoptera: Coccoidea: Diaspididae). *J. Ent.* **42**, 217-220.

WILLIAMS, D.J. (1980). *Andaspis dasi* Williams identical with *A. numerata* Brimblecombe (Hemiptera: Diaspididae), a species found on tea and associated with the fungus *Septobasidium*. *Bull. ent. Res.* **70**, 259-260.

WILLIAMS, D.J. & BUTCHER, C.F. (1987). Scale insects (Hemiptera: Coccoidea) of Vanuatu. *N.Z. Ent.* **9**, 88-99.

WILLIAMS, J.R. (1970). Studies on the biology, ecology, and economic importance of the sugar cane scale insect, *Aulacaspis tegalensis* (Zhnt.) (Diaspididae), in Mauritius. *Bull. ent. Res.* **60**, 61-95.

WILLIAMS, J.R., METCALFE, J.R., MUNGOMERY, R.W. & MATHES, R. (Eds) (1969). Pests of Sugar Cane. 568 pp. Amsterdam, London, New York, Elsevier.

WILLIS, J.C. (1973). A Dictionary of the flowering plants and ferns. 8th ed., revised by H.K.A. SHAW. London, Cambridge University Press. xii, 1245, lxv pp.

WILSON, F. (1960). A review of the biological control of insects and weeds in Australia and Australian New Guinea. *Tech. Commun. Commonw. Inst. biol. control* **1**, 1-102.

ZAHRADNÍK, J. (1959). *Borchseniaspis* novum genus, typus *Aspidiotus palmae* Morgan et Cockerell, 1893. *Sb. ent. Odd. nar. Mus. Praze* **5**, 65-67.

ZEHNTNER, L. (1897). Overzicht van de ziekten van het Suikerriet op Java. *Meded. Proefstn Oost-Java en Archf Java-Suiklnd.* **5**, 525-575.

ZEHNTNER, L. (1898). De plantenluizen van het suikerriet op Java. *Meded. Proefstn Oost-Java en Archf Java-Suiklnd.* **6**, 1085-1098.

ZIMMERMAN, E.C. (1948). Homoptera: Sternorhyncha. *Insects Hawaii* **5**: 1-464.

Index

Synonyms and *nomina nuda* are in *italics*.
Italicised numbers represent pages bearing illustrations.

Abgrallaspis 20, 22
Agrophaspis 19, 24, *25*
Andaspis 21, 27
Allantomytilus 21, 24
Aonidiella 19, 35
aonidum 9, *91*, 93
araucariae Genaparlatoria 127, *128*
araucariae Oceanaspidiotus 9, 184, *185*
articulatus 11, *240*, 241
Aspidiella 20, 44
Aspidiotus 20, 49
aspidistrae 212, *213*
astronidii 109, *110*
Aulacaspis 22, 69
aurantii 8, 35, *36*, 37
auriculata 168
australiensis 195, *196*

baluanensis 177, *178*
beckii 9, 143, *145*, 146, 148
biakana *113*, 115
biclavis 137, *138*
biformis 124, *126*
boisduvalii 100, *101*, 103
bromeliae 100, *102*, 103
broughae 80, *81*, 82
bugnicourti 17
buxi 212, *214*, 215
buxtoni Agrophaspis 24, *25*
buxtoni Lindingaspis *171*, 172

caledonicus 184, 186, *187*
Carulaspis 21, 79
casuarinae 100, 104, *105*
Chionaspis 22, 80
Chrysomphalus 20, 90
cinerea 9, 202, *203*
cinnamomi 75
citri Parlatoria 202, 204, *205*
citri Unaspis 7, 11, *246*, 247
citricola 146
citrina 35
Clavaspis 19, 98
claviger 107, *108*
clavigera 107

cochereaui 48, 49, 51
cockerelli Lopholeucaspis 9, 177, *179*, 180
cockerelli Pseudaulacaspis *221*, 222
cocotis 53
coloisuvae 222, *223*, 225
comperei 35, 37, *38*
comys 80, 82, *83*
coronata *114*, 115, 116
crotonis 202, *206*, 207
Cryptophyllaspis 49
cyanophylli 8, 22, *23*
cydoniae 132

dacryoides 26, 27
dasi 31
destructor 7, 8, 49, *50*, 52, 53, *54*, 55,
Diaspis 21, 100
dictyospermi *92*, 93, 94
dilatata 222
dubia 17
Duplaspidiotus 19, 107

echinocacti 100, 104, *106*
elmerrilleae 144, *147*, 148
eremocitri 35, *39*, 40
esakii 144, 148, *149*
eurychlidonis 144, 150, *151*
excisus 51, 56, *57*

fasciata 162
ficaria *242*, 243
ficus 93
fijiensis 115, 116, *117*
Fijifiorinia 21, 109
filiformis 142
Fiorinia 21, 112, 118
fioriniae 9, 115, 118, *119*
flacourtiae 227
freycinetiae 80, *84*, 85
Froggattiella 19, 123
Furcaspis 19, 124

Galeraspis 17
geigerae 112
Genaparlatoria 21, 127

geniostomae 144, *152*, 153
giffardi 79
gloverii 144, 153, *154*, 168
greenii 198

hartii 8, 44, *45*
hawaiiensis *28*, 29
hederae 64
Hemiberlesia 20, 130
Hemichionaspis 211
herculeana 98, *99*
Howardia 21, 137

inday 222
inornata 35, *41*, 42
Ischnaspis 21, 140

karkarica 144, 155, *156*
keravatana 80, 85, *86*

lasianthi 168
lataniae *131*, 132
laterochitinosa 143
Lepidosaphes 21, 143, 162
Leucaspis 17
leveri 222, *224*, 225
Lindingaspis 20, 170
longa 17
longirostris *141*, 142
longispina 9, *181*, 182
Lopholeucaspis 21, 176

maai 29, *30*, 31
macfarlanei 51, 58, *59*
maddisoni 51, 60, *61*
madiunensis 8, *68*, 70
major 220, *226*, 227
major 75
marginata 144, 157, *158*
martini 69, 70, *71*
maskelli 182
minima 79, 80
minor 212
moorsi 29
Morganella 19, 182
morrisoni 198
multiducta 220, *228*, 229
musae 51, 62, *63*

nendeanus 184, 188, *189*
neocaledonica 112
nerii 51, 64, *65*
numerata 8, 29, 31, *32*

Oceanaspidiotus 20, 183
oceanica 124
oconnori 109, *111*, 112
Octaspidiotus 20, 195
Odonaspis 19, 197
orientalis 35, 42, *43*

pacificus 49, 51, 66, *67*
palmae 132, *133*, 134
pandani 215
pandanicola 80, 87, *88*
pangoensis 184, *190*, *191*, 192
papulosa 220, *230*, 231
Parlatoria 21, 201
pellucida 118
penicillata 124, *125*
pentagona 9, 220, 231, *232*
pergandii 9, 202, *208*, 209
pinnaeformis 143
Pinnaspis 22, 211
pinnulifer 93, 95, 97
pinnulifera 97
pometiae 144, *159*, 160
ponticula 220, *234*, 235
proboscidaria *120*, 121
Prontaspis 247
propsimus 93, *96*, 98
proteus 202, 209, *210*
Pseudaonidia 19, 219
pseudaspidiotus 127, *129*, 130
pseudaspidistrae 215
Pseudaulacaspis 21, 220
pseudoaspidistrae 218

rageaui 17
rapax 132, 135, *136*
reducta 115, *122*, 123
rhaphidophorae 80, *89*, 90
rigidus 56
rosae 75
rosarum 70, 72, *73*
rossi 172, *173*, 176
rubrovittata 144, *161*, 162
rubrovittatus 162
ruebsaameni 17
ruthae 198, *199*
rutherfordi 76

sacchari 44, *46*, 47
saccharicaulis 198, 201
samoana Lindingaspis 172, *174*, 176
samoana Pseudaulacaspis 222, 235, *236*
Schizentaspidus 20, 237

secreta 197, *200*, 201
secretus 201
securicula 144, *163*, 164
Selenaspidus 20, 241
Silvestraspis 21, 243
silvicola *238*, 239
similis 172, *175*, 176
simmondsi 193
simulatrix 215
sinensis 245
spinosa 29, 33, *34*
spinosus 184, 193, *194*
stepta 144, *165*, 166
strachani 212, *216*, *217*, 218
stricklandi 137, *139*, 140
sumatrensis 70, *74*, 75
suvaensis 17, 183

tapleyi 144, *167*, 168
tegalensis 8, 69, 75, 77
Temnaspidiotus 49
temporaria 218

theae 115
tokionis 144, 168, *169*
townsendi 218
transparens 53
trilobitiformis 219
tubercularis 69, 75

uberifera 243, *244*, 245
ulapa 162
Unaspis 21, 247

veitchi 247
vitiensis 231
vitis 69, 76, *78*

wanatabei 56

yannonensis 247

zingiberi 47
ziziphi 202

CAB INTERNATIONAL INSTITUTE OF ENTOMOLOGY

CAB International Institute of Entomology (CIE) is a constituent institute of CAB International (CABI), formerly the Commonwealth Agricultural Bureaux, a co-operative, non-profit making organisation administered and financed by member governments. CABI provides scientific, information and development services for agriculture and allied disciplines throughout the world.

The Institute of Entomology provides a unique and expanding range of international services to entomologists in agriculture, horticulture, forestry, food storage, land management, public health, and related subjects. The main categories of service are:

Identification. A team of specialist taxonomists based in the Department of Entomology of the British Museum (Natural History) identify around 60,000 specimens annually, submitted mostly from scientists in local, national and international research organisations and universities. CIE taxonomists have access to the Museum's collections of over 20 million specimens and use of the Museum libraries and research facilities.

Research. CIE taxonomists publish many research papers each year and produce occasional monographs to describe new species, to clarify taxonomic and nomenclatural problems and to review the taxonomy and biology of major groups on a regional or world basis. Funded contract work is also undertaken and may include detailed studies of important groups of pests, parasitoids and predators, or assistance with surveys and quarantine inspection.

Training. A regular *International Course on Applied Taxonomy of Insects and Mites of Agricultural Importance* is provided in London by Institute staff with staff of the British Museum (Natural History) and other organisations. Practical instruction is given in the collection, curation, examination and identification of specimens, with field work and visits to relevant research establishments. Additional training is also provided overseas and in the UK as opportunities arise.

Information. *CABI Library Services* can provide information on all aspects of applied entomology and, by computerised literature searches of the *CAB ABSTRACTS* database and other appropriate databases, can retrieve information in seconds that might otherwise take weeks to find. The CIE Library has one of the largest collections of applied entomological literature in the world and provides the basis for the *Document Delivery Service* which can supply copies of most entomological and other relevant papers.

Publications. Two major entomological journals are published by CABI. The *Review of Applied Entomology, Series A (Agricultural)* and *Series B (Medical & Veterinary)* provide a unique abstract synopsis of world literature on every aspect of applied entomology. The *Bulletin of Entomological Research* is a quarterly research journal devoted to original papers on insects, mites and ticks of economic importance. CIE also produces a series of *Distribution Maps of Pests*, of special interest to those involved in plant quarantine. A new series of *Guides to Insects of Importance to Man* is being produced especially for laboratory and reference use.

Further details of these and other services may be obtained on request from:

Director
CAB International Institute of Entomology
56 Queen's Gate
London SW7 5JR
UK

Telephone: (01) 584 0067
Telecom Gold/Dialcom: 84:CAU006
Telex: 265871 (MONREF G) and quote CAU006
Fax: (01) 581 1676